Springer Tracts in Modern Physics
Volume 133

W0051026

Springer-Verlag Berlin Heidelberg GmbH

Springer Tracts in Modern Physics

Volumes 118–134 are listed at the end of the book

Covering reviews with emphasis on the fields of Elementary Particle Physics, Solid-State Physics, Complex Systems, and Fundamental Astrophysics

Manuscripts for publication should be addressed to the editor mainly responsible for the field concerned:

Gerhard Höhler
Institut für Theoretische Teilchenphysik
Universität Karlsruhe
Postfach 6980
D-76128 Karlsruhe
Germany
Fax: +49 (7 21) 37 07 26
Phone: +49 (7 21) 6 08 33 75
Email: hoehler@fphvax.physik.uni-karlsruhe.de

Johann Kühn
Institut für Theoretische Teilchenphysik
Universität Karlsruhe
Postfach 6980
D-76128 Karlsruhe
Germany
Fax: +49 (7 21) 37 07 26
Phone: +49 (7 21) 6 08 33 72
Email: johann.kuehn@physik.uni-karlsruhe.de

Thomas Müller
IEKP
Fakultät für Physik
Universität Karlsruhe
Postfach 6980
D-76128 Karlsruhe
Germany
Fax:+49 (7 21) 6 07 26 21
Phone: +49 (7 21) 6 08 35 24
Email: mullerth@vxcern.cern.ch

Roberto Peccei
Department of Physics
University of California, Los Angeles
405 Hilgard Avenue
Los Angeles, California 90024-1547
USA
Fax: +1 310 825 9368
Phone: +1 310 825 1042
Email: robertop@college.ucla.edu

Frank Steiner
Abteilung für Theoretische Physik
Universität Ulm
Albert-Einstein-Allee 11
D-89069 Ulm
Germany
Fax: +49 (7 31) 5 02 29 24
Phone: +49 (7 31) 5 02 29 10
Email: steiner@physik.uni-ulm.de

Joachim Trümper
Max-Planck-Institut
für Extraterrestrische Physik
Postfach 1603
D-85740 Garching
Germany
Fax: +49 (89) 32 99 35 69
Phone: +49 (89) 32 99 35 59
Email: jtrumper@mpe-garching.mpg.de

Peter Wölfle
Institut für Theorie
der Kondensierten Materie
Universität Karlsruhe
Postfach 69 80
D-76128 Karlsruhe
Germany
Fax: +49 (7 21) 69 81 50
Phone: +49 (7 21) 6 08 35 90/33 67
Email: woelfle@tkm.physik.uni-karlsruhe.de

H. Riffert
H. Müther
H. Herold
H. Ruder

Matter at High Densities in Astrophysics

Compact Stars and the Equation of State

With 86 Figures

Springer

Harald Riffert
Herbert Müther
Heinz Herold
Hanns Ruder

Universität Tübingen
Auf der Morgenstelle 10
D-72076 Tübingen
Germany

Cataloging-in-Publication Data applied for

Die Deutsche Bibliothek - CIP-Einheitsaufnahme

Matter at high densities in astrophysics : compact stars and the equation of state / H. Herold ... - Berlin ; Heidelberg ; New York ; Barcelona ; Budapest ; Hong Kong ; London ; Milan ; Paris ; Santa Clara ; Singapore ; Tokyo : Springer, 1996
(Springer tracts in modern physics ; Vol. 133)

NE: Herold, Heinz; GT

Physics and Astronomy Classification Scheme (PACS):
21.65.+f, 64.10.+h, 91.30.Ks, 96.60.Ly, 97.20.Rp, 97.60.Jd

ISBN 978-3-662-14090-1   ISBN 978-3-540-48454-7 (eBook)
DOI 10.1007/978-3-540-48454-7

Typesetting: Camera-ready copy from the authors using a Springer TeX macro package.
Cover design: Springer-Verlag, Design & Production
SPIN: 10493069   56/3144-5 4 3 2 1 0 – Printed on acid-free paper

In Honor
of Friedrich Hund's
100th Birthday

Friedrich Hund on his 100th Birthday

Friedrich Hund was born on the 4th of February 1896 in Karlsruhe. Thus we now have the rare opportunity to congratulate a distinguished physicist on his 100th birthday and to honor his achievements.

Let us first of all recall the major stages in his life as a physicist. Hund studied in Göttingen and Marburg, and in 1921 he took the state examination in Göttingen. He spent the following year as a teacher at a high school. During this time, Hund established contact with Max Born, who had just been appointed to the Chair of Theoretical Physics in Göttingen. This led to his working on a doctoral thesis for which he was awarded a PhD in 1922. At the same time Born offered him a position as scientific assistant, a post which Hund held until 1927. In 1925 he completed his *Habilitation* (the qualification needed to become a lecturer at a German university) and in 1926/27 he spent one semester in Copenhagen. In 1927 he was appointed to a professorship at the University of Rostock. From 1929 until 1946 Professor Hund was at the University of Leipzig, for the majority of this time together with Heisenberg. He then moved to the University of Jena, where he remained until 1951. He spent the years from 1951 until 1956 at the University of Frankfurt. In 1956 (after the death of Richard Becker) Hund accepted a chair at the University of Göttingen and thus returned to the place where he had begun his career as a physicist. Until only a few years ago, throughout the decades since his official retirement, Hund continued to hold regular lectures in Göttingen.

Anyone who hears the name Friedrich Hund immediately thinks of the discovery and development of quantum theory, events which he helped to shape during his time in Göttingen. In 1922 Hund attended the famous Göttingen Lectures of Niels Bohr. When, in 1925, the old quantum theory was cast aside in favor of Heisenberg's discovery of quantum mechanics, Hund experienced this at first hand. His work during the 1920s addressed the problem of understanding atomic and molecular spectra, based initially on the idea of the correspondence principle. He made essential contributions to explaining the observed multiplet structure of complicated spectra. "Hund's Rule" is to be found in all standard textbooks on quantum mechanics. In his first book, whose German title "Linienspektren und periodisches System der Elemente" translates as "Spectral Lines and the Periodic Table of the Elements" [1], he presented the results of this work.

Following the publication in 1926 of Schrödinger's equation, Hund recognized how it is possible to understand the connection between the electron system of a molecule and those of separate atoms. This important insight led to the development of the Hund–Mulliken method of molecular orbitals. Works were published on the interpretation of spectra and other properties of molecules.

We will have to be content here with these few keywords. Very much later, in his book on the history of quantum theory [2], Hund describes how he witnessed the creation of quantum theory. His own contributions to these developments are mentioned with his typical modesty.

During his years in Leipzig, he first of all pursued the theory of molecules and of chemical bonding. His contribution to the "Handbuch der Physik" (Encyclopedia of Physics) about the quantum mechanics of atoms and molecules was also written during this period. He then addressed his attention in particular to the quantum theory of solids. A series of works dealing with the states of electrons in crystal lattices bears witness to this interest. He was also active in the field of nuclear physics and, in 1937, he published a one-particle model of the atomic nucleus. This was a forerunner of the later shell model. Towards the end of the 1930s, Hund turned his attention to the field properties of matter. He was able to show that a qualitative understanding of chemical bonding and of the production of matter could already be achieved in the classical wave picture.

He began work on a monograph about the field theory of matter. However, due to the unfavorable circumstances that prevailed during and immediately after the Second World War, he was only able to complete this manuscript very much later. The book finally appeared in 1954 with the title "Materie als Feld" (Matter as Field) [3]. It contained a systematic description of the wave aspects of matter and also a particularly clear introduction to quantum field theory.

In the second half of his life, Hund was particularly concerned to communicate his knowledge and experience to younger physicists. During his last years in Leipzig he had already begun to write a textbook of theoretical physics based on his lectures there. The book appeared shortly after the Second World War, its final form being the three-volume work "Theoretische Physik. Eine Einführung" (Theoretical Physics. An Introduction) [4], which appeared in 1956/57. Shortly after this, in 1961, he published a book "Theorie des Aufbaues der Materie" (Theory of the Structure of Matter) [5]. In the preface to this book Hund expressed very clearly his desire to provide an understanding of the foundations of physics. Anyone who has read his books can confirm that he did indeed achieve his objective.

Hund has documented his thoughts about general questions of physics in very many articles and published lectures. In the book "Das Naturbild der Physik" (The Physical Concept of Nature) [6] one finds an impressive selection of these contributions.

Following his official retirement, Hund directed his attention in particular to matters concerning the history of physics, a subject that had always fascinated him. He wrote two further books: "Grundbegriffe der Physik" (Basic Concepts of Physics) [7] and "Geschichte der physikalischen Begriffe" (The History of Physical Concepts) [8], in addition to his history of quantum theory that was mentioned above.

A large portion of Hund's work was always dedicated to teaching. He was invariably exceptionally well prepared when presenting his lectures and talks. He got everything just right, even ensuring that he made best use of the available blackboards. His contributions to discussions in colloquia were of such high quality that all those who were present will remember them. His commitment was not confined to lectures. It is with pleasure that we recall how, in 1949, we took part (as guests from Berlin) in a vacation seminar in the Thuringian Forest. The seminar was organized by Hund and a mathematician for the benefit of Jena students. In the 1950s Hund regularly took part in the Oberwolfach meetings in the Black Forest, which at that time were organized by young theoreticians. Much energy was expanded not only in the scientific program but also in long hikes, one of Hund's favorite pastimes. The large number of his students confirms the fact that, throughout his career, he succeeded in gathering around himself a circle of highly motivated young physicists.

When it comes to personal affairs, Hund is rather reserved. What is conspicuous, however, is his distinctive personality, characterized by an upright and incorruptible nature, calmness, and deep sense of duty. He lived honorably throughout the evil times of Nazi rule (indeed, one of us recalls his kindness during this time with immense gratitude). In the few years for which he remained in the Soviet Zone, he retained his independence and made no secret of his critical opinions. Together with his wife and five children he left Jena in an unconventional manner, in other words, secretly. At that time there was no other way in which he could accept the professorship offered to him in Frankfurt.

His interest in physical research remains undiminished today. When one of us visited him in Göttingen shortly after his 99th birthday, he began the conversation with the question "What do we really know now about the mass of the top quark?"

Since 1936, together with E. Trendelenburg, Hund was the editor of the "Ergebnisse der exakten Naturwissenschaften" (Achievements of the Exact Natural Sciences), the forerunner of the "Springer Tracts in Modern Physics". It was here, in 1936, that his article "Materie bei sehr hohen Drucken und Temperaturen" (Matter Under Very High Pressures and Temperatures) was published. The scientific content of the present volume is written in connection with this article. Both the German original and an English translation of Hund's 1936 article are to be found starting with p. 175.

Together with the editors of the "Springer Tracts" and the authors of this monograph, we convey our warmest congratulations to Friedrich Hund on the occasion of his 100th birthday.

Gerhard Höhler

Harry Lehmann

Ernst Niekisch

References

1 Hund F. (1927) Linienspektrum und periodisches System der Elemente. Julius Springer, Berlin
2 Hund F. (1984) Geschichte der Quantentheorie, 3rd edn. Bibliographisches Institut, Mannheim
3 Hund F. (1954) Materie als Feld. Eine Einführung. Springer, Berlin
4 Hund F. (1956/57) Theoretische Physik. Eine Einführung, 3 Vols. Teubner, Stuttgart
5 Hund F. (1961) Theorie des Aufbaues der Materie. Teubner, Stuttgart
6 Hund F. (1975) Das Naturbild der Physik. Ausgewählte Schriften. (ed. by Hajdu J., Lüders G. with support of the Deutsche Physikalische Gesellschaft) Druckerei der Kernforschungsanlage Jülich GmbH.
7 Hund F. (1969) Grundbegriffe der Physik. Bibliographisches Institut, Mannheim
8 Hund F. (1972) Geschichte der physikalischen Begriffe. Bibliographisches Institut, Mannheim

Preface

This book would not exist were it not for some of Friedrich Hund's former students, whose idea it was to dedicate a special volume of the Springer Tracts in Modern Physics in honor of Hund's 100th birthday. For a long time Friedrich Hund was an editor of the "Ergebnisse der exakten Naturwissenschaften" (now the Springer Tracts) where in 1936 he published a review article on matter under extreme conditions, which contained a detailed discussion on astrophysical aspects and applications. In spite of the tremendous progress in laboratory experiments at high pressure and temperature, astrophysics of compact stars remained the realm of matter in genuinely extreme environments. In the 1930s when Hund wrote his article, white dwarf stars had already been observed and their stability was understood in terms of the properties of the degenerate electron gas. The existence of neutron stars had been predicted at about the same time but it took more than 30 years until rapidly pulsating radio and X-ray emitting sources were discovered and identified as rotating neutron stars, which possess in their interiors the largest densities in the universe. Due to the enormous progress in observational techniques and computer power, it is now possible to explore both experimentally and theoretically many details of the internal structure of planets, ordinary stars, white dwarf stars, and neutron stars. Thus, today the physics of compact astrophysical objects is more fascinating than ever, and it could serve as a common playground for astrophysicists, solid-state physicists, and nuclear physicists.

Some parts of this monograph have the character of short reviews on independent research areas and we are grateful to many friends and colleagues who are specialists in those fields for offering numerous suggestions and comments that have helped to improve the manuscript considerably. Amongst others we thank D. Koester, P. Kumar, G. Röpke, and M. Schneider, and we are particulary indebted to J. Ruoff for contributing to the chapter on neutron-star oscillations. In addition we are grateful to a legion of students for their assistance in our continuous battle with LaTeX macros, postscript files, scanners, and other hardware demons that had to be mastered in order to meet a deadline that had already been set one hundred years ago. We are also grateful to A. Wahl for re-typing Friedrich Hund's article, to A. Lahee for the English translation, and to J. Lenz and V. Wicks from Springer-Verlag for proof reading the manuscript.

Tübingen
January 1996

H. Riffert
H. Müther
H. Herold
H. Ruder

Contents

1. Introduction 1

2. The Equation of State
   and the Structure of Cosmic Objects 3
   2.1 Basic Structure Equations and Ideal Equations of State 4
   2.2 Non-ideal Equations of State 7
   2.3 Stellar Energy Sources 14
   2.4 Energy Transport 16

3. Nuclear Equations of State 21
   3.1 General Remarks 21
   3.2 Green's Function and Many-Body Theory 29
       3.2.1 Time-Dependent Perturbation Theory 29
       3.2.2 Perturbation Theory and Feynman Diagrams 33
       3.2.3 Single-Particle Green's Function 37
   3.3 Nucleon–Nucleon Interaction 43
       3.3.1 The One-Boson-Exchange Model 45
       3.3.2 NN Scattering 48
       3.3.3 Medium-Range Attraction 52
   3.4 Hole-Line Expansion and Other Approaches 55
       3.4.1 Brueckner–Hartree–Fock 55
       3.4.2 Beyond BHF 61
   3.5 Relativistic Effects 66
       3.5.1 Walecka Model for Nuclear Matter 66
       3.5.2 Dirac–Brueckner–Hartree–Fock 69
   3.6 Subnucleonic Degrees of Freedom 74
       3.6.1 Excitations of the Nucleons 74
       3.6.2 Pion Condensation 78
       3.6.3 Effective Quark Models 84

4. Neutron Stars: Spherically Symmetric
   and Rotating Models 93
   4.1 Spherically Symmetric Neutron Stars 93
       4.1.1 Relativistic Structure Equations 95

  4.1.2   Solution Method and Results........................ 98
  4.1.3   Stability and Maximum Mass 99
 4.2 Rapidly Rotating Neutron Stars 101
  4.2.1   Basic Formulation 103
  4.2.2   Numerical Solution Method 110
  4.2.3   Results... 113

5. Asteroseismology 121
 5.1 Oscillations of Spheres 121
 5.2 Free Oscillations of the Earth 136
 5.3 Helioseismology 142
 5.4 Asteroseismology of White Dwarfs 154
 5.5 Oscillations of Neutron Stars 162

Reprint of Friedrich Hund's Review Article................... 175

English Translation of Friedrich Hund's Review Article 217

References... 259

Subject Index.. 271

1. Introduction

Sixty years ago Friedrich Hund published a review article, in the German magazine *Ergebnisse der exakten Naturwissenschaften*, which later became the *Springer Tracts in Modern Physics*, about matter at extremely high pressure and temperature. Friedrich Hund celebrated his 100th birthday on February 4, 1996. We would like to take the occasion of this special event to discuss a number of modern aspects of the topics he reviewed in his 1936 article "Materie unter sehr hohen Drucken und Temperaturen" (Matter under Very High Pressure and Temperature). For the historically interested reader the German original and the translation are reprinted in full at the back of this book.

When we read this article today it is on the one hand impressive to recognize the enormous increase in knowledge that has occurred in physics over the last six decades, but on the other side it is equally impressive to see how Friedrich Hund drew a number of conclusions based on fundamental physical considerations that are still valid. A review article written today and covering the same subjects, including thermodynamics, atomic physics, solid-state physics, nuclear physics and the astrophysics of planets, and ordinary and compact stars would certainly go far beyond the scope of a Springer Tract.

In his introductory remark Hund emphasized the fundamental role of planets and stars as cosmic laboratories for investigating the properties of matter at high densities and temperatures:

> Extremely high pressures and temperatures exist in the interior of stars. Astronomy, however, is only able to give us very indirect information about what goes on inside stars; indeed, the situation today is that physics can make some relatively certain statements about the behavior of matter under such conditions, information which astronomy and astrophysics employs to interpret observations and to draw conclusions from these observations about the actual processes and states.

The modelling of the internal structure and evolution of stars based on well-established physical laws is still the standard procedure for interpreting the large variety of stellar observations. The enormous progress in observational instruments and techniques led among other things to the observation of neutron stars, which represent the most extreme example of matter at high densities. In a few cases the mass of such neutron stars can be determined with high accuracy, for example the famous binary pulsar PSR 1913+16 contains a neutron star of 1.4411 ± 0.0007 solar mass. Furthermore, vary rapidly rotating neutron stars have been discovered spinning as fast as 649 times per second. These observations led to certain constraints for the equation of state at densities above the density of atomic nuclei. Therefore neutron stars are on the edge of becoming laboratories for the nuclear physicist.

Chapter 2 of this monograph is a general introduction to the role of the equation of state in modelling the internal structure of cosmic objects. As a result of such calculations the temperature, pressure, and density at each point inside the star are obtained. In Chap. 3 we review recent developments in constructing equations of state for high-density matter in the interior of compact stars, with special emphasis on the conditions at nuclear densities and above. The next chapter is devoted to the relativistic structure of neutron stars, their mass–radius relation, and the effects of rotation. The last chapter starts with a discussion of small oscillations of spheres. The internal properties of an oscillating body determine the frequencies of the normal modes, which have been observed with incredible accuracy for the Earth, the Sun and some white dwarfs, and can thus be used to probe the entire interior. We will also discuss this fascinating method of asteroseismology in Chap. 5.

As is usual in the astrophysical literature we use cgs units throughout, and physical constants are denoted by their standard abbreviations.

2. The Equation of State
and the Structure of Cosmic Objects

This chapter is intended to provide the basic ideas about what determines the structure of planets, ordinary stars, white dwarfs, and neutron stars. Consequently, this will be by no means a complete introduction to this fundamental field of theoretical astrophysics; instead we refer the interested reader to a series of excellent texbooks by Chandrasekhar (1939), Schwarzschild (1958), Cox and Giuli (1968), Kippenhahn and Weigert (1990), or Hansen and Kawaler (1994) – just to name a few.

The matter in the interior of such objects is in a state of extreme conditions compared to terrestrial environments. In only a few laboratory experiments have temperatures and pressures that come close planetary or stellar magnitudes been achieved. In fusion reactors plasma temperatures larger than 10^8 K have been reached (Keilhacker and Watkins (1995)) which already exceed the values in the solar interior, although the density is much lower (electron density $n_e \approx 10^{14}$ cm^{-3} compared to $n_e \approx 10^{26}$ cm^{-3} in the center of the sun). High pressures of more than 10^{12} dyn cm^{-2} can be obtained dynamically through shock-wave experiments (Brown and McQueen 1986) or statically in diamond-anvil-cells (Jayaraman 1983); this corresponds to the pressure in the core of the Earth, about 3000 km below the surface. These values are quite impressive when compared with everyday conditions or even with the state of the art in experimental physics twenty years ago. However, compared with the interiors of celestial bodies such as giant planets, ordinary stars, white dwarfs, and neutron stars, the physical conditions are more extreme by orders of magnitude. Some properties of those objects are listed in Table 2.1. Unfortunately, with the exception of solar neutrinos, no information is obtained directly from those regions, but some insight can be gained through model calculations of the interior structure and evolution of stars based on the laws of physics. Each model contains a number of free parameters that can be adjusted to fit the observable properties such as mass, radius, surface gravity, surface temperature, total luminosty, spectral lines, or frequencies of stellar oscillations. It is in particular this latter information that, within the last decades, has opened an entire new window to the interior of the sun and a number of white dwarfs.

Table 2.1. Physical parameters for some celestial objects. The masses are given in units of solar mass (1 $M_\odot$ = 1.989 × 10^{33} g). The temperatures, densities, and pressures are values at the centers, except for Jupiter where the conditions in the metallic hydrogen core are given (Stevenson 1982). The data for the Earth are from Dziewonski and Anderson (1981), the solar model is taken from Guenther et al. (1992); the white dwarf (WD) and neutron star (NS) data are from model calculations by Wood (1994) and Herold and Neugebauer (1992), respectively.

Object	M $M_\odot$	R km	T K	ρ g cm^{-3}	p dyn cm^{-2}
Earth	3.0 × 10^{-6}	6.37 × 10^3	4 × 10^3	13.0	3.6 × 10^{12}
Jupiter	9.5 × 10^{-4}	6.98 × 10^4	2 × 10^4	4.5	4.5 × 10^{13}
Sun	1.0	6.96 × 10^5	1.6 × 10^7	1.5 × 10^2	2.3 × 10^{17}
WD	0.6	8.93 × 10^3	1.0 × 10^6	3.7 × 10^6	1.8 × 10^{23}
NS	1.4	9.0	--	2.4 × 10^{15}	5.5 × 10^{35}

2.1 Basic Structure Equations and Ideal Equations of State

Like every physical model that tries to mimic a certain aspect of nature, stellar-structure models are based on a number of simplifying assumptions, some of which are motivated by the problem under consideration, but others are introduced just to keep the entire model tractable. One basic assumption is a static spherically symmetric distribution of matter. Then all physical quantities depend on only the radial distance r from the center of mass. A static model ignores all time dependencies and velocity fields. Large-scale velocities such as those in rotation, stellar winds, or accretion flows are therefore not taken into account. The small-scale motion of convective flows, which, for example, are visible on the surface of the sun as the solar granulation, and which occur in some regions of most stars, are only treated in an average quasi-static way, i.e., there will be a transport of energy due to this process, the actual velocity field is not calculated. In addition, no magnetic fields are considered. With all this taken into consideration, the balance of forces is reduced to the equation of hydrostatic equilibrium:

$$\frac{1}{\rho}\frac{dp}{dr} = -\frac{Gm(r)}{r^2} , \qquad (2.1)$$

where ρ and p are the mass density and the pressure, respectively, G is the graviational constant, and the mass $m(r)$ inside the radius r is given by

$$m(r) = 4\pi \int_0^r r'^2 \rho(r')\, dr' . \qquad (2.2)$$

Here the effects of general relativity have been ignored. For the structure of neutron stars, however, these relations have to be replaced by the Tolman–Oppenheimer–Volkoff equations which will be discussed in Chap. 4.

The two equations (2.1) and (2.2) contain three functions ($p(r)$, $\rho(r)$, and $m(r)$), and are thus not sufficient to determine the structure of a spherical self-gravitating body. One essential additional relation is the equation of state (EOS)

$$p = p(\rho, T, \text{composition}) , \tag{2.3}$$

where T is the temperature, and "composition" stands for the various particle species that form the matter. In his article, Hund gave an example of such an EOS where the density has been calculated as a function of p and T covering many orders of magnitude. A three-dimensional representation of this global EOS structure is shown in the illustrative Fig. 2.1. It has been

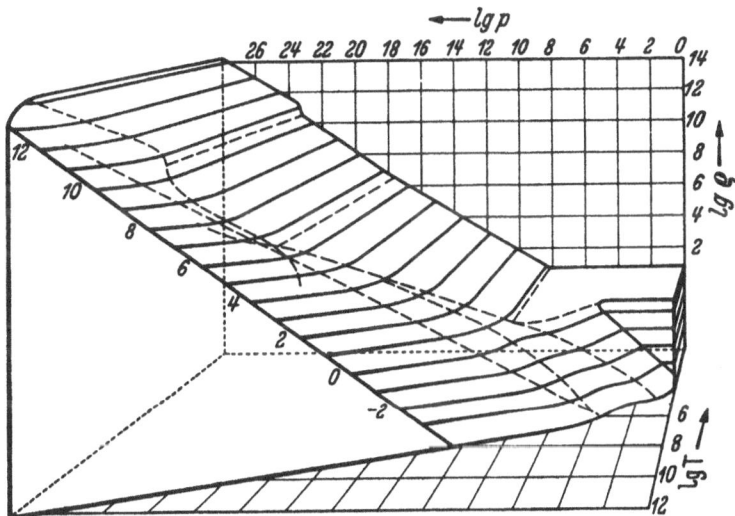

Fig. 2.1. A global overview of the equation of state $\rho = \rho(p,T)$ calculated by Hund in 1936 (densities are given in $\mathrm{g\,cm^{-3}}$, and pressures in bar). With the exception of the solid phase, a system of non-interacting particles has been assumed.

derived under a number of idealizing assumptions. Only two phases of the matter are distinguished: a gaseous phase that consists of an ideal gas mixture of electrons, ions, and – for high densities – protons and neutrons, and a solid phase of constant density. At high temperatures the contribution from the photons is taken into account.

An important regime of the EOS for the structure of compact stars is the line

$$p = p(\rho) , \tag{2.4}$$

in the $T=0$ plane. At pressures up to about 10^8 bar the density of matter is dictated by the laws of solid-state physics and rather insensitive to the pressure. At higher pressures the cristals collaps, the external pressure is balanced by the repulsion of the degenerate electron gas (see Sect. 3.1), and the density increases with pressure. At even higher densities protons turn into neutrons because of the inverse β decay, and the pressure at nuclear densities is dominated by the degenerate neutron gas, modified by the interaction between the baryons. This picture remains qualitatively valid for sufficiently low but finite temperatures, which can be seen as the temperature-independent part of the EOS surface in Fig. 2.1. This regime is limited at low densities by the evaporation of the solid phase. At higher densities temperature effects dominate where the thermal energy becomes comparable to the Fermi energy of the degenerate gas. At even higher temperatures the EOS can be represented by the ideal gas law

$$p = \frac{k}{\mu_m m_u} \rho T \ , \tag{2.5}$$

where μ_m is the mean molecular weight that implicitly depends on the temperature through the ionization state of matter. The constants m_u and k stand for the atomic mass unit and the Boltzmann constant, respectively. If the temperature is increased further until

$$T > \left[\frac{3k\rho}{a\,\mu_m m_u} \right]^{1/3} \ , \tag{2.6}$$

then the radiation pressure dominates the gas pressure, and the EOS is given by

$$p = \frac{a}{3} T^4 \ , \tag{2.7}$$

which is independent of density. Here a is the radiation constant. This idealized treatment of the EOS gives a valuable overview of the various physical mechanisms that determine the state of matter in different regimes of p, ρ, and T. For the calculation of stellar or planetary structures, however, this global picture has to be refined considerably, and we will discuss some of these aspects below. For the case of degenerate matter it is in most cases reasonable to consider the limit $T = 0$. Consequently, the EOS will be of the form (2.4), and it therfore allows the calculation of the entire stellar structure $p(r)$, $\rho(r)$, $m(r)$ by integraton of the hydrostatic equation (2.1) together with (2.2) and appropriate boundary conditions, one of which is already included in (2.2):

$$m(r=0) = 0 \ . \tag{2.8}$$

A second condition is usually specified at the surface of the sphere (at $r = R$)

$$p(r=R) = 0 \ . \tag{2.9}$$

These equations yield a one-parameter set of solutions. Specifying the central density then leads to definite values for the radius R and the total mass

$M = m(r = R)$. One well-known example of a $T = 0$ EOS is the case of a completely degenerate gas of non-interacting electrons in a positive ion background, which has been applied to model the structure of white dwarf stars by Chandrasekhar (1931) [see also Cox and Giuli (1968)]. This will be discussed in some detail in Chap. 3. There are two limiting regimes for which the EOS assumes the particulary simple form

$$p = K \rho^\gamma , \tag{2.10}$$

where $\gamma = 5/3$ for electrons having non-relativistic velocities, and $\gamma = 4/3$ for ultra-relativistic electrons. The relation (2.10) is known as a polytropic EOS with a polytropic exponent γ. Such equations of state have a long history of their own, which goes back to the work of Lane (1870), Emden (1907), or Chandrasekhar (1939). A polytropic EOS is sometimes used because it greatly simplifies the model calculations, some problems even admit analytic solutions, thus yielding in many cases a good qualitative picture of the underlying physical aspects of the problem. Because of this, polytropes have been applied not only in static stellar-structure calculations but also in some dynamical situations such as stellar-collapse models (Goldreich and Weber 1980) or accretion onto a compact star (Bondi 1952).

2.2 Non-ideal Equations of State

More realistic equations of state have to be determined from the microscopic properties of matter (and radiation) by employing the methods of quantum mechanics (or quantum field theory), statistics, and atomic, molecular, and nuclear physics. This is usually carried out by assuming a state of complete thermodynamic equilibrium between all particles involed, i.e., atoms, ions, photons, electrons, protons, neutrons, etc. The justification of this assumption is due to the small mean free paths of the microscopic particle interactions compared to the length scales for the change of macroscopic variables such as the pressure scale height $H_p = dr/d\ln(p)$. Depending on the choice of independent variables, the thermodynamic equilibrium is given by minimizing some appropriate thermodynamic function. For a given temperature T and volume V containing a fixed number N_s of particles of various species s the matter will adopt a state that minimizes the free energy

$$F = F(T, V, \{N_s\}) = E - TS , \tag{2.11}$$

where E and S denote the energy and the entropy, respectively. F is connected to the partition function $\mathcal{Z}$ of the entire system

$$F = -kT \ln \mathcal{Z} , \tag{2.12}$$

and the pressure follows from

$$p = \left(\frac{\partial F}{\partial V} \right)_{T, \{N_s\}} . \tag{2.13}$$

Thus, in order to derive an EOS it is most important to obtain an accurate calculation of the energy E. This is only trivial for non-interacting particles, and in general the energy of a complicated interacting many-body system has to be estimated. A detailed discussion for the regime of nuclear densities including nucleon–nucleon interactions is presented in Chap. 3. At somewhat lower densities ($\rho < 10^7$ g cm^{-3}) the pressure is determined by a (degenerate) electron gas interacting with positive ions. The exact solution of Schrödinger's equation for such a many-electron system is impossible to obtain. Thus, for practical purposes the electron interaction will be approximated by some effective potential acting on a single electron state. In the Thomas–Fermi method the electron density is expressed in terms of this average potential, which on the other hand has to be calculated from Poisson's equation with a source term that is due to the charge density of a nucleus with charge eZ surrounded by Z electrons. This method has been extended by Dirac to include the effects of the exchange term in the electron interactions. From that, Feynman, Metropolis, and Teller (1949) calculated equations of state at $T = 0$, and obtained electrostatic corrections to the pressure with respect to the ideal Fermi gas. This method delivers reasonable results for pressures that are not too low ($p > 10^{13}$ dyn cm^{-2}); the zero-pressure densities are, however, much too small. This can be partly corrected by including the contributions of electron correlation interactions in the total energy. Figure 2.2 shows the corresponding EOS calculations by Salpeter and Zapolsky (1967) for an assortment of elements together with experimental data.

Below $p \approx 10^{12}$ dyn cm^{-2} where outer electronic structure effects become important, the Thomas–Fermi–Dirac approach is definitely an oversimplification, and more sophisticated methods of modern molecular and solid-state physics have to be applied [see Bukowinski (1994) and the references therein]. An approach used widely for this pressure regime, which is essential for the internal structure of the Earth and other planets, is that by Kohn and Sham (1965) based on the density-functional theory. As in the Thomas–Fermi method the total energy of the many-electron system is expressed in terms of the electron density

$$\rho_e = \sum_j |\Phi_j(r)|^2 , \tag{2.14}$$

where Φ_j are single-electron wavefunctions. Kohn and Sham (1965) provided a total energy functional

$$E[\rho_e] = -\frac{1}{2} \sum_j \int \Phi_j^* \Delta \Phi_j \, dr + \sum_i \int \frac{Z_i \rho_e(r)}{|r - r_i|} \, dr$$

$$+ \frac{1}{2} \int\int \frac{\rho_e(r)\rho_e(r')}{|r - r'|} \, dr dr' + E_{\text{nuc}} + E_{\text{xc}}[\rho_e(r)] , \tag{2.15}$$

where E_{nuc} is the Coulomb interaction energy of the nuclei, $E_{\text{xc}}[\rho_e(r)]$ is a functional of the density ρ_e that includes the exchange and correlation energies. If this functional were known excactly, all exchange and correlation

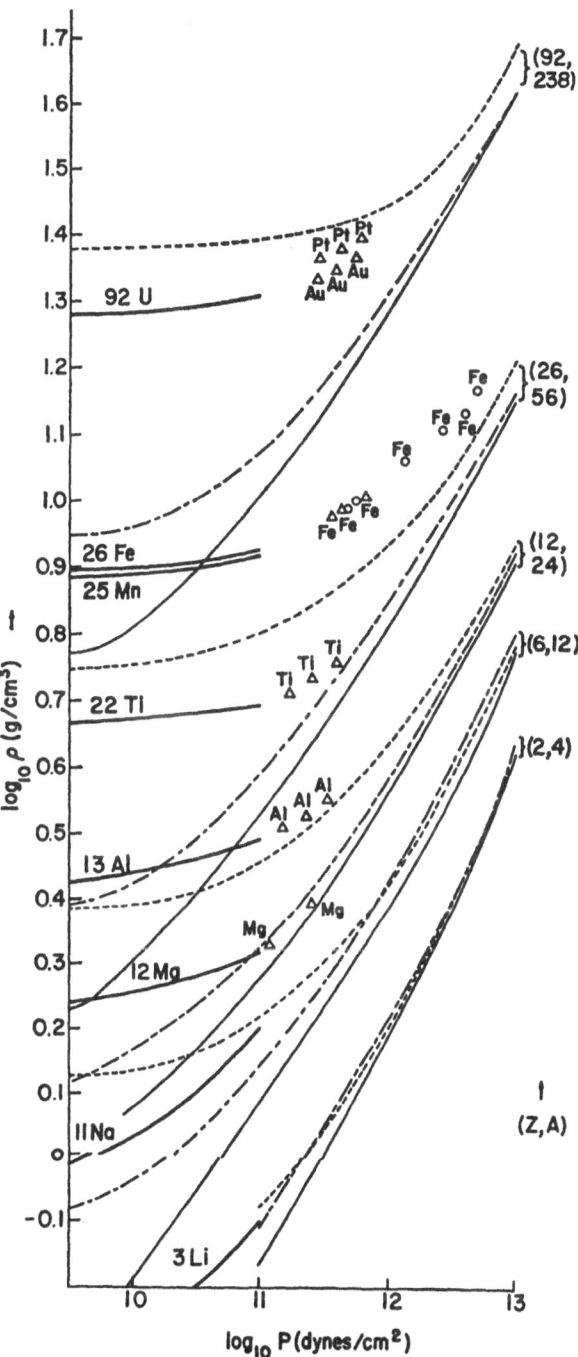

Fig. 2.2. Equations of state for various elements. Dots correspond to data from shock-wave experiments, solid curves on the left are from static-pressure experiments. Curves on the right are theoretical calculations based on a generalized Thomas–Fermi–Dirac model [from Salpeter and Zapolski (1967)].

effects would be fully taken into account. The problem is to find a reliable approximation for the exchange part $E_{ex}[\rho_e]$ in this energy functional. The standard procedure [proposed by Kohn and Sham (1965)] used to overcome this problem is the local density approximation, which is based on the idea that $E_{xc}[\rho_e]$ can be calculated for a uniform (interacting) electron gas and used subsequently as a local approximation in (2.15). This latter quantity has meanwhile been calculated very accurately with Monte Carlo simulations (Ceperley and Alder 1980, Perdew and Zunger 1981). With the aid of the local density approximation, a set of single-particle Schrödinger equations is obtained by variation of the total energy (2.15) with respect to Φ_j. These equations are solved by various well established techniques from atomic, molecular, and solid state physics such as LCAO (linear combination of atomic orbitals), APW (augmented plane waves), LMTO (linear muffin-tin orbitals), periodic Hartee–Fock, and so on. Applying these methods to EOS calculations of simple minerals yields results that are in very good agreement with experimental data (see Fig. 2.3). In order to obtain the EOS of more complex minerals

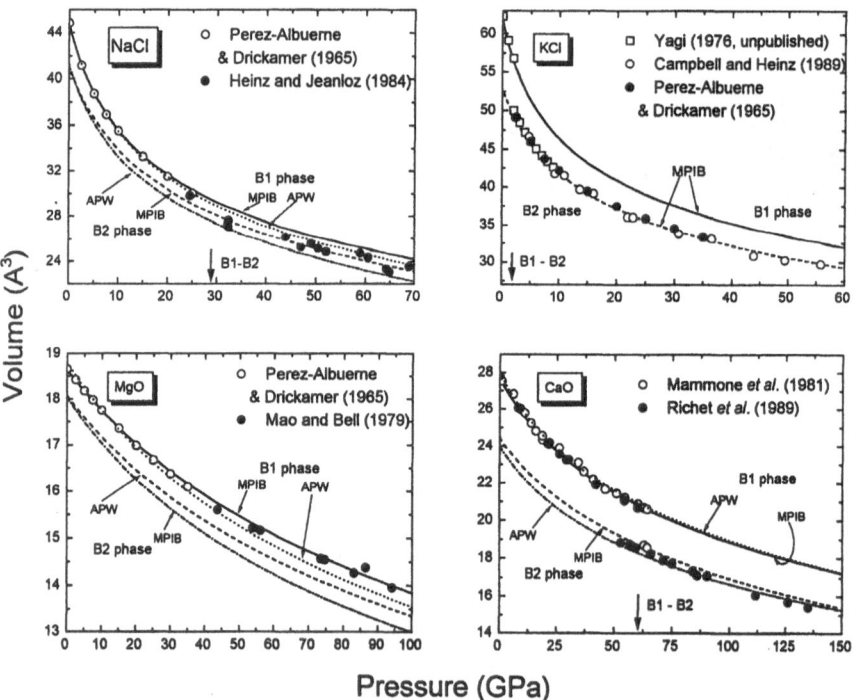

Fig. 2.3. Equations of state: experimental data (dots) and theoretical calculations (lines) from different models for band-structure calculations. The vertical arrows point at calculated pressures of crystal phase transitions. The unit of pressure is $1\ \text{GPa} = 10^{10}\ \text{dyn cm}^{-2}$. The figure is taken from Bukowinski (1994).

that are, for example, relevant for the lower Earth mantle, several specially designed methods have been developed within the last few years [for details see Zhang and Bukowinski (1991)]. Fig. 2.3 shows a comparison of calculated EOSs with laboratory high-pressure experiments. Again, the theoretical models produce quite accurate results.

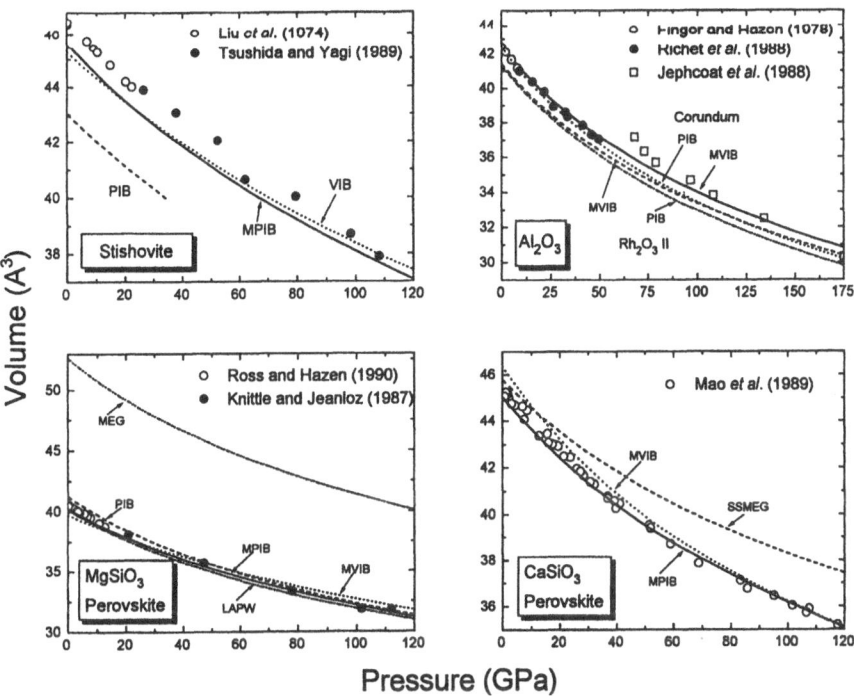

Fig. 2.4. Equation of state: experimental data (dots) and theoretical calculations (lines) for minerals under Earth mantle conditions. The various acronyms indicate the theoretical models used for the calculations [from Bukowinski (1994)].

The $T = 0$ approach for calculating the properties of high-density matter is sufficient for determining the structure of white dwarfs. However, for evolutionary scenarios connected to stellar cooling theories finite temperature effects have to be included. Thus, the free energy (2.11) is choosen to contain various additional contributions such as temperature corrections for the ideal Fermi gas of electrons, or the energy of phonons in an ion lattice (Segretain et al. 1994).

The physical conditions in stellar envelopes require the computation of the EOS at high temperatures and low densities, and a first approximation is given by the ideal gas law (2.5). The mean molecular weight μ_m can be expressed in terms of the numbers N_s of the various atomic and molecular species summed over all ionization states

$$\mu_{\mathrm{m}} = \frac{1}{m_{\mathrm{u}}} \frac{\sum_s m_s N_s}{N_{\mathrm{e}} + \sum_s N_s} , \tag{2.16}$$

where m_s is the particle mass, and N_{e} denotes the number of free electrons in the volume V. Due to global charge neutrality of the plasma, N_{e} is given by the numbers $N_{s,i}$ of ions of species s with ionization state i and charge $Z_{s,i}$ (in units of the electronic charge e):

$$N_{\mathrm{e}} = \sum_s \sum_i Z_{s,i} N_{s,i} . \tag{2.17}$$

Thus, the mean molecular weight (2.16) and the EOS (2.5) depend on the ionization state of the matter which can be expressed by a set of Saha equations

$$\frac{N_{s,i+1} N_{\mathrm{e}}}{N_{s,i} V} = \frac{2 U_{s,i+1}}{U_{s,i}} \left(\frac{2\pi m_{\mathrm{e}} kT}{h^2} \right)^{3/2} \exp(-\chi_{s,i}/kT) , \tag{2.18}$$

where m_{e} is the electron mass, $\chi_{s,i}$ is the ionization potential, and $U_{s,i}$ denotes the internal partition function for an ionization state i

$$U_{s,i} = \sum_j g_{s,i}^{(j)} \exp\left(-E_{s,i}^{(j)}/kT\right) . \tag{2.19}$$

The sum extends over all internal quantum states (bound states) of the ion with energy $E_{s,i}^{(j)}$ and statistical weight (degree of degeneracy) $g_{s,i}^{(j)}$. For a gas of non-interacting particles this sum diverges since $E_{s,i}^{(j)}$ approaches the finite ionization energy for large j, and $g_{s,i}^{(j)}$ generally increases with j. This problem is usually cured by truncating the series at some $j = j_{\max}$, because the highly excited states are sensitive to perturbations from the surrounding plasma particles, and are therefore easily depopulated. This effect results in a pressure induced ionization of the plasma and leads at high densities to a fully degenerate electron gas moving in an ion lattice, which has been discussed above. In order to render the sum (2.19) finite, atomic models have been proposed where the long-range Coulomb potential is modified to include interactions with other particles. The simplest method consists of an infinite potential wall at some characteristic radius, such as the interparticle separation (Graboske et al. 1969), to simulate hard-sphere collisions between the particles. This confined atomic model is a reasonable approach for neutral particles interactions. Charged-particle perturbations have been investigated by Rogers et al. (1970) using the static screened Coulomb potential

$$\frac{1}{r} e^{-r/\lambda_{\mathrm{D}}} , \qquad \lambda_{\mathrm{D}} = \sqrt{\frac{kTV}{4\pi e^2 \sum_{s,i} N_{s,i} Z_{s,i}^2}}$$

(with the Debye length λ_{D}) to solve Schrödinger's equation for hydrogen-like atoms. With this model the sum in the partition function (2.19) truncates itself at a maximum principal quantum number less than

$$\left(1.2701 \frac{Z_{s,i}\lambda_D}{a_0} - 0.1045\right)^{1/2} ,$$

where $a_0 = h^2/(4\pi^2 m_e e^2)$ denotes the Bohr radius (h is Planck's constant). Both the confined atom model and the screened Coulomb model can produce physically unreasonable discontinuities in the partition function because bound states are moved in and out of the sum (2.19) depending on the temperature and density. Hummer and Mihalas (1988) have therefore developed a method that leads to continuous and smooth changes in the free energy. The basic idea is that an atom with bound states is exposed to a fluctuating electric field generated by the surrounding charged particles. This will eventually lead to a field ionization of atoms in highly excited states, and the details will depend somewhat on the choice of the distribution function for the fluctuating microfield.

Any method that leads to a finite expression (2.19) takes into account many-particle interactions in one way or another, and therefore includes non-ideal effects in the free energy. EOS calculations for the low-density and high-temperature regime have been performed by Graboske et al. (1969) and Mihalas et al. (1988) using an expression for the free energy F that includes the translational energy of classical point particles, the contributions from bound states through an internal partition function that is kept finite by some appropriate formalism, the energy from an idealized partially degenerate electron gas, and a non-ideal term due to the interactions of charged particles via Coulomb collisions. Once F has been set up, the chemical potentials $\mu_{s,i}$ for each ionic species are obtained:

$$\mu_{s,i} = \left(\frac{\partial F}{\partial N_{s,i}}\right)_{T,V} , \tag{2.20}$$

and the equilibrium distribution between the various species follows from appropriate stoichiometric relations. For example, for an ionization process from state i to $(i+1)$ we have

$$\mu_{s,i} = \mu_{s,i+1} + \mu_e . \tag{2.21}$$

The minimization of F subject to these constraints in order to find the thermal equilibrium state of matter can only be done numerically. As a result, due to the non-ideal effects the various ionization fractions deviate strongly from the Saha equilibrium (2.18), especially for more dense plasmas (Mihalas et al. 1988). This approach has been applied successfully for densities up to $1\,\mathrm{g\,cm}^{-3}$.

For even higher densities a many-particle approach should be used which not only considers the plasma as a mixture of neutral and ionized atoms and electrons, but takes into account the collective degrees of freedom such as dynamic screening, phase space occupation, etc. From that, an interpolation between the region of the non-ideal plasma and the solid state is possible (Ebeling et al. 1991). A systematic quantum-mechanical approach to hot and dense plasmas (Kraeft et al. 1986) has been formulated by employing Green's function techniques which leads to a consistent description of the composition, ionization degree, pressure ionization, etc. of the plasma [for a review see Van Horn and Ichimaru (1993)]. Furthermore, phase instabilities connected with the change of the ionization degree (Mott transitions) have been investigated. In very dense plasmas the quasi-particle picture is usually replaced by spectral-function methods to calculate their thermodynamic, transport, and optical properties (Günter et al. 1991, Reinholz et al. 1995).

2.3 Stellar Energy Sources

We now return to the overall problem of stellar structure. As mentioned above, for white dwarfs and neutron stars the knowledge of the EOS at $T = 0$ is sufficient to solve the hydrostatic structure equations (2.1) and (2.2). In all other cases more relations are needed to determine the thermal state of the stellar interior. Obviously, a stationary state can only be maintained if the star is in a thermal balance between energy production and energy loss. Let F_r be the radial component of the energy flux density vector $\boldsymbol{F}$, then

$$\nabla \cdot \boldsymbol{F} = \frac{1}{r^2} \frac{\mathrm{d}(r^2 F_r)}{\mathrm{d}r} = \rho \varepsilon \;, \tag{2.22}$$

or in terms of the luminosity $\ell = 4\pi r^2 F_r$ through a spherical shell of radius r

$$\frac{\mathrm{d}\ell}{\mathrm{d}r} = 4\pi r^2 \rho \varepsilon \;. \tag{2.23}$$

Here ε stands for the energy released per unit mass per second. This energy source is due to nuclear reactions deep in the interior, and thus depends on the temperature, the density, and the abundance of the relevant nuclear species. For main sequence stars such as the sun, energy is delivered by the fusion of hydrogen to form helium. According to the basic picture of the hydrogen-burning process, four protons are converted into a helium nucleus, ^{4}He:

$$4\mathrm{p} \longrightarrow {}^4\mathrm{He} + 2\mathrm{e}^+ + 2\nu_{\mathrm{e}} + 26.731 \text{ MeV} \;, \tag{2.24}$$

where e^+ and ν_{e} denote the positron and the electron neutrino, respectively. Since the probabilty of a simultaneous collision of four protons is practically zero, the actual fusion will proceed through a series of binary collisions. There are two different possible fusion paths, the p–p chain and the CNO cycle, each of which consists of a number of alternative branches. The first step in the p–p chain reaction produces a deuterium nucleus ^{2}H from two protons:

$$\mathrm{p} + \mathrm{p} \longrightarrow {}^2\mathrm{H} + \mathrm{e}^+ + \nu_{\mathrm{e}} \;. \tag{2.25}$$

Because of the β decay involved, which is a weak interaction process, this is a very slow reaction having a time scale of about 10^{10} years for conditions encountered in the solar interior. It therefore prohibits the rapid consumption of stellar nuclear fuel and determines the lifetimes of stars. Next, deuterium–proton collisions produce ^{3}He nuclei, and then the p–p chain is completed along the main reaction branch by ^{4}He being produced from two ^{3}He nuclei. A completely different and more indirect path is the CNO cycle. It requires some carbon, nitrogen, and oxygen to act as catalysts during a sequence of collions with protons and the subsequent β decays [for details on the hydrogen burning processes see Rolfs and Rodney (1988), or Hansen and Kawaler (1994)]. The relative importance of the two fusion paths depends of course on the abundance of C, N, and O, but most noticable is the strong temperature dependence. A comparison of the energy release ε between the p–p chain and the CNO cycle is shown in Fig. 2.5, and it turns out that for the temperature at the center of the sun ($T = 1.6 \times 10^7$ K) the p–p chain is the dominant energy-producing process. Both processes drop with decreasing temperature because the thermal energy of the particles becomes insufficient to overcome their mutual Coulomb repulsion. Explicit expressions for $\varepsilon(\rho, T)$ can be found in Kippenhahn and Weigert (1990). Note that the hydrogen burning processes, including all reaction branches, produce neutrinos, which will escape from the star without further interactions. This leads to some energy loss that has to be subtracted from the production rate ε.

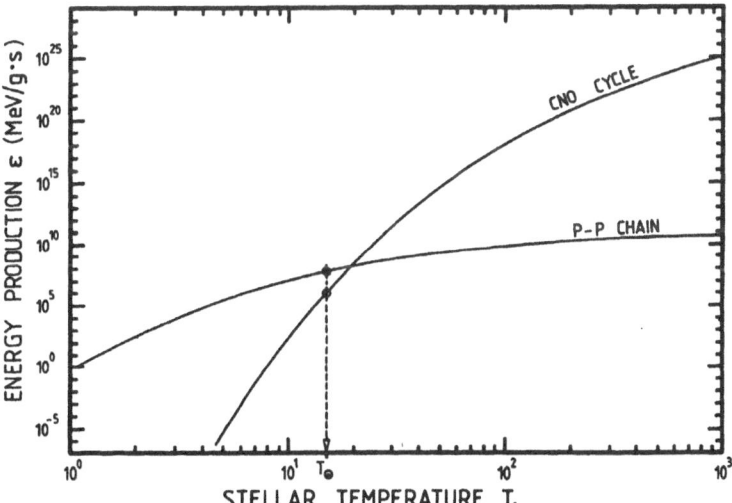

Fig. 2.5. Energy production rate as a function of temperature for the p–p chain and the CNO cycle [from Rolfs and Rodney (1988)].

2.4 Energy Transport

Now, we still do not have enough equations to determine the stellar structure, because even for a fixed composition and a given function $\varepsilon(\rho, T)$ there are only four equations (2.1), (2.2), (2.3), (2.23) to determine the five quantities $m(r)$, $p(r)$, $\rho(r)$, $T(r)$, and $\ell(r)$. The missing relation is an energy transport equation that describes how energy finds its way out from the interior to the surface, where it is radiated away. Several transport mechanisms have to be considered: radiative diffusion, thermal conduction, and convection.

Radiation and matter exchange energy and momentum through the emission, absorption, and scattering of photons, and the latter two processes can be desribed by the attenuation of the radiation intensity I_ν (at frequency ν) along some path length ds:

$$dI_\nu = -\kappa_\nu \rho I_\nu ds \ . \tag{2.26}$$

Here κ_ν is the opacity (with dimension $cm^2 g^{-1}$). One particularly simple source of opacity is the scattering of photons with free electrons, and in the low-energy limit ($h\nu \ll m_e c^2$) we have

$$\kappa^{es} = \sigma_T \frac{n_e}{\rho} \ , \tag{2.27}$$

where $n_e = N_e/V$ is the free electron density, and

$$\sigma_T = \frac{8\pi}{3} \left(\frac{e^2}{m_e c^2} \right)^2 = 6.65 \times 10^{-25} \ cm^2 \tag{2.28}$$

is the Thomson scattering cross section, which is independent of frequency (c indicates the speed of light). The following radiation processes are named according to the states of the electrons involved. Free-free transitions refer to scattering events where an electron in an initial free (unbound) state collides with an ion and is transfered to a final free state through the emission or absorption of a photon. The transition probability depends on the photon frequency, and for a thermal distribution of electrons the free-free opacity reads

$$\kappa^{ff} \propto \frac{Z^2}{\nu^3 \sqrt{T}} \left(1 - e^{-h\nu/kT} \right) \ . \tag{2.29}$$

If in either its initial or final state the electron is bound to an atom, ion, or molecule this process is called a bound-free transition: a photon can transfer its energy to a bound electron that then becomes a free particle by increasing the ionization state of matter, or a free electron can recombine with an ion, emitting a photon. The energy balance for this process is given by

$$h\nu = E_{s,i}^{(\infty)} - E_{s,i}^{(j)} + \frac{m_e}{2} v^2 \ , \tag{2.30}$$

where v is the electron velocity. If the photon energy is less than the amount needed to overcome the ionization energy of level j, the transition probabilty is zero for this energy state. It falls in a manner similar to the free-free expression (2.29) above the threshold. The overall opacity therefore shows a saw-tooth structure with peaks at the ionization energy of each level j.

Bound-bound transitions occur if an electron remains in a bound state before and after the interaction with a photon. Radiation is then absorbed or emitted only at discrete frequencies, or lines:

$$h\nu_{jk} = E_{s,i}^{(j)} - E_{s,i}^{(k)} \; . \tag{2.31}$$

These lines are, however, not infinitesimally narrow because of a variety of line-broadening mechanisms such as the finite lifetime of excited states, Doppler shifts from the thermal motion of the ions, collisions between particles, or broadening due to the Stark effect from the electric field of plasma particles.

It is now obvious that the most general calculation of opacities for the above radiation processes requires knowledge of the numbers of ions in each energy state, i.e., the occupation numbers for the energy levels, and the spectral properties of the radiation field. Such large-scale computations that take into account the spectral transfer of radiation coupled to the state of matter have to be done, for example, in the detailed modelling of stellar atmospheres or low-density winds from hot stars. The availiability of high-resolution spectrographs and the enormous increase in computer power over the last ten years have led to considerable improvements in quantitative spectroscopy, where observed spectra are fitted to model calculations including up to some 10^5 spectral lines from H, He, C, N, Si, or Fe in various ionization states, etc. (Kudritzki et al. 1991). The situation in stellar interiors is fortunately much easier. First of all, radiation and matter are in thermal equilibrium, at least locally, and second, the region is optically thick, i.e., the mean free paths of the photon-matter interactions are much smaller than the typical macroscopic length scales. This means photons undergo a large number of absorption, emission, and scattering processes, and they progress towards the stellar surface in a diffusive way. The energy flux then follows the local gradient of the radiation pressure:

$$F_{\text{diff}} = -\frac{c}{\kappa_{\text{R}}\rho}\frac{\mathrm{d}p_{\text{rad}}}{\mathrm{d}r} = -\frac{4ac}{3\kappa_{\text{R}}\rho T^3}\frac{\mathrm{d}T}{\mathrm{d}r} \tag{2.32}$$

where κ_{R} denotes the Rosseland mean opacity (Mihalas 1978),

$$\frac{1}{\kappa_{\text{R}}} = \frac{\int_0^\infty (\kappa_\nu + \kappa^{\text{es}})^{-1} \frac{\partial B_\nu}{\partial T} \, \mathrm{d}\nu}{\int_0^\infty \frac{\partial B_\nu}{\partial T} \, \mathrm{d}\nu} \; , \tag{2.33}$$

and $B_\nu(T)$ is the Planck spectrum, i.e., the frequency distribution of photons in thermal equilibrium,

$$B_\nu(T) = \frac{2h}{c^2}\frac{\nu^3}{\mathrm{e}^{h\nu/kT} - 1} \; . \tag{2.34}$$

The opacity κ_ν in (2.33) contains the sum over all bound-bound, bound-free, and free-free processes. As in the equation of state (2.3), κ_R is a function of ρ, T, and the composition of matter

$$\kappa_R = \kappa_R(\rho, T, \text{composition}) , \qquad (2.35)$$

and in fact, both function are based on a common set of atomic data, that is the various energy states $E_{s,i}^{(j)}$ and statistical weights $g_{s,i}^{(j)}$. In contrast to the EOS, however, for κ_R transition probabilities between the energy levels are needed. For practical stellar-structure calculations, opacities are access-able through a number of tables. The most widely used opacity tables were produced by the Los Alamos Scientific Laboratory (Huebner et al. 1977), but more recently scientists from various countries have been participating in "The Opacity Project" to re-calculate accurate bound-bound and bound-free transition probabilities by employing state-of-the-art theoretical calculations (Seaton 1987).

Another energy transport mechanism that has to be considered is thermal heat conduction, and the main contribution in a plasma is due to the scattering of electrons with ions. As in the case of photons in an optically thick medium, thermal conduction is a diffusive process, and the flux is again given by the temperature gradient

$$F_{\text{cond}} = -k_c \frac{dT}{dr} \qquad (2.36)$$

with the conductive coefficient k_c. From transport theory k_c can be expressed in terms of the mean free path λ_{coll} for the collision of an electron with an ion

$$k_c \approx c_p \, n_e \bar{v} \lambda_{\text{coll}} \qquad (2.37)$$

where c_p is the specific heat and $\bar{v}$ is the mean (thermal) electron velocity. For a high-temperature non-degenerate electron gas λ_{coll} is much smaller than the photon mean free path, and thermal conduction can be safely neglected. Defining a conductive opacity κ_C through

$$\kappa_C = \frac{4acT^3}{3k_c\rho} , \qquad (2.38)$$

allows us to give a unified descripton of radiative and conductive energy transport by replacing κ_R in equation (2.32) by κ_{tot}, where

$$\frac{1}{\kappa_{\text{tot}}} = \frac{1}{\kappa_R} + \frac{1}{\kappa_C} . \qquad (2.39)$$

For high densities when the electrons are degenerate, thermal conductivity be-comes the dominant transport mechanism. The mean velocity $\bar{v}$ is essentially given by the Fermi momentum ($\bar{v} \approx p_F/m_e$ for non-relativistic electrons), and λ_{coll} is large because in a scattering event most of the final states are already occupied. Thus, only the electrons close to the Fermi surface will contribute to the transport of heat. The thermal structure in the interior of white dwarfs is therefore dominated by a large thermal conductivity leading to an almost isothermal core, because even small temperature gradients result in a large heat flux.

A completely different energy transport process is related to the macroscopic motion of gas inside the star. Such an internal flow structure clearly violates our basic assumption of a hydrostatic equilibrium (2.1) and is caused by an unstable temperature distribution in some portions of the stellar interior. A simple stability criterion is due to Karl Schwarzschild (see Kippenhahn and Weigert 1990): consider a blob of gas that is displaced from its original location at radius r to a new position $r + \Delta r$. This has to be done slowly in order to maintain a pressure equilibrium with the environment. Further assume (for simplicity) that the displacement is carried out adiabatically, i.e., without exchange of heat. Now compare the density inside the blob ρ_b with the surrounding density ρ_s. If $\rho_b > \rho_s$ the blob will return towards its initial position, for $\rho_b < \rho_s$ there is a buoyancy force that will lift the blob further up, and the entire situation is unstable. The criterion of stability is usually written as

$$\nabla < \nabla_{ad} \tag{2.40}$$

where ∇ is defined as

$$\nabla = \frac{d \ln T}{d \ln p} = -\frac{r^2}{Gm(r)} \frac{p}{\rho T} \frac{dT}{dr} ; \tag{2.41}$$

the last expression follows from (2.1). ∇_{ad} corresponds to the value of ∇ for adiabatic conditions and follows from pure thermodynamic calculations including the EOS and the first law of thermodynamics. Once the actual temperature gradient is sufficiently large, and (2.40) is violated, the gas aquires a macroscopic convective motion that transports energy much more efficiently than radiative diffusion. For a quantitative analysis of this mechanism an expression of the convective flux F_{conv} is required. This is derived from the following picture of convective motion, called the mixing length theory. The basic idea is that the flow consists of rising blobs of various sizes and velocities that keep their identity over some characteristic distance l_m (the mixing length) and then blend into the surrounding matter. Analogous to the diffusive heat transfer in gases, the convection can be considered as a "gas" of blobs with a mean free path l_m, and an average velocity that can be estimated from the buoyancy force. The details of the calculations can be found in Cox and Giuli (1968), Kippenhahn and Weigert (1990), or Hansen and Kawaler (1994), and the result is

$$F_{conv} \propto \rho T \frac{l_m^2}{H_p^{3/2}} \left(\nabla - \nabla_{ad} \right)^{3/2} . \tag{2.42}$$

This can be improved further by taking into account that the blobs will not rise adibatically but rather exchange energy with the environment; then ∇_{ad} has to be replaced by some more appropriate gradient. In the dense inner parts of the stars the use of ∇_{ad} is a very good approximation. The difference

$\nabla - \nabla_{ad}$ turns out to be very small in this regime, i.e., convection establishes essentially an adiabatic temperature profile. The theory does not, however, deliver a rigorous estimate of the mixing length itself. It is assumed to be of the order of the pressure scale height H_p, but the exact value of the ratio l_m/H_p has to be taken as a free parameter in stellar-structure calculations. Note that the effects of convection have been considered only with respect to the energy transport. The associated transport of momentum would contribute to the hydrostatic equation (2.1). It can be shown, however, that this is a negligible correction as long as the fluid motion is much smaller than the sound velocity $\sqrt{dp/d\rho}$.

The set of equations needed to compute the (static) stellar structure is now complete, and it can be solved numerically in order to calculate stellar models in mechanical and thermal equilibrium. The properties of matter enter through the equation of state, the opacity, and the nuclear energy production rates, and all these functions depend on the chemical composition of the stellar material. The equations have to be solved together with appropriate boundary conditions. In addition to (2.8),

$$\ell(r=0) = 0 \tag{2.43}$$

is required at the center. In contrast to the zero-pressure condition at $r = R$, which is useful for compact stars, some other condition has to be taken here because it is not obvious where to put the stellar surface, and one has to include a stellar atmosphere in the outer layers of the star; for details on this photospheric boundary condition see Hansen and Kawaler (1994).

3. Nuclear Equations of State

3.1 General Remarks

As early as in 1936 F. Hund tried to obtain an estimate of the properties of baryonic matter at very high pressure and densities, as should exist in the interior of neutron stars. At that time any estimate of these properties had been rather courageous and even today, 50 years later, we are not yet able to determine the equation of state of baryonic matter at such densities in a doubtless way. However, during all these years physicists have achieved at least one important step of progress, as we are able now to phrase the basic ingredients for a microscopic theory of baryonic matter. It is generally accepted that such a microscopic theory should be based on quarks as the elementary building blocks of protons and neutrons and that the strong interaction of these quarks is described by quantumchromodynamics (QCD).

Knowing the basic theory of interacting baryons, however, does not imply that one can evaluate the energy of a system of interacting protons and neutrons. QCD represents a non-abelian gauge theory, and processes involving large energy and momentum transfers can be evaluated by applying the techniques of perturbation theory as have been developed for quantumelectrodynamics (QED). For systems at low energies (the systems of baryonic matter in stellar objects are systems of low energy densities on the scale of QCD) the perturbative treatment is not applicable. Therefore particle physicists have not yet succeeded in deriving the confinement of quarks in hadrons from QCD. This implies that one cannot yet describe properties of single nucleons from the basic theory of strongly interacting particles. Therefore it is currently also impossible to evaluate from the basic theory the properties of nucleons in the presence of external pressure.

A rather pragmatic way out of this problem would be to assume, on purely phenomenological grounds, an interaction between two nucleons, adjusting the parameters of these interactions in such a way that simple many-body calculations, such as the Hartree–Fock approximation, with these interactions would reproduce basic experimental properties such as the binding energies and the radii of nuclei. Typical examples of such phenomenological nucleon–nucleon (NN) interactions are the various versions of the so-called Skyrme interaction (Skyrme 1959, Friedrich and Reinhard 1986) or the interaction as defined in the relativistic Walecka model (Walecka 1974, Serot and Walecka 1986).

Within this phenomenological approach one obtains quite an accurate description for the bulk properties of nuclei. This implies that these models provide a description of nuclear matter at normal densities. Here and in the following the name *nuclear matter* identifies an artificial system invented by theoretical physicists. One considers a homogeneous infinite system of protons and neutrons, with the number of protons identical to the number of neutrons, interacting only by means of the strong interaction, which means that the Coulomb repulsion between protons is "switched off". This system corresponds to a nucleus without a surface and with no Coulomb interaction. Therefore the "experimental" value for the binding energy of this system can be deduced from the volume term in the empirical Bethe–Weizsäcker mass formula (von Weizsäcker 1935, Bethe and Bacher 1936, Ring and Schuck 1980) for the binding energy of nuclei. Furthermore one can derive the density of nuclear matter without any external pressure from the nucleon density in the interior of large nuclei. This means that one knows the minimum of the energy–versus–density plot of nuclear matter, which is sketched in Fig. 3.1 from empirical data. The energy of this saturation point saturation point of nuclear matter is around −16 MeV per nucleon at a density of 0.16 nucleons per fm^3.

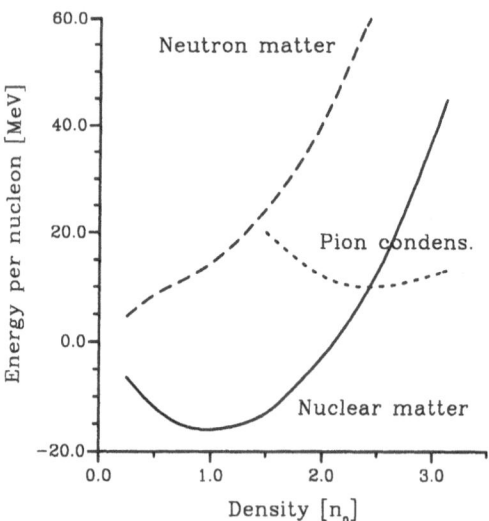

Fig. 3.1. Schematic plot of the energy per nucleon as a function of the density for nuclear matter (solid line) and neutron matter (dashed line). The only empirical information available is the saturation point of nuclear matter: the minimum of the corresponding curve at $n_0 = 0.16$ nucleon per fm^3. Also sketched is the energy of a pion condensate (dotted curve), which has been speculated to become the phase dominating nuclear matter at high densities.

Furthermore, one can try to get an estimate of the curvature of this energy–versus–density curve, which is related to the incompressibility of nuclear matter by

$$K = 9n_0^2 \frac{\partial^2 E}{\partial n^2}\bigg|_{n=n_0} , \tag{3.1}$$

from an investigation of the so-called "breathing-mode" vibrational excitations of nuclei. The excitation energies of the breathing-mode states are rather sensitive to the shell structure, i.e., the finite size, of real nuclei. Therefore the compressibility of nuclear matter cannot be deduced from these experimental data in a model-independent way and the estimates for K vary between 150 MeV and 300 MeV. Models that lead to a low value for the compressibility are said to produce a *soft* equation of state, whereas those predicting a large value for K produce a *hard* or *stiff* equation of state.

What we are interested in, however, are the properties of nuclear matter not at the saturation density n_0 without external pressure, but with external pressure due to the gravity. Such a large pressure may compress the matter up to five times this normal density. Furthermore one cannot "switch off" the Coulomb repulsion between protons, which means that compact stars at these densities will mainly consist of neutrons and we need to determine the equation of state for neutron matter rather than for symmetric nuclear matter (see also Fig. 3.1).

It is evident that phenomenological models for the NN interaction, which are simply adjusted to describe the properties of symmetric nuclear matter at normal saturation density n_0, cannot provide reliable estimates for the properties of neutron matter at much higher densities. Furthermore, it is obvious, e.g., from the results of nucleon knock-out experiments, that nuclei are rather complicated many-body systems that cannot be described in terms of a simple mean-field or Hartree–Fock description, as is typically associated with these phenomenological NN forces. As we will discuss more in detail below, a treatment of correlation beyond the Hartree–Fock approach is required to obtain a reliable description of the nuclear many-body problem.

For these reasons, we will concentrate our discussion of nuclear equations of state on those obtained with so-called "realistic" models for the NN interaction. Here the label "realistic NN force" is used to identify such NN interactions that reproduce all the observables of two isolated nucleons. These observables include the energy and electromagnetic properties of the bound two-nucleon system, the deuteron, as well as the phase shifts of NN scattering up to energies of a few hundred MeV. Such realistic NN forces can be based on a pure phenomenological ansatz for a local shape of a NN potential, a typical example being the potential defined by Reid (1968), on the assumption that the NN interaction is mediated by the exchange of mesons (see, e.g., Machleidt 1989) or on an effective quark model. The discussion here will concentrate to a large extent on the meson-exchange or on one-boson-exchange (OBE) models of the NN interaction since these models yield very accurate descriptions of the experimental NN scattering data and those of the deuteron with the adjustment of only a few parameters.

After the NN interaction has been determined from such a fit of the two-nucleon data, what remains to be done is a solution of the many-body problem of an infinite number of nucleons, forming a translationally invariant system of baryon density n, interacting with such a realistic NN interaction. The hope is that such a many-body calculation will reproduce the properties of nuclear matter around the saturation density without adjusting any additional parameters. This would imply that one has developed a tool which allows the prediction of baryon matter at normal density n_0 from the properties of nucleons interacting in the vacuum ($n = 0$). Such a theory should also provide reliable predictions for the properties of baryon matter up to densities that are a few times higher than the saturation density n_0.

The procedure just outlined defines the classical or *conventional* model of baryon matter. The basic assumption of this conventional model is that the nucleons are the basic building blocks of baryon matter and do not change their properties in the medium of baryon matter. One assumes that the bare interaction between the nucleons in the medium is identical to the interaction of two nucleons in the vacuum and that no three-nucleon or other many-body forces exist. In order to demonstrate that these basic assumptions of the conventional model for baryon matter could easily be violated, one should recall the typical length scales for the system. The radius of the nucleon, which is around 0.8 fm, is of the same order of magnitude as the average distance in between a nucleon and its next neighbor, which is around 1.9 fm for nuclear matter at the saturation density. This demonstrates that nuclear matter is a very dense system and one can easily imagine that the quark distribution in one nucleon is affected by the presence of the other nucleons. This means that one may not be allowed to consider nucleons as inert particles. Instead one has to consider possible excitation modes for the nucleons and their modifications at high densities. As we will see later in this chapter, this gives rise to many-body forces. The medium of baryonic matter could as well affect the properties and propagation of the mesons, which have radii quite similar to those of the baryons. The effective mass of the mesons in a nuclear system could be different from the mass in the vacuum. This would imply that the range of the NN interaction, described in terms of meson exchange, would be different in nuclear matter as compared to the vacuum, which leads to a density dependence of the NN interaction, which is again a mechanism beyond the conventional model.

It has been speculated that the modification of the meson masses in the medium could be so drastic that at densities larger than the saturation density of nuclear matter one obtains a condensation of pions. This would imply that the equation of state at high densities is no longer represented by the equation of state for normal nuclear matter, but that one obtains a phase transition to a pion condensation (see the schematic plot of Fig. 3.1). The phase transition may also occur for the system of neutron matter. As we will see below, such a phase transition would modify the cooling rate of a neutron star. Therefore a measurement of the temperatures and cooling rates of neutron stars provides very important information on the possible existence of non-conventional baryon matter.

At very high densities the nucleons will overlap with large probability. Presumably this implies that quarks are no longer confined to the quark bags of single nucleons. Therefore one expects for very high densities a phase transition from baryon matter to a quark–gluon plasma. The relevant degrees of freedom used to describe this phase transition are given in terms of the degrees of freedoms of interacting quarks and gluons and it is obvious that the description of this phase transition is beyond the scope of the model for baryonic matter as considered in this chapter.

In order to provide an appropriate nomenclature for the discussion of the meson-exchange model of the NN interaction as well as the presentation of various techniques to be used in the solution of the many-body problem, we present in the second section of this chapter a short introduction to the use of Green's functions in many-body theories. Section 3.3 is devoted to the determination of realistic NN forces. In Sect. 3.4 we discuss some initial results of non-relativistic nuclear matter calculations, employing realistic forces, and in Sect. 3.5 we describe some important effects, which are due to relativistic extensions of the many-body approach. The final section of this chapter is devoted to the discussion of sub-nucleonic degrees of freedom, where we consider possible modifications of the baryons as well as the meson in the nuclear medium and include a discussion of pion condensation.

However, before we turn to the discussion of the open questions about the properties of matter at such very high densities, we want to set up the nomenclature and repeat some well-established features of matter at lower densities. The status of such a system is well defined by specifying the number density in phase space for each species of particle s

$$\frac{\mathrm{d}\mathcal{N}_s}{\mathrm{d}^3 r \mathrm{d}^3 k} = \frac{\gamma_s}{(2\pi)^3 \hbar^3} f_s(\boldsymbol{r}, \boldsymbol{k}) \,, \tag{3.2}$$

where $(2\pi)^3 \hbar^3$ is the volume of a cell in phase space of particles with coordinates $\boldsymbol{r}$ and momenta $\boldsymbol{k}$. In the following we will employ "natural units" with $\hbar = c = 1$, where c is the velocity of light. In (3.2) γ_s denotes the degeneracy of states, which is $\gamma_s = 2S_s + 1$ for massive particles if S_s refers to the spin and f_s denotes the dimensionless distribution function, which for an ideal gas of fermions at temperature T takes the form of the Fermi–Dirac distribution

$$f_s = \frac{1}{1 + \exp\left(\frac{\epsilon - \epsilon_{\mathrm{F}}}{T}\right)} \,, \tag{3.3}$$

with ϵ the energy of a particle with phase-space coordinates $\boldsymbol{r}$ and $\boldsymbol{k}$ and ϵ_{F} the Fermi energy. For sufficiently low particle densities and/or high temperatures the Fermi–Dirac distribution can be very well approximated by the corresponding Maxwell–Boltzmann distribution. We will be more interested in the other extreme, the low temperature or $T = 0$ limit, which yields the distribution of a degenerate system of fermions

$$f_s(\epsilon) = \begin{cases} 1, & \epsilon \leq \epsilon_F, \\ 0, & \epsilon > \epsilon_F. \end{cases} \tag{3.4}$$

In the following we will assume that the energy of a particle increases monotonically with its momentum. In this case and in the limit of a degenerate gas of fermions the number density of each species of particle is given by

$$n_s = \int d^3k \, \frac{dN_s}{d^3r d^3k} = \frac{\gamma_s}{6\pi^2} k_{F_s}^3 , \tag{3.5}$$

with k_{F_s} referring to the Fermi momentum for the species s, i.e. the momentum of particles with the Fermi energy ϵ_F. As an example let us consider the case of a single species of non-interacting fermions, as discussed, e.g., by Chandrasekhar (1931). In this example the single-particle energy is related to the momentum by

$$\epsilon = \sqrt{k^2 + m_s^2} , \tag{3.6}$$

and the energy density is obtained as

$$\rho = \frac{\gamma_s}{2\pi^3} \int_0^{k_F} dk \, k^2 \sqrt{k^2 + m_s^2} , \tag{3.7}$$

while the pressure of a system with an isotropic distribution of momenta is given by

$$\begin{aligned} p &= \frac{1}{3} \frac{4\pi\gamma_s}{(2\pi)^3} \int_0^{k_F} dk \, k^2 \, kv \\ &= \frac{\gamma_s}{6\pi^2} \int_0^{k_F} dk \frac{k^4}{\sqrt{k^2 + m_s^2}} , \end{aligned} \tag{3.8}$$

with the velocity $v = p/\epsilon$ and a factor $1/3$ originating from the isotropy. It is convenient to define a dimensionless Fermi momentum by

$$x = \frac{k_F}{m_s} . \tag{3.9}$$

Note that this parameter x is also a measure for the importance of relativistic effects since these effects can be ignored for $x \ll 1$. With this parameter x, the pressure of (3.8) can be rewritten as

$$p = \frac{m_s^4}{8\pi^2} \left\{ x\sqrt{1+x^2} \left(\frac{2x^2}{3} - 1 \right) - \ln\left[x + \sqrt{1+x^2} \right] \right\} . \tag{3.10}$$

In the non-relativistic regime ($x \ll 1$) this expression can be expanded to

$$p \to \frac{m_s^4}{15\pi^2} \left\{ x^5 + \ldots \right\} , \tag{3.11}$$

whereas in the extreme relativistic domain ($x \gg 1$) one obtains

$$p \rightarrow \frac{m_s^4}{12\pi^2} \left\{ x^4 + \dots \right\} . \tag{3.12}$$

At low densities the pressure is dominated by the electrons and the mass density $\rho_0 = \sum_s n_s m_s$ is dominated by the mass of the baryons. If we assume that there are Y_B baryons per electron in the system, each of them with a mass m_B, this yields

$$\rho_0 = Y_B m_B n_e = Y_B m_B \frac{m_e^3}{3\pi^2} x^3 , \tag{3.13}$$

where we have used (3.5) and the definition of x in (3.9). For $Y_B = 2$, which is typical for light nuclei, this means that the non-relativistic approximation is valid for mass densities ρ_0 smaller than 10^6 g cm^{-3}. For such densities we can rewrite (3.12) as

$$p = \frac{1}{15\pi^2} \left(\frac{3\pi^2}{Y_B m_B} \right)^{5/3} \frac{1}{m_e} \rho_0^{5/3} . \tag{3.14}$$

On the other hand, in the extreme relativistic domain at densities larger than 10^6 g cm^{-3}, (3.12) can be transformed into

$$p = \frac{1}{12\pi^2} \left(\frac{3\pi^2}{Y_B m_B} \right)^{4/3} \rho_0^{4/3} . \tag{3.15}$$

These very simple estimates for the equation of state at lower densities have been corrected to account for the fact that the positive charges are not uniformly distributed but are concentrated in individual nuclei. A discussion of such electrostatic corrections can be found, e.g., in the book by Shapiro and Teukolsky (1983). Another correction, which is particularly important at higher densities, is due to the inverse β decay

$$e^- + p \rightarrow n + \nu_e , \tag{3.16}$$

a process which will occur whenever the electrons have enough energy to compensate for the mass difference between the protons and neutrons: $\delta m = 1.28$ MeV. If we ignore for the moment the fact that the nucleons are bound in nuclei and assume that we can regard the system as a mixture of free protons, neutrons, and electrons, the equilibrium is obtained if the chemical potential or the corresponding Fermi energies fulfill

$$\epsilon_F(p) + \epsilon_F(e) = \epsilon_F(n) ,$$
$$m_p \sqrt{1 + x_p^2} + m_e \sqrt{1 + x_e^2} = m_n \sqrt{1 + x_n^2} . \tag{3.17}$$

Charge neutrality implies that the number density for protons is identical to the one for electrons, or

$$x_e = \frac{m_p}{m_e} x_p . \tag{3.18}$$

This equation can be used to eliminate x_e from (3.17) and one can determine the ratio n_p/n_n. The fraction of protons decreases with increasing density, reaching a minimum of $n_p \approx 0.0032 \, n_p$ at a mass density $\rho_0 \approx 8 \times 10^{11}$ g cm^{-3} and then rises to $n_p/n_n = 1/8$ at higher densities. This simple estimate demonstrates how the inverse β decay converts the dense matter into a baryonic matter dominated by neutrons.

A more quantitative treatment has to account for the effects of the binding energy of protons and neutrons in nuclei. The most stable isolated nucleus is the one of ^{56}Fe. If, however, many of these nucleons are packed closely together and the effects of relativistic electrons becomes important, the inverse β decay shifts the equilibrium configuration to larger nuclei with a larger fraction of neutrons as compared to protons. For densities above $\rho_0 \approx 4 \times 10^{11}$ g cm^{-3}, the nuclei produced approach the neutron drip line and free neutrons coexist with the nuclei. Finally, when the mass density ρ_0 exceeds about 4×10^{12} g cm^{-3} the pressure is dominated by the free neutrons. An equation of state which accounts for the binding effects by employing a semi-empirical mass formula, has been evaluated by Baym, Pethick, and Sutherland (1971). This BPS equation of state is displayed in Fig. 3.2, supplemented by an equation of state determined from a relativistic Brueckner–Hartree–Fock calculation of Müther et al. (1987) at densities above 3×10^{13} g cm^{-3} (MPA). For a comparison we also show the prediction obtained from a non-relativistic (3.14) and a relativistic electron gas (3.15).

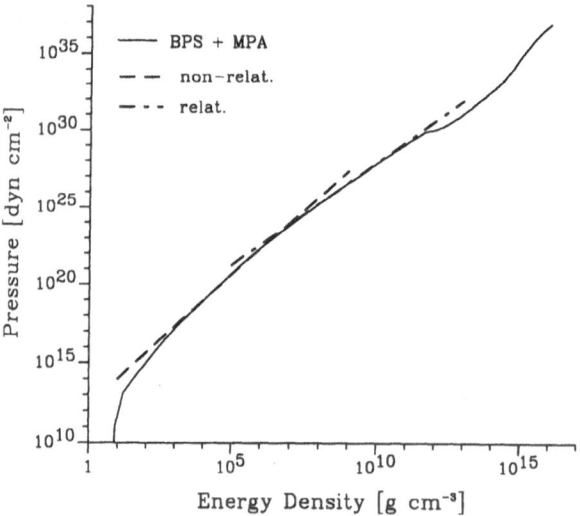

Fig. 3.2. Equation of state as evaluated by Baym, Pethick, and Sutherland (1971), supplemented by an equation of state determined from a relativistic Brueckner–Hartree–Fock calculation of Müther et al. (1987) at densities above 3×10^{13} g cm^{-3}. Also displayed are predictions from a non-relativistic (dashed line) and relativistic (dashed-dotted line) electron gas.

3.2 Green's Function and Many-Body Theory

3.2.1 Time-Dependent Perturbation Theory

It is the main aim of this section to introduce some basic concepts and approximations of the quantum theory for many-body systems. For that purpose we will mainly consider a system of interacting fermions (the nucleons), which can be described within non-relativistic quantum mechanics. We will sketch a perturbative expansion for the calculation of the energy of such a system at temperature $T = 0$ using the technique of many-body Green's functions. It is a nice feature of this expansion that it can easily be represented in terms of Feynman diagrams, a language we will introduce and use in this chapter. The perturbation expansion can easily be generalized to account for the effects of relativistic quantum mechanics, the inclusion of bosons (the mesons responsible for the NN interaction), and finite temperatures. A more detailed presentation of these techniques has been given, e.g., in the classical textbook by Fetter and Walecka (1971). New developments, in particular the connection to field theory, have been presented by Negele and Orland (1988). A rather intuitive interpretation of the Feynman diagrams can be found in the book by Mattuck (1976).

In order to set up the nomenclature we assume that the system is described by a Hamiltonian H_S, which does not explicitly depend on time t and can be split into a one-body part H_0 and a residual interaction V:

$$H_S = H_0 + V . \tag{3.19}$$

The one-body part contains the operator for the kinetic energy T_{kin} and a single-particle potential V_1 which can be chosen appropriately:

$$
\begin{aligned}
H_0 &= T_{\text{kin}} + V_1 \\
&= \sum_i \epsilon_i a_i^\dagger a_i .
\end{aligned}
\tag{3.20}
$$

Here H_0 is defined in terms of the single-particle energies ϵ_i and the fermion creation (annihilation) operators $a_i^\dagger$ (a_i), creating (annihilating) a fermion in the single-particle orbit identified by the quantum numbers i. Below we will consider a perturbation expansion in which H_0 is used for the unperturbed part, which can be treated in an exact way. We will assume a perturbative expansion for the residual part. This residual interaction V contains the original interaction between the fermions V_2, which we will consider to be a two-body interaction, reduced by the auxiliary potential V_1:

$$
\begin{aligned}
V &= V_2 - V_1 \\
&= \frac{1}{2} \sum_{i,j,k,l} \langle ij|V|kl \rangle a_i^\dagger a_j^\dagger a_l a_k ,
\end{aligned}
\tag{3.21}
$$

with $\langle ij|V|kl \rangle$ representing the matrix element of V calculated for the two-particle product states $|kl\rangle$ and $\langle ij|$.

Within the conventional Schrödinger picture, operators, such as the Hamiltonian H_S defined in (3.19) do not depend on time. The whole time dependence of the system is contained in the time dependence of the states, which satisfy the Schrödinger equation

$$i\frac{\partial}{\partial t}|\Psi^S(t)\rangle = H_S|\Psi^S(t)\rangle. \tag{3.22}$$

Note that we have dropped the $\hbar$ in this equation since we will assume units such that $\hbar = c = 1$. From this Schrödinger equation one can derive the time evolution operator $U_S(t, t_0)$ for the states within the Schrödinger scheme

$$\begin{aligned}|\Psi^S(t)\rangle &= U_S(t, t_0)|\Psi^S(t_0)\rangle \\ &= e^{-iH_S(t-t_0)}|\Psi^S(t_0)\rangle. \end{aligned} \tag{3.23}$$

In the so-called interaction scheme or interaction picture the time-dependence due to the unperturbed Hamiltonian H_0 is removed from the states by defining the states in the interaction picture as

$$|\Psi^I(t)\rangle = e^{iH_0 t}|\Psi^S(t)\rangle. \tag{3.24}$$

This implies that each operator, which is defined within the Schrödinger scheme as O_S, must be transformed in the interaction scheme to

$$O_I(t) = e^{iH_0 t}O_S e^{-iH_0 t} \tag{3.25}$$

in order to guarantee that the result for the calculation of matrix elements is independent of the scheme used

$$\begin{aligned}\langle\Psi^I(t)|O_I(t)|\Psi^I(t)\rangle &= \langle\Psi^S(t)|e^{-iH_0 t}O_I(t)e^{iH_0 t}|\Psi^S(t)\rangle \\ &= \langle\Psi^S(t)|e^{-iH_0 t}e^{iH_0 t}O_S e^{-iH_0 t}e^{iH_0 t}|\Psi^S(t)\rangle \\ &= \langle\Psi^S(t)|O_S|\Psi^S(t)\rangle. \end{aligned} \tag{3.26}$$

The Heisenberg picture can be understood as an extreme case of the interaction scheme in the sense that in the Heisenberg scheme H_0 is assumed to be identical to the total Hamiltonian H_S. This implies that the state in the Heisenberg picture [compare (3.24)],

$$|\Psi^H(t)\rangle = e^{iH_S t}|\Psi^S(t)\rangle = |\Psi^S(0)\rangle, \tag{3.27}$$

does not depend on time; the operators in the Heisenberg picture are defined by

$$O_H(t) = e^{iH_S t}O_S e^{-iH_S t}. \tag{3.28}$$

Using the definition of the state within the interaction scheme (3.24) and applying the Schrödinger equation (3.22) one derives the equation of motion for the state in the interaction scheme:

$$
\begin{aligned}
i\frac{\partial}{\partial t}|\Psi^I(t)\rangle &= -H_0|\Psi^I(t)\rangle + e^{iH_0t}\left(i\frac{\partial}{\partial t}|\Psi^S(t)\rangle\right) \\
&= -H_0|\Psi^I(t)\rangle + e^{iH_0t}H_S e^{-iH_0t}|\Psi^I(t)\rangle \\
&= V_I(t)|\Psi^I(t)\rangle .
\end{aligned}
\tag{3.29}
$$

In order to arrive at an equation for the evolution operator in the interaction scheme, $U_I(t,t_0)$, which is useful for computational purposes, one may rewrite (3.29) as

$$
i\frac{\partial}{\partial t}U_I(t,t_0)|\Psi^I(t_0)\rangle = V_I(t)U_I(t,t_0)|\Psi^I(t_0)\rangle ,
\tag{3.30}
$$

which yields

$$
i\frac{\partial}{\partial t}U_I(t,t_0) = V_I(t)U_I(t,t_0) .
\tag{3.31}
$$

This equation can be transformed into an integral equation

$$
U_I(t,t_0) = U_I(t_0,t_0) - i\int_{t_0}^t dt_1\, V_I(t_1)U_I(t_1,t_0) .
\tag{3.32}
$$

Using the fact that $U_I(t_0,t_0)$ is identical to the unit operator $\hat{1}$, we can iterate this integral equation to a perturbative expansion in powers of V_I:

$$
\begin{aligned}
U_I(t,t_0) &= \hat{1} + (-i)\int_{t_0}^t dt_1\, V_I(t_1) \\
&\quad + (-i)^2\int_{t_0}^t dt_1\, V_I(t_1)\int_{t_0}^{t_1} dt_2\, V_I(t_2) + \ldots
\end{aligned}
\tag{3.33}
$$

The integration variables used on the left-hand side are nested in a rather inconvenient way. Therefore one rewrites the term of order n in this expansion using

$$
\begin{aligned}
\int_{t_0}^t dt_1\, V_I(t_1)\ldots\int_{t_0}^{t_{n-1}} dt_n\, V_I(t_n) &= \frac{1}{n!}\int_{t_0}^t dt_1\ldots\int_{t_0}^t dt_n \\
&\quad \times \mathcal{T}\left(V_I(t_1)\ldots V_I(t_n)\right)
\end{aligned}
\tag{3.34}
$$

with the time ordering or chronological operator $\mathcal{T}$, which is defined for two operators by

$$
\mathcal{T}(A(t_1)B(t_2)) = \begin{cases} A(t_1)B(t_2), & \text{if } t_1 \geq t_2, \\ (-1)^m B(t_2)A(t_1), & \text{otherwise.} \end{cases}
\tag{3.35}
$$

Here m is the number of exchanges of fermion creation and annihilation operators contained in A and B, which are needed to bring A and B into chronological order. Note that for our present purpose (3.34) the factor $(-1)^m$ is always equal to 1, as the number of fermion operators defining V_I is even. The definition of $\mathcal{T}$ in (3.35) for two operators is easily extended to n operators. Applying (3.34) to the expansion of the time evolution operator in (3.33), one gets

$$U_I(t, t_0) = \sum_{n=0}^{\infty} \frac{(-i)^n}{n!} \int_{t_0}^t dt_1 \ldots \int_{t_0}^t dt_n \, \mathcal{T} \left(V_I(t_1) \ldots V_I(t_n) \right) \tag{3.36}$$

In order to arrive at a perturbation expansion for the calculation of matrix elements, one assumes that the perturbation V is "switched off" at times $t = -\infty$ and $t = +\infty$ and can be switched on in an adiabatic way for times $t \approx 0$. This can be achieved by a time-dependent Hamiltonian of the form

$$H_\alpha(t) = H_0 + e^{-\alpha|t|} V \tag{3.37}$$

where α is a small positive number that becomes infinitesimal in the adiabatic limit. This procedure implies that the eigenstates of the Hamiltonian H_α are identical to the eigenstates of the unperturbed Hamiltonian H_0 at times $|t| = \infty$ and should evolve to the corresponding eigenstates of the total H at $t \approx 0$ if we use the time evolution operator $U_{I\alpha}(t, t')$ for the Hamiltonian H_α. If we denote the (nondegenerate) ground state of H_0 by Φ_0, which is independent of time in the interaction picture, this means that we obtain an eigenstate of the exact H at time $t = 0$ by

$$|\Psi_0(t = 0)\rangle = \lim_{\alpha \to 0} U_{I\alpha}(0, -\infty)|\Phi_0\rangle. \tag{3.38}$$

It has been shown by Gell-Mann and Low (1951) that Ψ_0 is indeed an exact eigenstate of H if the perturbation expansion converges.

This means that a matrix element of an operator O can be calculated in the Heisenberg scheme, in which the state is time independent and identical to $\Psi_0(t = 0)$ [see (3.27) and (3.28)], as

$$\frac{\langle \Psi_0 | O_H(t) | \Psi_0 \rangle}{\langle \Psi_0 | \Psi_0 \rangle} = \lim_{\alpha \to 0} \frac{\langle \Phi_0 | U_{I\alpha}(\infty, 0) O_H(t) U_{I\alpha}(0, -\infty) | \Phi_0 \rangle}{\langle \Phi_0 | U_{I\alpha}(\infty, 0) U_{I\alpha}(0, -\infty) | \Phi_0 \rangle}$$

$$= \lim_{\alpha \to 0} \frac{\langle \Phi_0 | U_{I\alpha}(\infty, t) O_I(t) U_{I\alpha}(t, -\infty) | \Phi_0 \rangle}{\langle \Phi_0 | U_{I\alpha}(\infty, -\infty) | \Phi_0 \rangle}. \tag{3.39}$$

Using the explicit representation of the time evolution operator in (3.36) one can furthermore show (see, e.g., Fetter and Walecka 1971) that the matrix element for any time-ordered product of two Heisenberg operators can be calculated as

$$\frac{\langle \Psi_0 | \mathcal{T} \left(A_H(t) B_H(t') \right) | \Psi_0 \rangle}{\langle \Psi_0 | \Psi_0 \rangle}$$

$$= \lim_{\alpha \to 0} \left[\frac{1}{\langle \Psi_0 | \Psi_0 \rangle} \sum_{n=0}^{\infty} \frac{(-i)^n}{n!} \int_{-\infty}^{\infty} dt_1 \ldots \int_{-\infty}^{\infty} dt_n \, e^{-\alpha(|t_1| + \ldots |t_n|)} \right.$$

$$\left. \times \langle \Phi_0 | \mathcal{T} \left(V_I(t_1) \ldots V_I(t_n) A_I(t) B_I(t') \right) | \Phi_0 \rangle \right], \tag{3.40}$$

which means that one has to evaluate matrix elements of time-ordered products of operators in the interaction scheme for the ground state of the unperturbed Hamiltonian Φ_0.

3.2.2 Perturbation Theory and Feynman Diagrams

In order to demonstrate how the matrix elements occurring in (3.40) can be evaluated we will shortly sketch Wick's theorem for time-ordered products (see Fetter and Walecka 1971, Wick 1950) and demonstrate how the various contributions are visualized in terms of Feynman diagrams.

First we recall that the operator for the residual interaction as well as any other operator for the fermions considered can be expressed in terms of the basic single-particle creation $(a_i^\dagger)$ and annihilation operators $[a_i$, see (3.21)]. It is easy to verify that these operators in the interaction scheme are given by [see (3.25)]

$$a_{Ij}(t) = a_j \exp\left(-i\epsilon_j t\right),$$
$$a_{Ij}^\dagger(t) = a_j^\dagger \exp\left(+i\epsilon_j t\right), \tag{3.41}$$

with ϵ_j the single-particle energies defining the unperturbed Hamiltonian H_0 in (3.20). The ground state for this unperturbed Hamiltonian, Φ_0, is given by a Slater determinant in which all single-particle states i with an energy ϵ_i below the Fermi energy ϵ_F are occupied. The Fermi energy separates hole-states $(\epsilon_{\text{hole}} \le \epsilon_F$, occupied in the unperturbed ground state) from particle states $(\epsilon_{\text{part.}} > \epsilon_F$, unoccupied in the unperturbed ground state). If $M, N, O, P, \dots$ represent creation or annihilation operators, in the Schroedinger or in the interaction picture, one can define the normal product of such operators by

$$\mathcal{N}(MNOP\dots) = (-1)^\gamma OP \dots MN \dots \tag{3.42}$$

where the sequence of operators on the right-hand side of this equation is such that all creation operators for particle states $(a_{\text{part.}}^\dagger)$ and annihilation operators for hole states (a_{hole}) are moved to the left (O, P), whereas all creation operators for hole states $(a_{\text{hole}}^\dagger)$ and annihilation operators for particle states $(a_{\text{part.}})$ are moved to the right (M, N) and γ counts the number of exchanges of these operators required to obtain the normal ordering. This normal ordering guarantees that the matrix element of such an ordered product calculated for the unperturbed state Φ_0 vanishes:

$$\langle\Phi_0|\mathcal{N}(MNOP\dots)|\Phi_0\rangle = 0. \tag{3.43}$$

Furthermore we define a "contraction" of two operators, using the chronological operator of (3.35) for two such operators M, N in the interaction scheme, as

$$\underbrace{MN} = \langle\Phi_0|\mathcal{T}(MN) - \mathcal{N}(MN)|\Phi_0\rangle$$
$$= \langle\Phi_0|\mathcal{T}(MN)|\Phi_0\rangle, \tag{3.44}$$

which is just a complex number given by

$$\underbrace{a_{Ij}(t)a_{Ik}^\dagger(t')} = \begin{cases} \delta_{jk}e^{-i\epsilon_j(t-t')} & \text{if } j \text{ is a particle state and } t > t', \\ -\delta_{jk}e^{-i\epsilon_j(t-t')} & \text{if } j \text{ is a hole state and } t' > t, \\ 0 & \text{otherwise}, \end{cases} \tag{3.45}$$

$$\underbrace{a_{Ik}^\dagger(t')a_{Ij}(t)} = -\underbrace{a_{Ij}(t)a_{Ik}^\dagger(t')} , \tag{3.46}$$

and

$$\underbrace{a_{Ij}(t)a_{Ik}(t')} = \underbrace{a_{Ij}^\dagger(t)a_{Ik}^\dagger(t')} = 0 . \tag{3.47}$$

If we furthermore define the normal-ordered product with the inclusion of one contraction by

$$\mathcal{N}\left(\underbrace{MN\ldots O}P\ldots\right) = (-1)^\beta \underbrace{MO}\,\mathcal{N}\left(\not\!M N\ldots \not\!O P\ldots\right) , \tag{3.48}$$

where β is the number of permutations of operators needed to bring the operators M and O next to each other and $\not\!M$ indicates that the corresponding operator should be eliminated from the list. In a similar way one defines normal-ordered products with more contractions. With this nomenclature Wick's theorem can be formulated as

$$\mathcal{T}(MNOP\ldots Z) = \mathcal{N}(MNOP\ldots Z) \tag{3.49}$$

$$+ \ \mathcal{N}\left(\underbrace{MN}OP\ldots Z\right) + \text{other terms with one contraction}$$

$$+ \ \mathcal{N}\left(\underbrace{MN}\,\underbrace{OP}\ldots Z\right) + \text{other terms with two contractions}$$

$$+ \ \ldots + \text{all terms with all operators contracted.}$$

From (3.43) it is evident that in calculating the various contributions to the perturbation expansion in (3.40) one has to consider only the completely contracted terms in applying Wick's theorem.

The use of Feynman diagrams provides an easy control of all the contributions. To demonstrate this, we consider as an example the matrix element occurring in the first-order term of (3.40)

$$\tfrac{1}{2}\sum_{i,j,k,l}\langle ij|V|kl\rangle$$

$$\times \langle \Phi_0|\mathcal{T}\left(a_{Ii}^\dagger(t_1)a_{Ij}^\dagger(t_1)a_{Il}(t_1)a_{Ik}(t_1)a_{I\alpha}(t)a_{I\beta}^\dagger(t')\right)|\Phi_0\rangle , \tag{3.50}$$

where we have made the substitution $A_I \to a_{I\alpha}$, $B_I \to a_{I\beta}^\dagger$, and used the explicit representation of V according to (3.21). For this example we will furthermore assume that $t' < t_1 < t$. The operator of the residual interaction is represented in the Feynman diagram by a horizontal dashed line, with an outgoing and incoming arrow at each end. The outgoing arrows refer to creation operators contained in V ($a_{Ii}^\dagger$, $a_{Ij}^\dagger$) in our notation, and the incoming ones refer to annihilation operators. The external operators are represented by a dot with an ingoing and an outgoing arrow for the annihilation and creation operator, respectively. These objects are displayed in Fig. 3.3a. With the assumption that the vertical axis represents a time axis the objects are ordered according to the choice of our example: $t' < t_1 < t$.

Any contraction implies that two of the lines with arrows must be paired. In the graphical representation this is achieved by connecting them. Looking

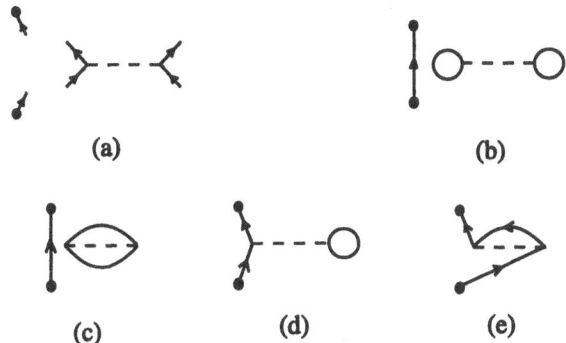

Fig. 3.3. Building blocks for Feynman diagrams (a) and completely contracted contributions as discussed in the text.

at the contractions which yield results different from zero [see (3.45)–(3.47)], one finds that only those pairs of lines in which the arrows point into the same direction must be connected. Furthermore we can distinguish connected lines with an arrow pointing upwards, which refers to a particle state, i.e., the corresponding summation indices in (3.50) can be restricted to particle states, whereas connected lines with an arrow pointing downwards refer to hole lines.

All diagrams representing the completely contracted terms of the expression shown in (3.50) are displayed in of Fig. 3.3b–e. Note that we show only those diagrams that are topologically distinct in the sense that a diagram obtained from another one by just mirroring the ends of an interaction line is not displayed again. Recalling Wick's theorem, it is evident that all non-vanishing contributions to the perturbation calculation of the nth-order term in the matrix element of a time-ordered product of operators in (3.40) can be obtained in the following way.

– Draw n interaction lines (dashed lines in Fig. 3.3) and mark on the creation and annihilation operators for the external operators to be calculated (dots in Fig. 3.3).
– Construct all diagrams by connecting the "arrows" linked to these basic building blocks according to the rules given in the example of Fig. 3.3.
– Keep in mind that all possible time orderings of the interaction vertices relative to the external operators must be considered [see time integrations in (3.40)].
– Each of the resulting diagrams represents a non-vanishing contribution to the evaluation of (3.40) and there exist well-established Feynman rules that translate the contribution of the diagram into a calculable expression [see, e.g., Mattuck (1976)].

In the example discussed above and represented in Fig. 3.3 we obtained two kinds of diagrams: the linked diagrams (d) and (e), in which all lines are connected to the points representing external operators; and the unlinked diagrams (b) and (c), in which parts are completely disconnected from the external operator. We also notice that until now we have completely ignored the denominator $\langle \Psi_0 | \Psi_0 \rangle$ on the right-hand side of (3.40) and discussed only the expansion of the numerator. The linked cluster theorem [see, e.g., Fetter and Walecka (1971)] now proves that the contributions from all unlinked diagrams occurring in the expansion of the numerator in (3.40) can be cancel with the corresponding expansion of the denominator. This means that we can restrict the discussion simply to the expansion of the numerator and add a new rule to the construction of the Feynman diagrams:

– Consider only the contribution of linked diagrams.

Up to this point the diagrams have been used only as a kind of book-keeping tool to identify all non-vanishing contributions in the perturbation expansion. However, we have seen already that each connecting line in those diagrams represents a contraction, and a line with an arrow pointing upwards stands for [see (3.44)]

$$\underbrace{a_{Ij}(t)a^{\dagger}_{Ik}(t')} = \langle \Phi_0 | \mathcal{T} \left(a_{Ij}(t)a^{\dagger}_{Ik}(t') \right) | \Phi_0 \rangle , \qquad (3.51)$$

with the creation of a particle taking place before the annihilation $(t > t')$. This means that we can ignore the operator $\mathcal{T}$ and rewrite this contraction as

$$\langle \Phi_0 | a_{Ij}(t)a^{\dagger}_{Ik}(t') | \Phi_0 \rangle = \sum_{\beta} \langle \Phi_0 | a_{Ij}(t) | \beta \rangle \langle \beta | a^{\dagger}_{Ik}(t') | \Phi_0 \rangle$$

$$= \delta_{jk}e^{-i\epsilon_j(t-t')} \quad \text{for } j \text{ a particle state.} \qquad (3.52)$$

In the first line of this equation we have inserted a summation over a complete set of states $|\beta\rangle$ with one particle in addition to the number of fermions in $|\Phi_0\rangle$, in order to show that this contraction describes the product of a probability amplitude to create a particle at a time t' and producing a state β and the probability that it is annihilated at the later time t reproducing the unperturbed ground state Φ_0. In the second line of this equation we have copied the result for the contraction from (3.45) that such a propagation of a particle on top of the unperturbed state is only possible if $j = k$ refers to a state above the Fermi energy, in order not to violate the Pauli principle.

In a similar way one can convince oneself that a line with an arrow pointing down represents the propagation of a hole state, i.e., a particle must be removed first from a state h below the Fermi energy before it is put back at a later time. With this interpretation of the contractions visualized in the diagrams one can easily interpret the Feynman diagrams in terms of a time-dependent processes.

As an example we consider the two diagrams displayed in Fig. 3.4. Part (a) of this figure describes a process with a particle created at time t' which propagates until the time t_1 when it interacts with another particle in the Fermi sea of Φ_0 producing a two-particle one-hole configuration. This configuration continues until the time t_2 when the nucleons interact again to return to the ground state with one extra particle which continues to propagate until t. In Fig. 3.4b a particle is created at t' but propagates without any interaction

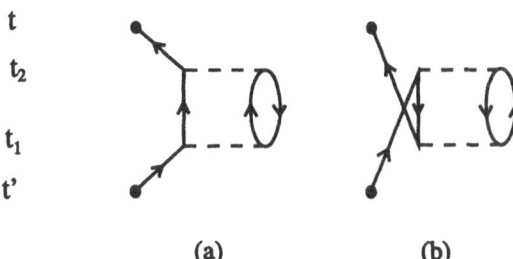

t

t_2

t_1

t'

(a)   (b)

Fig. 3.4. Examples of the diagrams discussed in the text

until t_2. At the time $t_1 < t_2$ two nucleons in the Fermi sea interact forming altogether a three-particle two-hole excitation. The second interaction at t_2 restores the state Φ_0 with an extra nucleon.

This physical interpretation of the Feynman diagrams can also be used to stress the importance of the linked cluster theorem. Unlinked diagrams like those displayed in parts (b) and (c) of Fig. 3.3 describe processes in which the motion of the extra particle is completely independent of the unlinked interaction of two other nucleons. In a neutron star or even an infinite system these two processes can occur at completely different places and the contribution of the unlinked diagrams should diverge. Therefore the restriction to linked clusters is important in order to obtain a reasonable expansion.

3.2.3 Single-Particle Green's Function

The single-particle Green's function can be considered as a special example of an expectation value for the time-ordered product of two operators calculated for the exact ground-state in (3.40). It is defined by

$$ig(\alpha t, \beta t') = \langle \Psi_0 | \mathcal{T}\left(a_{H\alpha}(t) a_{H\beta}^\dagger(t') \right) | \Psi_0 \rangle . \tag{3.53}$$

Note that here and in the following we have dropped the denominator $\langle \Psi_0 | \Psi_0 \rangle$, assuming that the exact ground state is properly normalized. The creation and annihilation operators in the Heisenberg representation are defined in an appropriate basis, characterized by quantum numbers α and β. If we assume that α refers to a position r' and β to a position r of the considered fermion in r space, the single-particle Green's function $ig(r't', rt)$ describes the propagation of this fermion from the space-time point (rt) to $(r't')$. In contrast to the discussion of the contractions in the previous section, it is in this case the propagation with respect to the exact ground state and the complete Hamiltonian.

For a system that is invariant under translation, such as the infinite nuclear or neutron matter, which we want to consider, the appropriate set of basis states is that of the momentum eigenstates; the Green's function is diagonal in this representation. Rewriting the effect of the chronological operator $\mathcal{T}$ defined in (3.35) in terms of step functions $\Theta(t - t')$, we find the Green's function is given as

$ig(k, t - t')$

$$= \Theta(t - t') \quad \langle \Psi_0 | a_{Hk}(t) a_{Hk}^\dagger(t') | \Psi_0 \rangle - \Theta(t' - t) \langle \Psi_0 | a_{Hk}^\dagger(t') a_{Hk}(t) | \Psi_0 \rangle$$

$$= \Theta(t - t') \sum_\gamma e^{-i(E_\gamma^{(A+1)} - E_0^A)(t - t')} \left| \langle \Psi_\gamma^{(A+1)} | a_k^\dagger | \Psi_0 \rangle \right|^2$$

$$- \Theta(t' - t) \sum_\delta e^{-i(E_0^A - E_\delta^{(A-1)})(t - t')} \left| \langle \Psi_\delta^{(A-1)} | a_k | \Psi_0 \rangle \right|^2 \qquad (3.54)$$

To arrive at the second part of this equation we have inserted a complete set of eigenstates for the system with $A + 1$ particles ($\Psi_\gamma^{(A+1)}$) and $A - 1$ particles ($\Psi_\delta^{(A-1)}$), as appropriate, and replaced the Hamiltonian in the exponential functions of the definition for the Heisenberg operators [see (3.28)] by the corresponding eigenvalues. This means that the energies E_0^A, $E_\gamma^{(A+1)}$, and $E_\delta^{(A-1)}$ refer to the exact energies for the ground state of our reference system, and the exact energies of eigenstates with $A + 1$ and $A - 1$ particles, respectively. Note that the step function can be represented by its integral form:

$$\Theta(t) = -\lim_{\eta \to 0} \frac{1}{2\pi i} \int_{-\infty}^{\infty} d\omega \frac{e^{-i\omega t}}{\omega + i\eta}. \qquad (3.55)$$

Thus the Fourier transformed Green's function, transforming the time difference $t - t'$ to an energy variable ω, can be written as

$$g(k, \omega) = \lim_{\eta \to 0} \left(\sum_\gamma \frac{\left| \langle \Psi_\gamma^{(A+1)} | a_k^\dagger | \Psi_0 \rangle \right|^2}{\omega - \left(E_\gamma^{(A+1)} - E_0^A \right) + i\eta} \right.$$

$$\left. + \sum_\delta \frac{\left| \langle \Psi_\delta^{(A-1)} | a_k | \Psi_0 \rangle \right|^2}{\omega - \left(E_0^A - E_\delta^{(A-1)} \right) - i\eta} \right). \qquad (3.56)$$

This is the so-called spectral or Lehmann representation of the single-particle Green's function [Lehmann (1954)]. Inspecting this equation one finds that the single-particle Green's function is represented in terms of quantities that are measurable. It shows poles at energies that correspond to energies of the system with one particle added ($A + 1$) and one particle removed ($A - 1$) relative to the energy of the ground state for the reference system. The residua of these poles are given by the spectroscopic factors, i.e., the probabilities of adding and removing one particle with momentum k to produce the specific state γ (δ) of the residual system. The infinitesimal quantity η shifts those poles below the Fermi energy (the states of the $A - 1$ system) to slightly above the real axis and those above the Fermi energy (the states of the $A + 1$ system) to slightly below the real axis (see Fig. 3.5).

In our spectral representation of the single-particle Green's function (3.56) we assumed that the spectra of the many-body system are defined in terms of

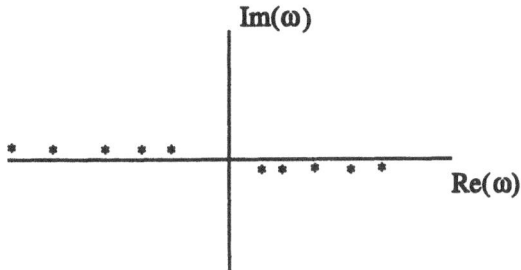

Fig. 3.5. Schematic picture of the position of the poles for the single-particle Green's function in the spectral or Lehmann representation

discrete energies $E_\gamma^{(A+1)}$. This is true of course only for systems confined to a finite area. For infinite systems, such as we want to consider here, the energy spectra are continuous and it is more appropriate to introduce the so-called hole and particle spectral functions defined by

$$
\begin{aligned}
S_\mathrm{h}(k,\omega) &= \frac{1}{\pi}\mathrm{Im}g(k,\omega), \quad \text{for } \omega \le \epsilon_\mathrm{F} \\
&= \sum_\gamma \left| \langle \Psi_\gamma^{(A-1)} | a_k | \Psi_0 \rangle \right|^2 \delta\left(\omega - (E_0^A - E_\gamma^{(A-1)}) \right), \\
S_\mathrm{p}(k,\omega) &= \frac{1}{\pi}\mathrm{Im}g(k,\omega), \quad \text{for } \omega > \epsilon_\mathrm{F} \\
&= \sum_\gamma \left| \langle \Psi_\gamma^{(A+1)} | a_k^\dagger | \Psi_0 \rangle \right|^2 \delta\left(\omega - (E_\gamma^{(A+1)} - E_0^A) \right) \quad (3.57)
\end{aligned}
$$

where we have made reference to the case of discrete spectra in the second and fourth lines. These definitions imply that the single-particle Green's function is given in terms of the spectral functions by

$$
g(k,\omega) = \lim_{\eta \to 0} \left(\int_{-\infty}^{\epsilon_\mathrm{F}} d\omega' \frac{S_\mathrm{h}(k,\omega')}{\omega - \omega' - i\eta} + \int_{\epsilon_\mathrm{F}}^{\infty} d\omega' \frac{S_\mathrm{p}(k,\omega')}{\omega - \omega' + i\eta} \right). \quad (3.58)
$$

The single-particle Green's function or the spectral functions defining the single-particle Green's function can be used to evaluate observables of the system. As a first example we mention the momentum distribution or momentum density

$$
n(k) = \langle \Psi_0 | a_k^\dagger a_k | \Psi_0 \rangle = \int_{-\infty}^{\epsilon_\mathrm{F}} d\omega\, S_\mathrm{h}(k,\omega). \quad (3.59)
$$

For a system of interacting fermions this momentum distribution will in general deviate from that of a free Fermi gas, in which all states with momenta k less than a Fermi momentum k_F are completely occupied:

$$
n_0(k) = \begin{cases} 1 & \text{for } |k| \le k_\mathrm{F}, \\ 0 & \text{otherwise.} \end{cases} \quad (3.60)
$$

Therefore the deviation of the momentum distribution $n(k)$ of (3.59) from the free one provides information about the correlations between the interacting fermions. Modern electron accelerators such as NIKEF in Amsterdam and MAMI in Mainz are a tool for performing nucleon knock-out experiments by inelastic electron scattering [(e,e')p] with high precision. Therefore, there has been quite some effort, especially recently, from experimental (see Bobeldijk et al. 1994, Blomqvist et al. 1995) and theoretical groups (Müther et al. 1995, and references therein) to explore the momentum distribution, in particular at very high momenta.

The single-particle Green's function allows the evaluation of the expectation value for any single-particle operator O. If we return to the non-diagonal nomenclature used, e.g., in (3.53) and generalize the definition of spectral functions in a corresponding way, such a calculation is performed via

$$\langle \Psi_0 | O | \Psi_0 \rangle = \sum_{\alpha,\beta} \int_{-\infty}^{\epsilon_F} d\omega \, S_h(\alpha\beta, \omega) \langle \alpha | O | \beta \rangle , \tag{3.61}$$

with $\langle \alpha | O | \beta \rangle$ denoting the single-particle matrix element calculated in the basis $\alpha, \beta \ldots$. Furthermore, if the interaction between the fermions is a two-body interaction, one can even calculated the energy of the correlated ground state via

$$
\begin{aligned}
E_0^A &= \langle \Psi_0 | H | \Psi_0 \rangle \\
&= \frac{1}{2} \sum_{\alpha,\beta} \int_{-\infty}^{\epsilon_F} d\omega \, S_h(\alpha\beta, \omega) \left(\langle \alpha | T_{\text{kin}} | \beta \rangle + \omega \delta_{\alpha,\beta} \right) ,
\end{aligned} \tag{3.62}
$$

with T_{kin} representing the single-particle operator for the kinetic energy.

What remains to be discussed are the tools and approximations used to determine the single-particle Green's function. The simplest, non-trivial approximation is the Hartree–Fock (HF) approximation, assuming that the fermions move independently from each other in a single-particle potential U_{HF}. This independent particle picture of HF implies of course that the ground state of the reference system, as well as the states of the $(A + 1)$ and $(A - 1)$ systems, are described by Slater determinants, and the spectroscopic factors in the Lehmann representation (3.56) of the single-particle Green's function are identical to 1 for one special state and zero otherwise. Therefore the spectral representation of the single-particle Green's function is given by

$$g_{\text{HF}}(k, \omega) = \frac{\Theta(|k| - k_F)}{\omega - \epsilon_k^{\text{HF}} + i\eta} + \frac{\Theta(k_F - |k|)}{\omega - \epsilon_k^{\text{HF}} - i\eta} . \tag{3.63}$$

The single-particle energies ϵ_k^{HF} are defined as the sum of the kinetic energies of the particles with momentum k plus the energy of the HF potential U, which is given by

$$
\begin{aligned}
U_{\text{HF}}(k) &= \sum_{\alpha,\beta} (\langle k\alpha | V | k\beta \rangle - \langle k\alpha | V | \beta k \rangle) \lim_{\eta \to 0} \int \frac{d\omega}{2i\pi} e^{i\omega\eta} g_{\text{HF}}(\alpha\beta, \omega) \\
&= \sum_{\alpha < F} \langle k\alpha | V | k\alpha \rangle_A .
\end{aligned} \tag{3.64}
$$

The antisymmetrized matrix element of the two-body interaction in the second line, labeled by the index A, is just the sum of the two matrix elements in the first line for $\alpha = \beta$. The sum over all single-particle states below the Fermi energy ($\alpha < F$) refers to an integral over momenta below k_F and includes a summation on spin or other quantum numbers as well. The first line of (3.64) has been written so as to indicate that the HF approximation requires a solution of a non-linear self-consistency problem. In order to define the potential U_{HF} one should know the Green's function g_{HF}, which is defined in terms of single-particle energies by using U_{HF}. The translational symmetry of infinite systems, however, yields a Green's function diagonal in the basis of plane waves and therefore the self-consistency is taken into account in a trivial manner.

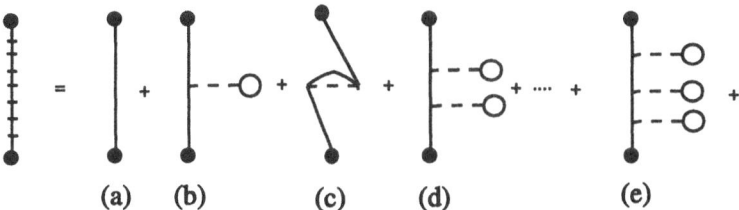

(a)   (b)   (c)   (d)   (e)

Fig. 3.6. Diagrammatic representation of the expansion in (3.65). The line on the left represents g_{HF}.

If one now expands the contribution of the potential energy to ϵ^{HF} from the Green's function in (3.63), one observes that this expansion can be written in terms of Feynman diagrams as displayed in Fig. 3.6. One can see from this expansion that the HF Green's function corresponds to a power expansion

$$g_{HF}(k, \omega) = g_0(k, \omega) \left[1 + \sum_{n=1}^{\infty} (U_{HF}(k)g_0(k, \omega))^n \right], \tag{3.65}$$

where g_0 represents the Green's function of (3.63) with the assumption $U_{HF} = 0$. When g_{HF} is given as diagrams, one sees that each order n yields an additional contraction, which corresponds to g_0, and the terms representing U_{HF}. This concept is generalized by grouping in a special way all Feynman diagrams that represent the exact single-particle Green's function. The irreducible self-energy or mass operator $\mathcal{M}$ is an extension of the pieces representing U_{HF} in Fig. 3.6 insofar that it contains all diagrams of any order in V without an intermediate state, which is just a single-particle Green's function. This means that (b) and (c) of Fig. 3.6 (without the external lines representing g_0) as well (a) and (b) of Fig. 3.4 are part of $\mathcal{M}$, whereas the diagram (d) of Fig. 3.6 does not belong to the definition of the irreducible self-energy.

If we represent this self-energy by the oval in Fig. 3.7, it is obvious that all diagrams for the exact g can be classified as indicated in the upper part of that figure. In the algebraic representation this is translated into the Dyson equation

$$\begin{aligned}
g(\alpha\beta,\omega) &= g_0(\alpha\beta,\omega) + \sum_{\gamma\delta} g_0(\alpha\gamma,\omega)\mathcal{M}_{\gamma\delta}(\omega)\Big[g_0(\delta\beta,\omega) \\
&\quad + \sum_{\mu\nu} g_0(\delta\mu,\omega)\mathcal{M}_{\mu\nu}(\omega)g_0(\nu\beta,\omega) + \ldots\Big] \\
&= g_0(\alpha\beta,\omega) + \sum_{\gamma\delta} g_0(\alpha\gamma,\omega)\mathcal{M}_{\gamma\delta}(\omega)g(\delta\beta,\omega).
\end{aligned} \tag{3.66}$$

The self-energy calculated beyond the HF approximation is itself energy dependent (or time dependent if one employs the time-dependent representation). Therefore the solution of the Dyson equation (3.66) requires the solution of a non-trivial set of non-linear equations, even if the translational symmetry requires $\mathcal{M}$ to be diagonal in the momentum representation. The various approximations discussed below can now be defined in terms of the approximation used for $\mathcal{M}$.

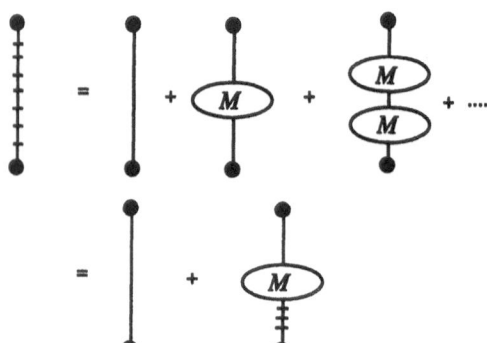

Fig. 3.7. Classifying the diagrams for the single-particle Green's function in terms of the mass operator $\mathcal{M}$ and graphical representation of the Dyson equation. The railed line represents the exact single-particle Green's function.

Up to this point we have discussed only the single-particle Green's function for a system of fermions described in terms of non-relativistic quantum mechanics. Below we will discuss the evaluation of the NN interaction within the framework of a relativistic meson exchange model. For that purpose we will also need the single-particle Green's function describing the propagation of a free fermion in the vacuum within a relativistic theory. This is given by the Feynman propagator

$$\begin{aligned}
g_{\text{Fe}}(k) &= \frac{1}{\not k - m + i\eta} \\
&= \frac{m}{E_k}\left[\frac{\Lambda^+(\mathbf{k})}{k_0 - E_k + i\eta} + \frac{\Lambda^-(\mathbf{k})}{k_0 + E_k - i\eta}\right],
\end{aligned} \tag{3.67}$$

with $\not k = \gamma^\mu k_\mu$, m the mass of the nucleon, E_k its free energy $E_k = \sqrt{\mathbf{k}^2 + m^2}$, and Λ^+ (Λ^-) the projector on states with positive (negative) energies [see also for the nomenclature of Bjorken and Drell (1964)].

$$
\begin{aligned}
\Lambda^+(\mathbf{k}) &= \sum_\lambda u(\mathbf{k}, \lambda)\bar{u}(\mathbf{k}, \lambda) \\
&= \frac{\gamma^0 E_k - \gamma\mathbf{k} + m}{2m},
\end{aligned}
\tag{3.68}
$$

where $u(\mathbf{k}, \lambda)$ refers to the positive energy solution of the free Dirac equation for a nucleon with spin projection (or helicity) λ,

$$
u(\mathbf{k}, \lambda) = \sqrt{\frac{E_k + m}{2m}} \begin{pmatrix} 1 \\ \frac{\sigma\cdot\mathbf{k}}{E_k+m} \end{pmatrix} |\lambda\rangle,
\tag{3.69}
$$

which is normalized by $\bar{u}u = u^\dagger\gamma^0 u = 1$. Inspecting (3.67) one observes that the poles of the Green's function, which refer to the states of negative energy $\omega = -E_k$ occupied in the Dirac sea, are slightly above the real axis, whereas those for positive energies are below [compare (3.56) and Fig. 3.5]. In order to obtain the unperturbed relativistic Green's function for a system at finite density, one should modify the Green's functions g_{Fe} in such a way that the states of the Fermi sea are also represented by poles slightly above the real axes. This is achieved by defining

$$
g_{0\rho}(k) = g_{\mathrm{Fe}}(k) + \frac{m}{E_k}\Lambda^+(k)2i\pi\delta(k_0 - E_k)\Theta(k_{\mathrm{F}} - |\mathbf{k}|),
\tag{3.70}
$$

in which only the second part depends on the density ρ.

3.3 Nucleon–Nucleon Interaction

The determination of a realistic NN interaction, which reproduces the NN scattering data and the properties of the bound two-nucleon system, the deuteron. The most conspicuous approach is to assume a local ansatz for the NN interaction of the form

$$
V = \sum_\alpha v_\alpha(r)O_\alpha(r),
\tag{3.71}
$$

with O_α defining a set of two-body operators, which account for the spin and isospin quantum numbers of the interacting nucleons and are compatible with basic symmetry properties of the NN interaction. Examples of such operators are the scalar product of the two spin operators $\sigma_1\sigma_2$ represented by the Pauli spin matrices σ_i acting on the spin part of the wave function for nucleon i, the corresponding product of the two isospin operators, or the tensor operator

$$
S_{12} = 3(\sigma_1\hat{r})(\sigma_2\hat{r}) - \sigma_1\sigma_2
\tag{3.72}
$$

in which $\hat{r}$ defines the unit vector in the direction of the relative vector. This tensor operator is a scalar product of two tensor operators of rank two, acting in spin and coordinate space, respectively. The local functions $v_\alpha(r)$ in (3.71), depending on the distance of the interacting nucleons, are adjusted in such a way that the phase shifts calculated from these potential for NN at energies up to the threshold for the production of pions fit the empirical data in the various partial waves. Examples of such empirical potentials are the Argonne potential (Wiringa et.al. 1984), the Urbana V_{14} potential (Lagaris and Pandharipande 1981), where the index 14 is used to indicate that fourteen operators O_α are considered in the expansion of (3.71), or the Reid soft-core potential (Reid 1968). It is evident from the large number of operators that must be considered that also the number of parameters to be adjusted is quite large (typically around 50). The radial dependences for such local interactions for two protons or two neutrons (total isospin $T_{iso}=1$) with antiparallel spins (total spin $S=0$) in a partial wave with orbital angular momentum for the relative motion $l = 0$, the partial wave $^{2S+1}l_J = {}^1S_0$, are displayed in Fig. 3.8. Very similar shapes are obtained if one tries to determine the local potential by scattering inversion methods as is done, e.g., by von Geramb (1993).

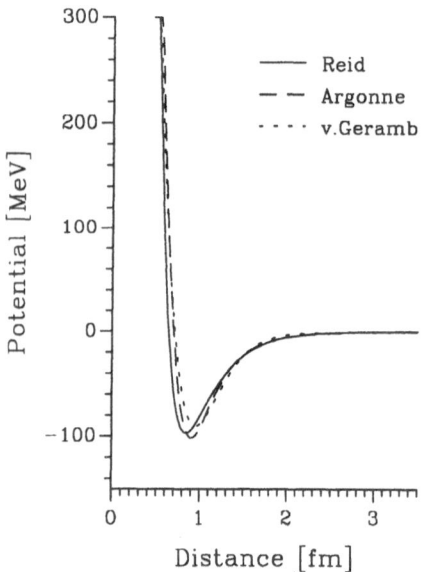

Fig. 3.8. Local NN interaction in the $^1S_0$ partial wave as defined in the Reid (1968) potential, the Argonne interaction of Wiringa et al. (1984), and that evaluated by von Geramb (1993) using scattering inversion techniques.

It is typical for these interactions that they exhibit a strong repulsion for distances around 0.5 fm or smaller. This short-range repulsion is required to obtain the negative phase shifts at scattering energies above 200 MeV as deduced from the analysis of the empirical data (see Arndt et al. 1987). For various empirical potentials such as that from Hamada and Johnston (1962), one even assumes a hard-core that is infinite for distances smaller than 0.4 fm.

The various potentials essentially reproduce the same experimental data: they are phase-shift equivalent. This means that they provide identical solutions for two nucleons interacting in the vacuum. Within a nuclear medium, however, the interacting nucleons are not on-shell any longer in the sense that the relation between momentum and energy is different from the one in the vacuum. Various models for the NN interaction, although they yield identical results for "on-shell" nucleons, will give different answers for the interaction of "off-shell" nucleons. Therefore one would like to get rid of this model dependence and use models for the NN interaction that have a microscopic foundation and less parameters to adjust than those empirical descriptions.

Since a direct solution of QCD in the low-energy regime seems to be impossible, one follows the arguments of t'Hooft (1974) and Witten (1979), who demonstrated that in this limit QCD should be dominated by the features of a meson theory. For the model of the NN interaction this means that it is dominated by the exchange of the "collective" quark–antiquark excitations, i.e., the exchange of mesons as displayed in Fig. 3.9. More detailed discussions of the meson exchange model have been presented, e.g., by Erkelenz (1974), Holinde (1981), Machleidt (1989), Brown and Jackson (1976).

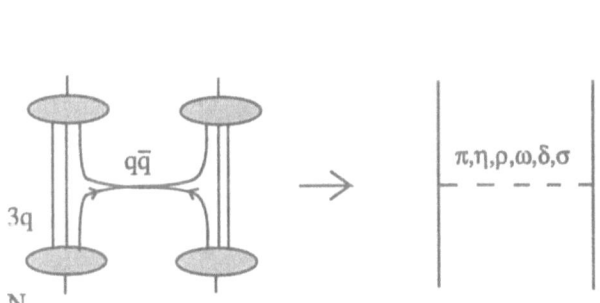

Fig. 3.9. Nucleon-Nucleon interaction in the quark- and meson-exchange picture. The lines on the left refer to propagating quarks, whereas those in the right part refer to nucleons and mesons. Note that the exchange of quark-antiquark (left) is replaced by various mesons (right).

3.3.1 The One-Boson-Exchange Model

As early as in 1935 Yukawa suggested that a new particle with "intermediate" mass, eventually called a meson, should be responsible for the strong interaction between nucleons. In order to exchange such a meson the interacting nucleons have to supply the energy for its creation ($\Delta E \approx \mu c^2$ with μ the mass of the mesons). In a process of elastic scattering the nucleons can provide this energy only virtually, i.e., only for a time period compatible with the uncertainty principle, $\Delta t \approx \hbar / \Delta E$. Assuming that the meson is exchanged at the speed of light c, the range of the interaction is around

$$\Delta r = c\Delta t = \frac{\hbar c}{\mu c^2} = \frac{197}{\mu \,[\text{MeV}]}\,[\text{fm}]. \tag{3.73}$$

This means that the meson with the lowest mass, the pion ($\mu = 139$ MeV), should be responsible for the long-range part of the interaction. The attraction at medium ranges is dominated by the exchange of two pions described in terms of an effective meson, called the σ meson, with a mass around 500 MeV, whereas the short-range repulsion should be provided by the exchange of a vector meson, the ω, with a mass of 783 MeV, which yields repulsion like the exchange of a massive photon between particles with identical charges.

In the one-meson or one-boson-exchange (OBE) model of the interaction one assumes that the basic NN interaction is described by the exchange of all mesons displayed in the right-hand part of Fig. 3.9. The diagram of that figure can be interpreted as a Feynman diagram representing the simplest non-trivial contribution to the two-nucleon Green's function. Such contributions can be evaluated by using essentially the same rules as discussed for the single-particle Greens's function in the Sect. 3.2. The basic building blocks for evaluating these contributions are the operators for the meson-nucleon vertices, which are given by

$$
\begin{aligned}
\Gamma_{\rm s} &= & \sqrt{4\pi}g_{\rm s} & \qquad \text{for a scalar meson,} \\
\Gamma_{\rm ps} &= & i\sqrt{4\pi}g_{\rm ps}\gamma^5 & \qquad \text{for a pseudoscalar meson,} \\
\Gamma_{\rm v} &= & \sqrt{4\pi}\left[g_{\rm v}\gamma^\mu + \frac{f_{\rm v}}{2m}i\sigma^{\mu\nu}k_\nu\right] & \qquad \text{for a vector meson.}
\end{aligned} \tag{3.74}
$$

As above, γ^μ are the usual matrices using the conventions of Bjorken and Drell (1964), the tensor operator is defined by the commutator $\sigma^{\mu\nu} = i[\gamma^\mu,\gamma^\nu]$ and k_μ refers to the 4-momentum of the exchanged meson. Instead of using the pseudoscalar coupling $\Gamma_{\rm ps}$ one also employs the pseudovector coupling

$$\Gamma_{\rm pv} = i\sqrt{4\pi}\,\frac{f_{\rm pv}}{m_{\rm mes}}\gamma^\mu\gamma^5 k_\mu. \tag{3.75}$$

Both couplings yield equivalent results for on-shell nucleons, if one makes $m_{\rm mes}g_{\rm ps} = 2mf_{\rm pv}$. For nucleons described by Dirac spinors different from those of a free particle, however, the pseudovector coupling suppresses the enhancements due to the antiparticle admixture as compared to pseudoscalar coupling.

If now one considers the interaction of two nucleons in the center-of-mass frame with momenta q and $-q$ before and the momenta q' and $-q'$ after the interaction, the matrix element for the exchange of a meson of the kind α is given by

$$V_\alpha(q',q) = \left(\bar{u}(-q')\Gamma_\alpha u(-q)\right)P_\alpha(k)\left(\bar{u}(q')\Gamma_\alpha u(q)\right), \tag{3.76}$$

with the Dirac spinors u as defined in (3.69) and $\bar{u} = u^\dagger\gamma^0$. Momentum conservation requires that the 4-momentum of the exchanged meson is $k = q - q'$ and the meson propagators are defined by

$$P_{\mathrm{s}} = \frac{1}{k^2 - m_{\mathrm{s}}^2} \quad \text{for scalar and pseudoscalar mesons,}$$

$$P_{\mathrm{v}} = \frac{-g_{\mu\nu} + k_\mu k_\nu / m_{\mathrm{v}}^2}{k^2 - m_{\mathrm{v}}^2} \quad \text{for vector mesons.} \tag{3.77}$$

A very efficient way for the evaluation of the OBE matrix elements in (3.76) using the helicity representation for the Dirac spinors has been presented by Erkelenz (1974). The two-particle states can be expanded in terms of eigenstates with respect to the total angular momentum J. Since the OBE amplitudes are invariant under rotation the angular momentum is a good quantum number and one obtains matrix elements

$$\langle \lambda_1' \lambda_2' q' | V | \lambda_1 \lambda_2 q \rangle_J = 2\pi \int_0^\pi d\theta \, \sin\theta \, d_{\lambda\lambda'}^J(\theta) \, \langle \lambda_1' \lambda_2' q' | V | \lambda_1 \lambda_2 q \rangle , \tag{3.78}$$

where λ_i denotes the helicity of nucleon i, i.e. the projection of its spin onto its momentum, θ is the angle between the momenta q and q', and the $d_{\lambda\lambda'}^J$ are the reduced rotation matrices with $\lambda = \lambda_1 - \lambda_2$ and $\lambda' = \lambda_1' - \lambda_2'$. Inspection of the symmetries for the matrix elements in (3.78) shows that there are six independent matrix elements between the various helicity states for each J. These matrix elements in the helicity representation can easily be transformed into the conventional basis of partial waves for two nucleons

$$|\lambda_1 \lambda_2 q\rangle_J \implies |^{2S+1} L_J q\rangle , \tag{3.79}$$

where S identifies the total spin of the nucleons, L is the orbital angular momentum of the relative motion, which is usually labelled by the letter S, P, D... for $L = 0, 1, 2...$, and $J = L + S$ is the total angular momentum. The Pauli principle for the interacting nucleons requires that the total isospin T_{iso} is related to these quantum numbers by the requirement that the sum $L + S + T_{\mathrm{iso}}$ is an odd number.

In order to account for the composite structure of the mesons and baryons, which is reflected by their finite size, one may consider the coupling constants in (3.74) not as universal constants but as depending on the 4-momenta of the interacting hadrons. A simple choice for such a form factor, which is commonly used, is to assume that it depends only on the momentum transfer k, the momentum of the meson, and takes the form

$$g_\alpha(k) = g_\alpha \left(\frac{\Lambda_\alpha^2 - m_\alpha^2}{\Lambda_\alpha^2 - k^2} \right)^\nu , \tag{3.80}$$

with a cut-off parameter Λ and an exponent ν, which is 1 for the so-called monopole form factor.

To demonstrate some features of the meson exchange model we will briefly discuss the one-pion-exchange amplitude. Since the π is a pseudoscalar and isovector meson, with electric charges π^+, π^-, and π^0, we have to consider the vertex for a pseudoscalar meson in (3.74) multiplied by the vector of Pauli matrices acting on the isospin part of the nucleon states, τ, and the corresponding propagator from (3.77). If we assume elastic scattering, $|q'| = |q|$, the zero-component of the momentum transfer vanishes and the operator takes the form

$$V_\pi = -4\pi \frac{g_\pi^2}{4m^2} \frac{(\sigma_1 k)(\sigma_2 k)}{k^2 + m_\pi^2} \tau_1 \cdot \tau_2 \,. \tag{3.81}$$

Using some simple manipulations we can write this expression as

$$
\begin{aligned}
V_\pi &= -\frac{4\pi g_\pi^2}{12m^2} \left[\left(3\sigma_1 \cdot \hat{k}\sigma_2 \cdot \hat{k} - \sigma_1 \cdot \sigma_2 \right) \frac{k^2}{k^2 + m_\pi^2} \right. \\
&\quad \left. +\sigma_1 \cdot \sigma_2 - \sigma_1 \cdot \sigma_2 \frac{m_\pi^2}{k^2 + m_\pi^2} \right] \tau_1 \cdot \tau_2 \,,
\end{aligned}
\tag{3.82}
$$

with $\hat{k}$ the unit vector in the direction of momentum transfer. From this representation we can see that the one-pion-exchange potential can be separated into three terms. The first one shows the structure of a tensor operator S_{12}, as already discussed in (3.72). The second term with the spin operator $\sigma_1 \cdot \sigma_2$ is independent of the momentum transfer k, which means that it is represented in r space by a δ function with respect to the distance between the interacting nucleons. The Fourier transformation of the third term yields the well known Yukawa potential:

$$\frac{4\pi m_\pi^2}{k^2 + m_\pi^2} \xrightarrow[\text{Fourier}]{} \quad m_\pi \frac{e^{-m_\pi r}}{r} \,. \tag{3.83}$$

It must be kept in mind, however, that this local representation of the one-pion-exchange potential has been possible only because we have ignored an energy transfer k^0 and the effects of a form factor. In general, more so for other mesons, the OBE model yields a non-local NN interaction.

3.3.2 NN Scattering

The parameters of the OBE model of the NN interaction, coupling constants, form factors, and meson masses, or any other model should be adjusted by solving the problem of two nucleons interacting in the vacuum. In the terminology of the Green's function approach discussed above, this means that one should evaluate the exact two-particle Green's function. As is also the case for the single-particle Green's function discussed above, we can classify the various diagrams contributing to the evaluation of the Green's function as displayed in Figs. 3.10 and 3.11.

The leading terms in Fig. 3.10 represent the products of two uncorrelated single-particle Green's functions denoting the direct and the exchange part of the propagation without any interaction. In the following discussion we will generally distinguish no longer between these two contributions, but represent them by one diagram. It is worth noting that for single nucleons in the vacuum the exact single-particle Green's function is identical to the free one.

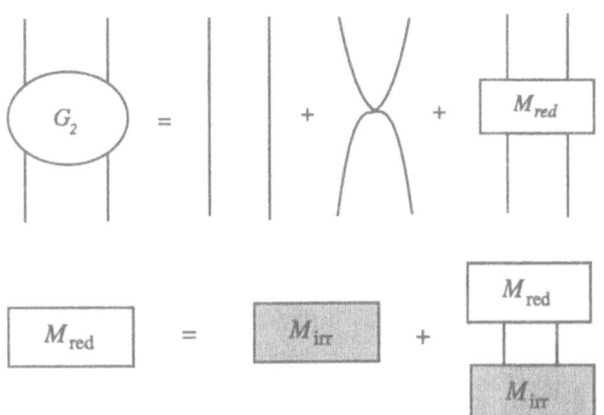

Fig. 3.10. The two-particle Green's function G_2 contains the direct and exchange term without any interaction plus all terms with an interaction M_{red} connecting the two lines.

Fig. 3.11. The Bethe–Salpeter equation (3.84) in terms of diagrams

The diagram on the very right of Fig. 3.10 represents all diagrams containing an interaction between the two nucleons. All the diagrams represented by M_{red} are reducible in the sense that they can be classified further into irreducible parts M_{irr}; those diagrams that cannot be further separated by a horizontal cut, which just cuts two lines, represent the Green's functions of the interacting nucleons. The reducible interaction term M_{red} is then obtained from the irreducible pieces by solving the Bethe–Salpeter equation

$$M_{red}(q',q;P) = M_{irr}(q',q;P) + \int d^4k \frac{i}{(2\pi)^4} M_{irr}(q',k;P)$$
$$\times \mathcal{G}_{BS}(k;P) M_{red}(k,q;P), \qquad (3.84)$$

which is displayed in Fig. 3.11 via the language of diagrams. Examples for irreducible diagrams contributing to M_{irr} are given in Fig. 3.12.

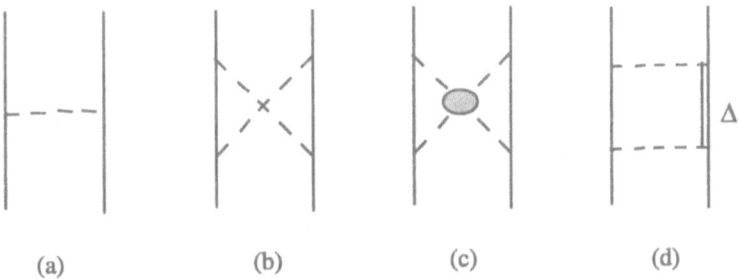

(a)   (b)   (c)   (d)

Fig. 3.12. Examples of diagrams contributing to the irreducible NN interaction M_{irr}. The one-meson term (a), the cross-box diagrams without (b) and with (c) the interaction of two mesons being exchanged simultaneously, and a contribution (d) in which one of the interacting nucleons is excited (indicated by the double line) are displayed.

In (3.84) and in the following, the momenta of the interacting nucleons are defined in terms of the center-of-mass momentum P, which is conserved, and the relative momenta q, q', and k in such a way that the momenta of the particles are, e.g., $p_i = 1/2 P \pm k$. In the center of mass frame the total momentum P has a time like component, which is identical to the total energy $\sqrt{s}$, with s referring to the corresponding Mandelstam variable (see e.g. Itzykson and Zuber, 1980), whereas the space component of P is equal to zero. The uncorrelated two-particle Green's function $\mathcal{G}_{BS}$ occurring in (3.84) can be written as a product of single-particle Green's functions, to give, using the relativistic form of (3.67),

$$\mathcal{G}_{BS}(k; P) = \left(\frac{1}{\frac{1}{2}\not{P} + \not{k} - m + i\eta} \right)^{(1)} \left(\frac{1}{\frac{1}{2}\not{P} - \not{k} - m + i\eta} \right)^{(2)}, \qquad (3.85)$$

where the superscript (1) or (2) refers to the corresponding nucleon. It is common practice to ignore the propagation of the solutions of negative energy and furthermore reduce the 4-dimensional Bethe–Salpeter equation (3.84) to an integral equation in three dimensions by fixing the time component of k in a covariant way. One of the possible choices is the approach suggested by Blankenbecler and Sugar (1966):

$$\mathcal{G}_{BBS}(k; P) = \delta(k_0) \frac{i}{2\pi} \frac{m^2}{E_k} \frac{\Lambda^{+(1)}(k)\Lambda^{+(2)}(-k)}{\frac{1}{4}s - E_k^2 + i\eta}, \qquad (3.86)$$

with $\Lambda^{+(i)}$ referring again (3.68) to the projector on Dirac states with positive energy for particle i. Assuming that the time-like component of k_0 vanishes means that for the propagation of the intermediate states both nucleons are considered to be off-shell by the same amount and the irreducible interaction terms do not transfer energy between the interacting nucleons. This implies that no energy transfer should be assumed for the meson propagators if one considers the contributions of OBE in the Blankenbecler–Sugar approximation. Replacing the Bethe–Salpeter propagator $\mathcal{G}_{BS}$ in (3.84) by the Blankenbecler–Sugar approximation one obtains

$$\begin{aligned} M_{red}(q', q) &= M_{irr}(q', q) + \int \frac{d^3 k}{(2\pi)^3} M_{irr}(q', k) \frac{m^2}{E_k} \\ &\quad \times \frac{\Lambda^{+(1)}(k)\Lambda^{+(2)}(-k)}{q^2 - k^2 + i\eta} M_{red}(k, q), \end{aligned} \qquad (3.87)$$

where we have replaced s by $4E_q^2$, using $E_q = \sqrt{q^2 + m^2}$. Assuming that the matrix elements of M_{irr} are calculated between spinors, which correspond to the solution of the Dirac equation with positive energy (to account for the projectors Λ^+), we may define

$$V(q', q) = \sqrt{\frac{m}{E_{q'}}} M_{irr}(q', q; P) \sqrt{\frac{m}{E_q}},$$

$$T_{\text{scat}}(q', q) = \sqrt{\frac{m}{E_{q'}}} M_{\text{red}}(q', q; P) \sqrt{\frac{m}{E_q}}. \qquad (3.88)$$

This allows us to rewrite the Blankenbecler–Sugar equation (3.87) as

$$T_{\text{scat}}(q', q) = V(q', q) + \int \frac{\mathrm{d}^3 k}{(2\pi)^3} V(q', k) \frac{m}{q^2 - k^2 + i\eta} T_{\text{scat}}(k, q), \qquad (3.89)$$

which has the form of the non-relativistic Lippmann–Schwinger equation for the scattering T_{scat} matrix. This means that if we evaluate the relativistic matrix elements of M_{irr} and apply the so-called "minimal relativity" factors of (3.88), the (relativistic) Blankenbecler–Sugar equation (3.87) becomes identical to the non-relativistic scattering equation. The states of this scattering equation can be rewritten in the usual partial-wave basis and the integral (3.89) and can be solved with the techniques as described by Haftel and Tabakin (1970).

In addition to the Blankenbecler–Sugar approximation, which we have just discussed, various other three-dimensional reductions of the Bethe–Salpeter equation have been discussed in the literature (see, e.g., Kadychevsky 1968, Gross 1969, Thompson 1970, Schierholz 1972, Erkelenz 1974). A detailed discussion of the various approaches has been presented by Brown and Jackson (1976).

The OBE model for the realistic NN interaction is now defined by assuming that the irreducible part of the relativistic interaction term is defined by considering just one-meson-exchange contributions for the various mesons. With this ansatz one solves a scattering equation like, e.g., the Blankenbecler–Sugar equation, calculates phase shifts, and adjusts the parameter of the OBE ansatz until the experimental data are reproduced. A typical set of parameters resulting from such a fit are displayed in Table 3.1. As an example we consider in this table parameters of a recent version of the "Bonn" potential [potential B defined in Table A.1 by Machleidt (1989)]. Some of the OBE parameters, such as the masses of the π, η, ρ, ω, and δ mesons, are not free parameters but are taken from the mass table of the Particle Data Group (1994). Other parameters, such as the cut-off parameters Λ and the contributions from η and δ exchange, do not effect the fit very much but are used as a fine tuning. The coupling constant for the π is very well constrained by the πN scattering data.

Also the coupling constants for the ρ meson, in particular the large tensor coupling f_ρ are deduced from a dispersion analysis of πN data by Höhler and Pietarinen (1975). As we will see below, this strong coupling for the ρ is of some significance for the nuclear structure calculation. A non-relativistic reduction for the ρ exchange, similar to the one performed for the π in the preceding subsection, yields a tensor component for the NN interaction with a sign opposite to the one deduced from one-pion-exchange in (3.82). Therefore a strong ρ exchange reduces the tensor force originating from the π exchange significantly.

Table 3.1. Parameters of a realistic OBE potential. The second column displays the type of meson: pseudoscalar (ps), vector (v) and scalar (s) and the third its isospin T_{iso}.

Meson	Type	T_{iso}	m_α [MeV]	$g_\alpha^2/4\pi$	Λ_α [MeV]
π	ps	1	138.03	14.4	1700
η	ps	0	548.8	3	1500
ρ	v	1	769	0.9[a]	1850
ω	v	0	782.6	24.5[a]	1850
δ	s	1	983	2.488	2000
σ	s	0	550[b]	8.9437[b]	2000

[a] The tensor coupling constants are $f_\rho = 6.1\, g_\rho$ and $f_\omega = 0$.
[b] The σ parameters apply for NN channels with isospin 1, for isospin 0 values of 720 MeV and 18.3773 were used for the mass and the coupling constant, respectively.

The ω coupling constant used in the OBE potential displayed here but also in other realistic OBE models is rather large. A simple quark model with SU(3) flavor predicts the ω coupling to be nine times as large as that for the corresponding isovector vector meson, the ρ. The strong ω exchange contribution, however, is required to obtain sufficient repulsion for the NN interaction at short distances. A possible reason for this discrepancy may be that the ω exchange in the OBE model contains an effective parametrization of short-range repulsion originating from quark effects (see, e.g., Faessler et al. 1983) or irreducible multi-meson exchange, terms which we will discuss in the next subsection.

The only part of the OBE model, that has a purely phenomenological origin is the σ exchange, which describes medium-range attraction of the NN interaction. Nevertheless, it must be considered as a great success that the OBE model yields a very accurate description of the NN scattering experiments by adjusting fewer parameters than, e.g., the parametrizations in terms of local interactions [see (3.71)]. Even more, as we have just discussed, most of these parameters are adjusted within the constraints given by independent observables.

3.3.3 Medium-Range Attraction

Since the medium-range attraction of the NN interaction has so far been described in a very phenomenological way only, by employing the σ exchange, we are going to discuss the microscopic origin of this part in more detail. From our discussion of the irreducible parts of the NN interaction M_{irr} to be used in the Bethe–Salpeter equation (3.84) we know already that the one-meson exchange part describes only a part of M_{irr}. Since the mass of the π is so low compared to the mass of the other mesons, it seems quite plausible that 2π-exchange terms, which are irreducible with respect to the NN states, contribute significantly to the NN interaction at medium range.

As a first example of such irreducible diagrams we will discuss the contribution displayed in Fig. 3.12d. This diagram represents the process in which the interacting nucleons exchange a meson (here a π), one of them gets excited (we will assume the mode lowest in energy, the $\Delta(3,3)$ excitation at 1232 MeV) and gets deexcited by a second exchange of a π. The contribution from this diagram can be evaluated from

$$M_{\mathrm{irr}}^{\mathrm{N}\Delta}(q',q) = \langle q'|V_{\mathrm{N}\Delta}^{\dagger}P_{\mathrm{N}\Delta}V_{\mathrm{N}\Delta}|q\rangle \tag{3.90}$$

in terms of a transition amplitude for the process NN→NΔ, $V_{\mathrm{N}\Delta}$, its hermitian conjugate, and the propagator $P_{\mathrm{N}\Delta}$ for the intermediate NΔ state. Relativistic expressions for these terms have been discussed, e.g., by Holinde et al.(1978) and Machleidt et al. (1987). Here we want to simplify the discussion and simply consider the non-relativistic reduction of the π contribution to the transition potential $V_{\mathrm{N}\Delta}$ in analogy to (3.81) and (3.82). Using the coupling constants for pseudovector coupling at the NNπ vertex ($f_{\mathrm{NN}\pi}$) and the N$\Delta\pi$ vertex ($f_{\mathrm{N}\Delta\pi}$) it is given by

$$V_{\mathrm{N}\Delta}^{\pi} = -\frac{f_{\mathrm{NN}\pi}f_{\mathrm{N}\Delta\pi}}{m_{\pi}^2}\frac{(\sigma_1 k)(S_{\mathrm{N}\Delta,2}k)}{k^2 + m_{\pi}^2}\tau_1 \cdot T_{\mathrm{N}\Delta,2}\,, \tag{3.91}$$

where $S_{\mathrm{N}\Delta}$ denotes the three vector components of a transition operator between a two-component spinor for a nucleon (spin 1/2) and a four-component spinor for the Δ, which has a spin of 3/2. The matrix elements of the spherical components of this vector operator are defined by the reduced matrix element

$$\langle\Delta||S_{\mathrm{N}\Delta}||\mathrm{N}\rangle = 2\,. \tag{3.92}$$

The operator $T_{\mathrm{N}\Delta}$ is defined in straight analogy, acting on the isospin part of the states, as the isospins are also 1/2 and 3/2 for N and Δ, respectively. The value for the coupling constant $f_{\mathrm{N}\Delta\pi}$ is very well determined from the decay width of the Δ. In a way quite similar to the steps leading from (3.81) to (3.82), the transition potential of (3.91) can also be decomposed into a zero-range, a central and a tensor term. Within the non-relativistic approach the propagator for the NΔ system is given by

$$P_{\mathrm{N}\Delta} \rightarrow \int \frac{\mathrm{d}^3 k}{(2\pi)^3}\frac{|\mathrm{N}\Delta\,k\rangle\langle\mathrm{N}\Delta\,k|}{2E(q) - E(k) - E_{\Delta}(k) + i\eta}\,, \tag{3.93}$$

with a projector on intermediate NΔ states with momentum k and $-k$ for the two baryons. Here E_k denotes the energy of a nucleon with momentum k and $E_{\Delta}(k)$ is the energy of a Δ including the NΔ mass difference. When we put the propagator and the transition potential together, it is evident that the $M_{\mathrm{irr}}^{\mathrm{N}\Delta}$ defined in (3.90) provides an attractive contribution to the NN interaction.

Since the isospin of the Δ is $3/2$, terms with intermediate $N\Delta$ states can contribute only to the NN interaction in partial waves with total isospin $T_{iso} = 1$. An attractive contribution for both $T_{iso} = 0$ and $T_{iso} = 1$ partial waves is obtained from terms with intermediate $\Delta\Delta$ states, which can be described in a similar way as just discussed. These terms provide a significant contribution to the medium-range attraction, which is simulated by the σ meson in OBE. Note that the OBE displayed in Table 3.1 uses a larger mass for the σ meson in the NN $T_{iso} = 0$ than in the $T_{iso} = 1$ partial waves. This may reflect that the Δ terms in $T_{iso} = 0$ (two intermediate Δ) lead to a shorter range than those in $T_{iso} = 1$, where one has terms with intermediate $N\Delta$ as well.

Until now we have considered only the contribution of the π exchange to the transition potentials such as $V_{N\Delta}$. If we take into account the spin–isospin selection rules, the ρ meson can also contribute to these transition potentials. Also the ρ contribution can be approximated in a local form. One finds that the tensor part of the ρ exchange contribution to the transition potential has the opposite sign as compared to the tensor term in the π exchange. This is similar to the role of π and ρ in the OBE part of the NN interaction discussed above. As the tensor contributions play a significant role, this implies that by including the ρ, part of the attraction just discussed is canceled. It is obvious that the terms with ρ exchange in the transition potential are of shorter range than the corresponding terms with π only. This suggests that this reduction in the medium-range attraction is simulated to some extent by the strong ω exchange used in pure OBE models of the NN interaction.

The irreducible terms with intermediate Δ excitations discussed so far represent just one specific kind of multi-meson term, which should be considered for M_{irr}. Others such as the cross-box diagrams of Fig. 3.12b,c with and without interaction of the meson exchanged also play a very important role in describing the NN interaction at medium and short ranges. A discussion of these various contributions has been given by Machleidt et al. (1987).

It is of course very tedious work to evaluate all the higher-order contributions to M_{irr} and one can never be sure that one has considered all important parts. Therefore we will shortly sketch another technique which has been used, e.g., by Lacombe et al. (1980) and Durso et al. (1977) to determine the parts of shorter range in the NN interaction within a meson model. In this scheme one relates the 2π-exchange contributions to the NN interaction to empirical data from nucleon-antinucleon ($N\bar{N}$) scattering, πN scattering and $\pi\pi$ scattering by means of dispersion relations.

Assuming that the NN scattering amplitude $M(s,t)$ is given as an analytic function depending on the Mandelstam variables s (the square of the NN center-of-mass energy, see above) and t (the square of the invariant momentum transfer), one can calculate the 2π-exchange contribution by means of the dispersion relation

$$M_{2\pi}(s,t) = \frac{1}{\pi} \int_{4m_\pi^2}^\infty \frac{\mathrm{Im}\{M_{2\pi}(s,t')\}}{t'-t} \mathrm{d}t' , \tag{3.94}$$

where $\text{Im}\{M_{2\pi}(s,t')\}$ denotes the imaginary part of the scattering amplitude M for positive and real momentum transfers due to 2π poles. This imaginary part is given by the cross section for $N\bar{N} \rightarrow 2\pi$. By using the dispersion relation above one makes a mathematical connection between two processes described by the same scattering amplitude but for different regimes of the variables s and t. One of these processes takes place in the particle–particle channel for positive values of s, i.e. time-like 4-momenta in the NN channel, while the other one can be observed for time-like 4-momenta in the cross channel $N\bar{N}$ for values of t above the threshold for 2π.

Within kinematical factors, the $\text{Im}\{M_{2\pi}(s,t')\}$ for the values of t' required above is related to the square of the scattering amplitude for $N\bar{N} \rightarrow 2\pi$. Also for this amplitude one can assume analyticity and relate it again by means of a dispersion relation to the scattering amplitude $\pi N \rightarrow \pi N$. The resulting amplitude $M_{2\pi}(s,t)$ contains the information from all contributions to M_{red} with two intermediate π. In order to obtain the parts that are irreducible with respect to NN cuts and therefore belong to M_{irr}, one has to subtract the contribution of the so-called 2π box terms, as these are accounted for by the solution of the Bethe–Salpeter equation (3.84).

This dispersion technique has been used in particular by Lacombe et al. (1980), leading to the so-called Paris potential. The Paris potential contains a very detailed description of the medium-range part of the NN interaction. For this part it is much more sophisticated than the OBE model and its extension is accounting for some diagrams beyond OBE. Since, however, it accounts only for terms with two pions, it cannot provide a microscopic description for the components of shorter range. Therefore, the Paris potential contains a very phenomenological description for the short-range components, which is adjusted by fitting NN scattering data. Due to the restriction to 2π contributions, the dispersion technique cannot account for the possible cancellations between π and ρ exchange as we have discussed above for the example of the terms with intermediate Δ excitations.

3.4 Hole-Line Expansion and Other Approaches

3.4.1 Brueckner–Hartree–Fock

We have reviewed various descriptions of a realistic NN interaction and will now discuss a few features of many-body calculations employing such realistic NN interactions for nuclear matter. The simplest non-trivial approach used to describe the properties of a many-fermion system like nuclear matter is the Hartree–Fock (HF) approximation. The translational invariance of this system ensures that the single-particle wave functions are plane waves. If we assume that the single-particle energies are given by a function that increases monotonically with the momentum of the nucleon k, the ground state of nuclear matter in the HF approximation is given by the Slater determinant, in which all states with momenta k smaller than the Fermi momentum k_F are occupied. This Fermi momentum is related to the density n of nuclear matter by

$$n \; = \; 4 \int_0^{k_F} \frac{d^3 k}{(2\pi)^3} \; = \; \frac{2k_F^3}{3\pi^2} \qquad (3.95)$$

with a factor 4 on the right-hand side of the first equation to account for the spin and isospin degeneracy of the plane-wave states in symmetric nuclear matter. As the empirical saturation density of nuclear matter is around 0.17 nucleons per fm^3, (3.95) yields a Fermi momentum k_F of 1.36 fm^{-1}. The average kinetic energy per nucleon for this Fermi gas can be calculated very simply by

$$\langle T_{\text{kin}} \rangle \; = \; \frac{1}{n} 4 \int_0^{k_F} \frac{d^3 k}{(2\pi)^3} \frac{k^2}{2m} \; = \; \frac{3}{5} k_F^2 2m \,. \qquad (3.96)$$

This yields a kinetic energy per nucleon of 23 MeV when $k_F = 1.36$ fm^{-1} (see Table 3.2). If now we calculate the single-particle energy in the HF approximation [see (3.64)] as

$$\epsilon_k^{\text{HF}} = \frac{k^2}{2m} + \sum_{\alpha < F} \langle k\alpha | V | k\alpha \rangle_{\text{A}} \qquad (3.97)$$

and evaluate the total energy using (3.62) as

$$E_{0,\text{HF}} \; = \; \frac{1}{A} \frac{1}{2} \sum_{\alpha < F} \left[\frac{k_\alpha^2}{2m} + \epsilon_\alpha^{\text{HF}} \right]$$

$$= \; \frac{1}{n} \frac{1}{2} 4 \int_0^{k_F} \frac{d^3 k}{(2\pi)^3} \left[\frac{k^2}{2m} + \epsilon_k^{\text{HF}} \right] \,, \qquad (3.98)$$

we obtain positive total energies rather than any binding energy. Using as an example for a realistic NN interaction the soft-core potential defined by Reid (1968), the result for the total energy per nucleon calculated at the empirical saturation density is about 176 MeV, which is far away from the empirical value of -16 MeV [see Table 3.2 and Müther (1986)].

The origin of this disaster in the HF approximation can easily be understood. If one calculates the single-particle energies according to (3.97), one has to evaluate matrix elements of the realistic NN potential V for a two-particle state, which is just a product of two uncorrelated plane waves. This product state can be rewritten in the center-of-mass frame for the two interacting nucleons, and the wave function for the relative motion is given in terms of a Bessel function $j_l(kr)$ with momenta k ranging from zero to the Fermi momentum k_F. Therefore the evaluation of the matrix elements of V occuring in (3.97) involves an integration on the distance of interacting nucleons r with an integrand $V(r) r^2 j_l^2(kr)$. The different contributions to the integrand are displayed in the schematic Fig. 3.13. The uncorrelated wavefunction $r j_l(kr)$ is quite large for distances r smaller than 0.5 fm, for which the NN potential

Table 3.2. Results for the kinetic energy per nucleon $\langle T_{\text{kin}} \rangle$ and the total energy per nucleon $E_{0,\text{HF}}$ and $E_{0,\text{BHF}}$ Also listed are the results for the energy corrections due to two-body correlations ($\Delta E_2 = E_{0,\text{BHF}} - E_{0,\text{HF}}$), three-body correlations (ΔE_3), as well as the effects of three- and four-body correlations evaluated in the particle-hole ring-diagram approximation (ΔE_3^{ring}, ΔE_4^{ring}) All entities are in MeV per nucleon. The results are displayed for densities of nuclear matter characterized by $k_F = 1.2$ fm^{-1} and 1.36 fm^{-1}. For the NN interaction the Reid soft-core potential has been used.

	$k_F = 1.2$	$k_F = 1.36$
$\langle T_{\text{kin}} \rangle$	17.9	23.0
$E_{0,\text{HF}}$	119.8	176.2
$E_{0,\text{BHF}}$	−9.0	−10.2
ΔE_2	−128.8	−186.4
ΔE_3	−3.5	−5.4
ΔE_3^{ring}	−3.4	−5.0
ΔE_4^{ring}	−0.5	−1.0

is very repulsive (see also Fig. 3.8). This leads to positive matrix elements for V and consequently to a positive energy in the HF approximation.

A more realistic nuclear wave function would be optimized in the sense that it yields a small probability whenever two nucleons are as close as the repulsive core and yields a probability larger than the one obtained from the HF wave function whenever two nucleons are at distances for which the interaction provides attraction. For large relative distances the correlated wave function should merge with the uncorrelated one ("healing property"). These features are indicated by the schematic wave functions displayed in Fig. 3.13.

This discussion demonstrates that it is essential for nuclear structure calculations employing realistic NN forces that one uses an approach that accounts for the effects of the correlations induced by the strong short-range components of the interaction. One possibility to achieve this goal is to solve the problem of two interacting nucleons in the presence of the others by solving the Bethe–Goldstone equation

$$G(Z) = V + V \frac{Q_{\text{Pauli}}}{Z - H_0} G(Z) . \tag{3.99}$$

This Bethe–Goldstone equation is very similar to the Lippmann–Schwinger equation (3.89) for the scattering matrix T_{scat}, which is the solution for the problem of two nucleons in the vacuum. One difference is due to the Pauli operator Q_{Pauli}, which is a projector on two-particle states with both interacting nucleons above the Fermi sea. Furthermore, the energy denominator in

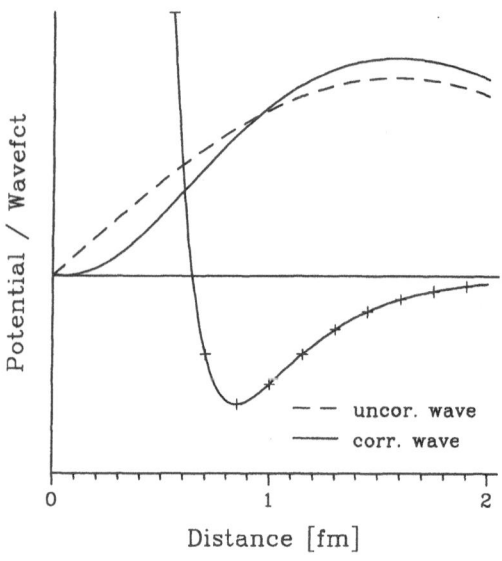

Fig. 3.13. Schematic plot of a local NN interaction V as a function of distance r. A typical wave function for the relative motion without correlations ($rj_0(kr)$, dashed line) is also displayed and compared to one that accounts for NN correlations.

(3.99) is defined in terms of energies of the many-body system rather than by using pure kinetic energies as in (3.89). This means that the propator $Q_{\text{Pauli}}/(Z - H_0)$ represents the part of the unperturbed two-particle Green's function in the nuclear medium, which describes the propagation of intermediate states with positive energy (or forward in time). In straight analogy with the scattering matrix T_{scat}, the G matrix deduced from the Bethe–Goldstone equation (3.99) can also be understood as an effective two-body interaction, which applied to the uncorrelated two-particle state yields the same result as does the bare interaction V applied to the correlated wave function.

Replacing the bare NN interaction V in (3.97) by this G matrix, one obtains the definition of the single-particle energy in the Brueckner–Hartree–Fock (BHF) approximation

$$\epsilon_k^{\text{BHF}} = \frac{k^2}{2m} + \sum_{\alpha < F} \langle k\alpha | G(Z = \epsilon_k^{\text{BHF}} + \epsilon_\alpha^{\text{BHF}}) | k\alpha \rangle_{\text{A}} . \tag{3.100}$$

Note that the definition of the starting energy Z in this equation, which is due to the BBP theorem of Bethe, Brandow, and Petschek (1963), requires a self-consistent solution of (3.99) and (3.100). Strictly speaking this BBP theorem can be justified only for states with momenta k below the Fermi momentum. Therefore, there has been a controversy for quite a long time [see, e.g., Jeukenne et al., (1976)] about the appropriate definition of the single-particle energies for the particle states, which also defines H_0 in (3.99). The difference between the "conventional choice", which ignores any single-particle potential for the particle states, and the "continous choice", which extends (3.100) to particle states as well, are not negligible and we will come back to this below.

Having evaluated the ϵ_k^{BHF} one can calculate the energy per nucleon for nuclear matter at a given density, using (3.98) and replacing ϵ_k^{HF} by ϵ_k^{BHF}. Repeating this procedure for various densities or Fermi momenta k_F one obtains the energy per nucleon as a function of density in the BHF approximation as indicated by the corresponding curve in Fig. 3.14 (using the Reid soft-core potential with "conventional choice" for H_0). The first point to be noticed is that the inclusion of NN correlations in the BHF approximation has cured the problems of the HF approximation to quite an extent. The predicitions for the energies are around -10 MeV per nucleon, which means that the inclusion of NN correlations provides more than 100 MeV per nucleon of binding energy as compared to the HF approach (see Table 3.2).

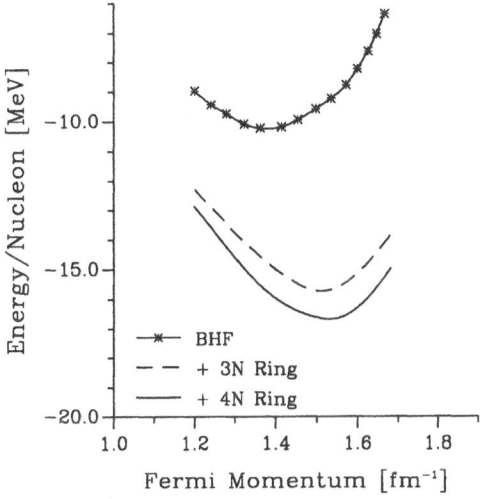

Fig. 3.14. Predictions for the energy of nuclear matter as a function of the Fermi momentum evaluated for the Reid soft-core potential. The results are displayed using the BHF approximation (with conventional choice for H_0) and including the effects of particle–hole ring diagrams of third and fourth order in G.

The BHF approach provides a first estimate for the energy of nuclear matter as a function of density and therefore for the equation of state for cold nuclear matter, which is derived from a realistic NN interaction and provides results that are in rough agreement at the normal density of nuclear matter. Inspecting the prediction for the saturation point of nuclear matter, which is determined as the minimum in the energy-versus-density (or Fermi momentum) curve of Fig. 3.14, one finds that the BHF approximation predicts -10.3 MeV at $k_F = 1.4$ fm^{-1}, which means a lack of binding energy and a saturation density that is too large as compared to the empirical value (-16 MeV at $k_F = 1.36$ fm^{-1}).

For some time physicists had hoped that using a different model for the NN interaction would lead to an improvement in the description of the saturation properties of nuclear matter. Therefore the evaluation of the saturation point has been repeated within the BHF approximation for various realistic NN interactions, which all reproduce the NN scattering data. Some results for the saturation points obtained in this way are presented in Fig. 3.15. One finds that for certain NN interactions (like, e.g., the Reid soft-core potential discussed above) one obtains the minimum at a density that is close to the empirical saturation density but that these potentials predict an energy per nucleon which is around 6 MeV, or even more, below the experimental value. Other potentials yield approximately the correct result for the binding energy but at a Fermi momentum around 1.68 fm^{-1} rather than 1.36 fm^{-1}. Notice that this means that the calculated density is almost twice as large as the empirical one. Other models for a realistic NN interaction yield results in between or predict larger energies at even larger saturation densities.

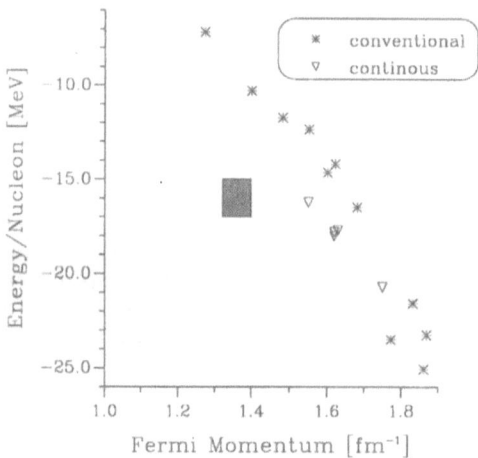

Fig. 3.15. Saturation points of nuclear matter calculated in the BHF approximation for various realistic NN interactions. The results are presented using the conventional as well as the continous choice for H_0 and are compared to the empirical data (crossed area).

This feature was first observed by Coester et al. (1970) and therefore the band of calculated saturation points is called the "Coester band". A similar phenomenon has also been obtained by Sauer and Tjon (1973) for the three-particle system and in the calculation of ground-state properties of finite nuclei (see, e.g., Schmid et al. 1991). Also for these finite nuclear systems it seems to be impossible to derive the binding energy and the radius in one microscopic calculation.

The intriguing question is, of course: what is the typical difference between NN interactions that yield a large binding energy for nuclear matter and those that give a small binding energy? One finds (see, e.g., Müther 1985) that those producing a large binding energy at a large saturation density typically contain a weak tensor force. Modern OBE potentials with a strong ρ exchange (see Sect. 3.3 on NN interactions) belong to this kind. The importance of the tensor force in determining the calculated binding energy is easily understood from the following argument. All the different NN interactions considered in Fig. 3.15 yield a good description for the NN phase shifts, which means that the NN scattering matrix $T_{\rm scat}$ is essentially the same. Depending on the potential used the contribution of the terms of second or higher order in V is more or less important than the Born term in (3.89). Especially a strong tensor force enhances the attraction obtained from the terms of higher order. In the Brueckner G matrix, however, this attraction originating from the higher-order terms is quenched due to the Pauli operator Q_{Pauli} in the Bethe–Goldstone equation (3.99) and the larger energy denominmators as compared to the Lippmann–Schwinger equation (3.89). If, for NN interactions with a strong tensor force, this attractive contribution is more important, the quenching will have stronger effects as well and therefore the G matrix will be less attractive for interactions with strong tensor components than for those with a weaker one.

A similar argument also explains why the "continous" choice for H_0 yields more binding energy than the conventional one discussed so far (see Fig. 3.15). The continous choice for H_0, which considers an attractive single-particle potential also for the particle states above the Fermi sea, leads to smaller energy denominators in the Bethe–Goldstone equation (3.99), which means the dispersive quenching discussed above is weaker. Nevertheless, also the continous choice produces saturation points along the "Coester" band, although one may observe a slight shift towards the empirical data in Fig. 3.15.

3.4.2 Beyond BHF

Since a careful treatment of NN correlation turns out to be very important, one is led to assume that correlations beyond those accounted for in the BHF approximation could change the results in a significant way. This suspicion had been supported by variational calculations for the binding energy of nuclear matter, which were found to be a few MeV below the results of the BHF approximation. This has been called a "crisis in nuclear matter theory" by Clark (1979). Such a variational calculation assumes an ansatz for the correlated many-body wave function of the form

$$|\Psi\rangle = \left(S \prod_{i<j} \hat{C}_{ij} \right) |\Phi\rangle , \qquad (3.101)$$

with Φ denoting the uncorrelated ground state of the free Fermi gas. The pair correlation operators $\hat{C}_{ij}$ are in general non-commuting operators, so the product must be symmetrized by S. These operators can be expanded into a basis set of two-body operators like the NN interaction in (3.71). The simplest non-trivial choice of $\hat{C}_{ij}$ has been the Jastrow ansatz, in which $\hat{C}_{ij}$ is a state-independent function depending on the distance of the pair of nucleons ij. Such a Jastrow correlation function could be chosen to modify the relative wave functions as displayed in Fig. 3.13. The optimal form of the pair correlation operators is determined by the variational principle

$$\delta \frac{\langle \Psi|H|\Psi\rangle}{\langle \Psi|\Psi\rangle} = 0 . \qquad (3.102)$$

There are two main problems in such a variational calculation. The first one is the reliable evaluation of the matrix elements in (3.102) for a given trial function of (3.101). Note that the integrals to be evaluated in (3.102) are of dimension $3A$, if A is the number of particles. Very efficient cluster expansion techniques have been developed, such as the Fermi-hypernetted-chain scheme (see Clark 1979), which account for the antisymmetrization of the many-body state. These cluster expansion techniques converge very well if the density of the system is not too high. The second problem is the optimal choice of the correlation operator, which can be rather complicate due to the rich operator structure of the NN interaction. Here, we do not want to go into the technical details of variational calculations but just mention that the comparison of the variational calculations with those obtained in the BHF approach has encouraged calculations in nuclear matter that go beyond the BHF approach.

The BHF approximation is the lowest-order approach in the hole-line expansion. This means that the definition of the irreducible self-energy in BHF accounts for all and only those diagrams that contain one internal hole line (these include, e.g., diagrams like the HF contributions displayed in Fig. 3.3d,e and Fig. 3.4a but not Fig. 3.3b). The reasoning behind this ordering of contributions with respect to the number of hole lines involved, is the assumption that nuclear matter is a system of low density in the sense that the average distance to the next neighbor is smaller than the range of the strong components in the NN interaction. Therefore the probability of observing a process in which n particles are excited in a coherent way, described by an irreducible diagram with n hole lines, is suppressed as compared to processes involving $n-1$ hole lines. The BHF approximation accounts for all terms with one hole line in the definition of the self-energy and two hole lines in the calculation of the energy. These are represented by the diagram of first order in the Brueckner G matrix (wiggly line at the left-hand side of the equation displayed in Fig. 3.16), which includes the whole series of diagrams displayed on the right-hand side of the equation in Fig. 3.16 (plus all exchange terms, which are not displayed).

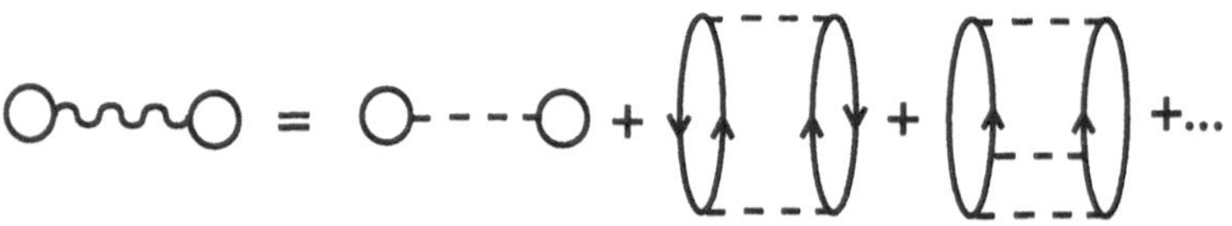

Fig. 3.16. Diagram representing the energy of the BHF approximation. The wiggly interaction line represents the G matrix and accounts for all ladders of NN interactions indicated on the left-hand side of the equation.

In order to account for all diagrams including up to three hole lines, one has to sum up all ladder diagrams for three particles. This means one has to solve the Bethe–Faddeev equations as defined, e.g., by Day et al. (1972), which corresponds to the solution of the problem of three interacting particles in the presence of all others. The contribution of these three-particle ladder diagrams can be evaluated as an expectation value of the form

$$\Delta E_3 = \sum_{i,j,k<F} \langle ij, k | F_{03}^{\dagger} F_3 (1 + \mathcal{X}) | ij, k \rangle \tag{3.103}$$

with a three-body operator

$$F_{03} = \mathcal{X} \frac{Q_{\text{Pauli}}}{Z_3 - H_0} G \tag{3.104}$$

acting on three-body states $|ij, k\rangle$ in such a way that the two-body operators of the G matrix and the Pauli operator Q_{Pauli} refer to the nucleons labeled by quantum numbers ij, while the third particle k is a spectator. Z_3 denotes the starting energy for all three particles and the exchange operator $\mathcal{X}$ is defined by

$$\mathcal{X} |ij, k\rangle = |jk, i\rangle + |ki, j\rangle. \tag{3.105}$$

Finally, the Bethe–Faddeev amplitude F_3 in (3.103) is obtained by solving an integral equation similar to the Faddeev equation for three particles in the vacuum of the form

$$F_3 = G \left[F_{03} + \mathcal{X} \frac{Q_{\text{Pauli}}}{Z_3 - H_0} F_3 \right]. \tag{3.106}$$

The effects of these three-hole-line contributions on the binding energy of nuclear matter have been accurately calculated for the Reid soft-core potential by Day (1981). At the empirical value of the saturation density one obtains an additional binding energy ΔE_3 of about -5 MeV per nucleon (see Table 3.2). This gain in binding energy would be large enough to move the saturation point calculated for the Reid soft-core potential to the empirical result. Repeating this calculation for various densities, one finds, however, that the minimum of the energy-versus-density curve is also shifted to higher densities. Therefore the inclusion of three-body correlations is not negligible (at least for the Reid potential) but simply moves the saturation point along the Coester band.

The evaluation of the effects of three-body correlations can also be used as a test for the convergence of the hole-line expansion. The gain in binding energy due to the three-body correlations (-5.4 MeV) is indeed very small compared to the effect of two-body correlations (-186.4 MeV, see Table 3.2), indicating a very good convergence of the hole-line expansion. To confirm this conclusion one may try to estimate the importance of four-nucleon correlations, which have not yet been evaluated accurately. For that purpose we consider the effects of the ring diagrams, examples of these diagrams are displayed in Fig. 3.17. It has been observed by Day (1981) that the ring diagrams with three hole lines yield by far the dominant contribution among all three-hole-line terms (see also Table 3.2). Therefore one can assume that the convergence of the ring diagrams with an increasing number of hole lines is a good indicator for the convergence of the hole-line expansion. The particle-hole ring diagrams describe the effect of ground-state correlations induced by collective particle-hole excitations within the RPA approximation. Note, that the ring diagram with n hole lines contains n G matrix interaction vertices. The series of ring diagrams may therefore be interpreted as a summation of some special diagrams classified according to the number of G interactions involved. However, it is not a perturbative expansion in terms of the bare NN interaction V.

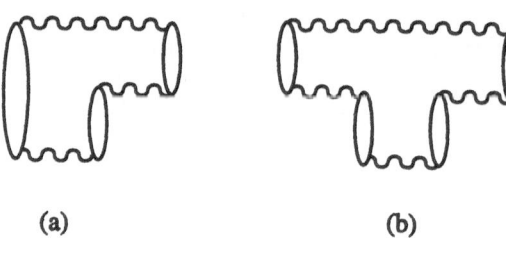

(a)   (b)

Fig. 3.17. Example for ring diagrams with three **(a)** and four **(b)** hole lines.

The effects of third- and fourth-order ring diagrams have been evaluated by Dickhoff et al. (1982). The results for the Reid soft-core potential were in very good agreement with those obtained by Day (1981) and are displayed in Table 3.2 and Fig. 3.14. One finds that the contribution due to four-hole ring diagrams is only about -1 MeV per nucleon; indeed a good indication for the convergence of the hole-line expansion. The results displayed in Fig. 3.14 furthermore show that the inclusion of these ring diagrams not only yields more binding energy but also shifts the saturation point to large densities, which means that it is moved again along the Coester band.

The conclusion about the convergence of the hole-line expansion gets strong support from a comparison with variational calculations mentioned above (3.102). Using the Argonne potential of Wiringa et.al. (1984), Day and Wiringa (1985) demonstrated that the hole-line expansion, after the effects of three-hole line terms are included, yields results for the energy that are below but fairly close to the upper bounds obtained in the variational calculation.

Unfortunately, such comparisons have been made possible until now only for NN interactions of the form of (3.71) defined in coordinate space. Modern OBE potentials, which are typically defined in a momentum space and would correspond to non-local interactions, cannot yet be handled by the tools developed for variational methods. The estimate of ring diagrams by Dickhoff et al. (1982) gives rise to the hope that the convergence of the hole-line expansion for these modern OBE interactions discussed in Sect. 3.3 is even better than for the Reid potential due to the weaker tensor force.

Of course there exist also schemes other than the hole-line expansion for truncating an expansion for the nuclear many-body problem. We mention the self-consistent Green's function (SCGF) approach, which we have sketched in Sect. 3.2 and which has been discussed in more detail by

Dickhoff and Müther (1992). The main difference between the hole-line expansion and SCGF is the symmetric treatment of particle and hole states. This implies that a low-order calculation would not only account for particle–particle ladders as in the G matrix, but consider hole–hole states, propagating backward in time as well. Calculations by Jiang et al. (1988) indicate that the inclusion of all these particle–particle hole–hole ring diagrams tends to move the calculated saturation point away from the Coester band towards the empirical data. However, a complete calculation with a real self-consistent treatment of the single-particle Green's function is still needed to confirm this result. It is interesting to note that recent investigations for finite nuclei by Müther et al. (1995) also yield such an improvement.

Concluding this part, one may say that the strong short-range components of a realistic NN interaction require a many-body calculation of nuclear systems, which account at least for two-nucleon correlations. One very popular way to achieve this is the BHF approximation. It yields results for the saturation properties of nuclear matter, which depend on the model for the NN interaction, and form the so-called Coester band in the energy-versus-density plot (see Fig. 3.15), which does not meet the empirical data. Higher-order corrections lead to modifications, but the results for the saturation point are typically shifted along the Coester band. Although this conclusion may not be final (see, e.g., the SCGF approach), one may say that the conventional model of nuclear physics discussed so far yields quite reasonable predictions for the properties of normal nuclear matter but some fine tuning might be required to obtain a good agreement with the empirical data.

One possibility for such fine tuning is to assume a three-body force, as has been done by Friedman and Pandharipande (1981). They adjust parameters of a three-nucleon interaction in such a way that a variational calculation [see (3.102)] for a Hamiltonian that contains this three-body force in addition to the realistic Urbana potential of Lagaris and Pandharipande (1981) yields a result for the energy as a function of density $E(n)$ that saturates at the empirical value for nuclear matter. From $E(n)$ one can evaluate the energy density $\rho(n)$ and the pressure $p(n)$ as

$$\rho(n) = n\big(E(n) + m\big),$$
$$p(n) = n^2 \frac{\partial E(n)}{\partial n}. \tag{3.107}$$

The equation of state $p(\rho)$ for cold nuclear matter is obtained by eliminating the density n from these two equations. In a similar way one can calculate the equation of state for cold neutron matter or any ratio of protons and neutrons in asymmetric nuclear matter.

In variational calculations for matter at temperatures different from zero, one replaces the Slater determinant Φ in (3.101) by a corresponding Fermi-gas state with occupations of single-particle states between zero and one and performs a variational calculation for the free energy. In a rather similar way one can also perform BHF calculations for a Fermi gas at a temperature different from zero. In this case one replaces the Pauli operator in the Bethe–Goldstone equation (3.99) by one which accounts for these occupations.

The equation of state evaluated by Friedman and Pandharipande (1981) has been used very frequently in studying phenomena of stellar objects. One must keep in mind, however, that the three-body force has been introduced in a very phenomenological way. This is a rather delicate approximation, in particular because many-body forces have large effects at high densities. Of course there is no reason why such three-nucleon forces should not be present within a nuclear medium. However, as we will see below, it is very important to treat many-body forces and many-nucleon correlations in a consistent way.

3.5 Relativistic Effects

3.5.1 Walecka Model for Nuclear Matter

Up to this point we have considered a non-relativistic many-body theory to describe the properties of nuclear systems. At first sight this can be justified from the following consideration. The depth of the single-particle potential, or in the nomenclature of Sect. 3.2 of the self energy $\mathcal{M}$, is around 60 MeV for finite nuclei and 80 MeV for nuclear matter. This is quite small compared to the rest mass of the nucleons, $m = 938$ MeV, and therefore relativistic effects should be negligible.

This argument has been called in question by phenomenological models accounting for nucleonic as well as mesonic degrees of freedom such as the so-called Walecka model (see Walecka 1974, Chinn 1977, Serot and Walecka 1986). The starting points in these models are relativistic covariant Lagrangians for a field theory of nucleons, represented by the Dirac field Ψ, a scalar meson σ (mass m_σ), and a vector meson ω (with a mass m_ω), which are of the form

$$
\begin{aligned}
\mathcal{L}_{\sigma\omega} = {} & \bar{\Psi}\left(i\gamma_\mu\partial^\mu - m - g_\sigma\sigma - g_\omega\gamma_\mu\omega^\mu\right)\Psi \\
& + \frac{1}{2}\partial_\mu\sigma\partial^\mu\sigma - \frac{1}{2}m_\sigma^2\sigma^2 - \frac{1}{3}g_{3\sigma}\sigma^3 - \frac{1}{4}g_{4\sigma}\sigma^4 \\
& - \frac{1}{4}\left(\partial_\mu\omega_\nu - \partial_\nu\omega_\mu\right)\left(\partial^\mu\omega^\nu - \partial^\nu\omega^\mu\right) + \frac{1}{2}m_\omega^2\omega_\mu\omega^\mu.
\end{aligned}
\tag{3.108}
$$

The g_σ and g_ω denote the meson–nucleon coupling constants and the coefficients $g_{3\sigma}$ and $g_{4\sigma}$ are self-coupling constants for the scalar field, which we will ignore in the first part of the following discussion. One can derive the Euler–Lagrange equations for this Lagrangian, which yields a coupled set of equations: a Dirac equation for the nucleon field,

$$
\left(i\gamma_\mu\partial^\mu - m - g_\sigma\sigma - g_\omega\gamma_\mu\omega^\mu\right)\Psi = 0\,;
\tag{3.109}
$$

and a Klein–Gordon equation and a Proca equation for the scalar and vector fields, respectively,

$$\partial_\mu \partial^\mu \sigma - m_\sigma^2 \sigma \;=\; -g_\sigma \bar{\Psi}\Psi,$$
$$\partial_\mu \left(\partial^\mu \omega^\nu - \partial^\nu \omega^\mu\right) + m_\omega^2 \omega^\nu \;=\; g_\omega \bar{\Psi}\gamma^\nu \Psi. \tag{3.110}$$

The meson fields are approximated in the mean-field approximation by the expectation value, which is for our static infinite system just a number independent of space and time. Therefore the solution of (3.110) is trivial and the Dirac equation (3.109) can be rewritten as

$$\left(i\gamma_\mu \partial^\mu - m + \frac{g_\sigma^2}{m_\sigma^2}\langle\bar{\Psi}\Psi\rangle - \frac{g_\omega^2}{m_\omega^2}\langle\bar{\Psi}\gamma^0\Psi\rangle\right)\Psi = 0, \tag{3.111}$$

The same Dirac equation can also be obtained by considering the contribution of a σ and ω meson to the OBE potentials described in Sect. 3.3, constructing a self-energy $\mathcal{M}$ in the Hartree approximation, which keeps track of the transformation properties under a Lorentz transformation, and adding this self-energy to the Dirac equation of the nucleon. In a microscopic many-body calculation one would like to account for the Pauli effects and therefore we prefer to discuss the Hartree–Fock approximation to the Dirac theory of nucleons interacting by the exchange of ω and σ mesons.

For systems that are invariant under translation, the self-energy for nucleons with momentum k can be split into three contributions, depending on the transformation properties:

$$\mathcal{M}(k) = \mathcal{M}^s(k) - \gamma^0 \mathcal{M}^0(k) + \gamma \cdot k \mathcal{M}^v(k). \tag{3.112}$$

If we insert this self-energy into the Dirac equation, the Dirac equation can be written in momentum-space representation:

$$[\gamma \cdot k + m + \mathcal{M}(k)]\,\tilde{u}(k, \lambda) \;=\; \gamma_0 E(k)\tilde{u}(k, \lambda)\,,$$
$$[\gamma \cdot k^* + m^*(k)]\,\tilde{u}(k, \lambda) \;=\; \gamma_0 E^*(k)\tilde{u}(k, \lambda)\,, \tag{3.113}$$

with the definitions

$$k^* \;=\; k\,(1 + \mathcal{M}^v(k))\,,$$
$$m^*(k) \;=\; m + \mathcal{M}^s(k)\,,$$
$$E^*(k) \;=\; E(k) + \mathcal{M}^0(k)\,. \tag{3.114}$$

The formal similarity between (3.113) and a free Dirac equation allows the immediate determination of the nucleon spinor $\tilde{u}$ in nuclear matter in the form of (3.69):

$$\tilde{u}(\mathbf{k}, \lambda) = \sqrt{\frac{2m^*}{E^*(k) + m^*}}\left(\begin{array}{c} 1 \\ \frac{\sigma \cdot k}{E^*(k)+m^*} \end{array}\right)|\lambda\rangle, \tag{3.115}$$

which are normalized in such a way that $\tilde{u}^\dagger \tilde{u} = 1$. The on-shell condition in nuclear matter is now

$$E^*(k)^2 = m^*(k)^2 + k^{*2}\,. \tag{3.116}$$

On the level of the Hartree–Fock approximation, if we assume the exchange of a scalar σ and a vector ω meson the different contributions to the self-energy are given by

$$\mathcal{M}^{\text{s}}(k) \;\doteq\; -\left(\frac{g_\sigma}{m_\sigma}\right)^2 n_{\text{s}} \tag{3.117}$$

$$+ \frac{1}{(4\pi)^2}\frac{1}{k}\int_0^{k_{\text{F}}} q\,dq\,\frac{m^*(q)}{E^*(q)}\left[g_\sigma^2 \Theta_\sigma(k,q) - 4g_\omega^2 \Theta_\omega(k,q)\right],$$

$$\mathcal{M}^0(k) \;=\; -\left(\frac{g_\omega}{m_\omega}\right)^2 n_0 \tag{3.118}$$

$$- \frac{1}{(4\pi)^2}\frac{1}{k}\int_0^{k_{\text{F}}} q\,dq\left[g_\sigma^2 \Theta_\sigma(k,q) + 2g_\omega^2 \Theta_\omega(k,q)\right],$$

$$\mathcal{M}^{\text{v}}(k) \;=\; -\frac{1}{(4\pi k)^2}\int_0^{k_{\text{F}}} q\,dq\,\frac{q^*}{E^*(q)} \tag{3.119}$$

$$\times \left[2g_\sigma^2 \Gamma_\sigma(k,q) + 4G_\omega^2 \gamma_\omega(k,q)\right].$$

The first term in $\mathcal{M}^{\text{s}}(k)$ and $\mathcal{M}^0(k)$ correspond to the Hartree contribution with

$$n_0(k_{\text{F}}) \;=\; \frac{4}{(2\pi)^3}\int_0^{k_{\text{F}}} d^3q\,\bar{\tilde{u}}\gamma^0\tilde{u} \;=\; \frac{2}{3\pi^2}k_{\text{F}}^3, \tag{3.120}$$

which is identical to the baryon density n, and

$$n_{\text{s}}(k_{\text{F}}) \;=\; \frac{4}{(2\pi)^3}\int_0^{k_{\text{F}}} d^3q\,\bar{\tilde{u}}\tilde{u} \;=\; \frac{2}{\pi^2}\int_0^{k_{\text{F}}} q^2\,dq\,\frac{m^*(q)}{E^*(q)}, \tag{3.121}$$

the so-called scalar density. The remaining expressions are due to the Fock (exchange) contributions where we have used the abbreviations

$$A_i(k,q) \;=\; k^2 + q^2 + m_i^2 - \left(E(k) - E(q)\right)^2 \tag{3.122}$$

$$\Theta_i(k,q) \;=\; \ln\left(\frac{A_i(k,q) + 2kq}{A_i(k,q) - 2kq}\right) \tag{3.123}$$

$$\Gamma_i(k,q) \;=\; \frac{A_i(k,q)\Theta_i(k,q)}{4kq} - 1, \tag{3.124}$$

for the mesons $i = \{\sigma, \omega\}$. The parameters of this Dirac–Hartree–Fock model g_σ and g_ω [assuming that the masses m_σ and m_ω correspond to the values of OBE potentials (see Table 3.1) and keeping the constants for the self-coupling of the scalar field $g_{3\sigma}$ and $g_{4\sigma}$ in the Lagrangian of (3.108) at zero] can be adjusted to obtain the empirical saturation point of nuclear matter.

These fits lead to a strong cancellation in the various contributions to the self-energy. The scalar part $\mathcal{M}^{\text{s}}$ is mainly due to the exchange of the σ meson [see the corresponding Hartree contribution in (3.117)] and is very attractive leading to values as large as -400 MeV at the saturation density. The zero component of the vector term $\mathcal{M}^0$, which is dominated by the ω exchange, on the other hand yields a repulsive contribution (around 320 MeV) to the energy $E(k)$, as can be seen from (3.114). Therefore the net effect of the NN interaction on the energy of a nucleon in nuclear matter ($\mathcal{M}^s - \mathcal{M}^0 = -80$ MeV for $k=0$, with the numbers just quoted) is small as compared to the rest mass of the nucleons. This is in agreement with the experimental observation and led to the argument that relativistic effects can be ignored.

The results of the Walecka model demonstrate, however, that the small binding energy is due to a cancellation of strong attractive and repulsive components, as we have seen already in the discussion of realistic NN interactions in Sect. 3.3. The various contributions to the self-energy $\mathcal{M}$ are comparable in size to the rest mass of the nucleons. As they have quite a different transformation properties under Lorentz transformation, they enter the Dirac equation in different ways. The large scalar contribution $\mathcal{M}^s$ can be expressed in terms of an effective mass m^* [see (3.114)] which is substantially smaller than the bare mass. This leads to solutions of the Dirac equation $\tilde{u}$ in (3.115) that show an enhancement of the small component as compared to the Dirac spinors of nucleons with the same momentum in the vacuum.

Recently the Walecka model in various versions has become very popular for evaluating an equation of state for stellar objects (see, e.g., Glendenning et al. 1983, Weber and Weigel 1989, Sumiyoshi and Toki 1994, and references therein). A simple fit of just the nuclear matter saturation point by adjusting the coupling constants g_σ and g_ω produces a rather stiff equation of state. Attempts have been made to include the self-interaction terms of the scalar field as well and adjust the constants $g_{3\sigma}$ and $g_{4\sigma}$ in the Lagrangian (3.108) to describe properties of finite nuclei (see, e.g., Gambhir et al. 1990). Various extensions of this model have been considered to account for additional degrees of freedom like fields for pions, hyperons, and Δ excitations. All these approaches, however, should be considered as purely phenomenological ones in the sense that the parameters are adjusted to describe properties of nuclear systems at normal densities. Therefore the predictive power for the equations of state at higher densities or for systems with large proton–neutron asymmetries is rather limited. It is comparable to those obtained with effective forces like the Skyrme force between nucleons (see discussion in Sect. 3.1).

3.5.2 Dirac–Brueckner–Hartree–Fock

It is the aim of the Dirac–Brueckner–Hartree–Fock approach (DBHF) to include the relativistic features exhibited in the Walecka model in a microscopic many-body calculation, starting from a realistic meson-exchange interaction as discussed in Sect. 3.3. This implies that the evaluation of the matrix elements for the bare potential requires the knowledge of the Dirac spinors, evaluated for the nucleons in the medium. Employing these matrix elements one can solve the Bethe–Salpeter equation, or a three-dimensional reduction of the Bethe–Salpeter equation such as the Blankenbecler–Sugar equation (3.87), modified to account for the Pauli and medium effects as is done in the Bethe–Goldstone equation (3.99), which yields a relativistic G matrix. This G matrix allows the determination of the self-energy $\mathcal{M}$, which should be included in the Dirac equation (3.113). Finally, the solution of the Dirac equation provides the Dirac spinors and energies, which are required for the evaluation of matrix elements and the solution of the Bethe–Goldstone equation.

If we reduce the scattering or Bethe–Goldstone equation to a three-dimensional one, it is not a trivial problem to extract the Dirac structure of the G interaction, which is needed to determine the Dirac structure of the self-energy $\mathcal{M}$. One possibility is to represent the G matrix operator, similar to the non-relativistic ansatz of (3.71), in terms of a complete set of two-body operators forming a Lorentz scalar:

$$
\begin{aligned}
\hat{G} \;=&\; \sum_m {}^m G \, \kappa_m^{(1)} \cdot \kappa_m^{(2)} \\[4pt]
=&\; {}^{\text{scalar}}G \mathbb{1}^{(1)} \mathbb{1}^{(2)} + {}^{\text{vector}}G \gamma^{\mu(1)} \gamma_\mu^{(2)} + {}^{\text{tensor}}G \sigma^{\mu\nu(1)} \sigma_{\mu\nu}^{(2)} \\[4pt]
&+ {}^{\text{pscalar}}G \gamma_5^{(1)} \gamma_5^{(2)} + {}^{\text{Avector}}G \gamma^{\mu(1)} \gamma_5^{(1)} \gamma_\mu^{(2)} \gamma_5^{(2)}
\end{aligned}
\tag{3.125}
$$

where the nomenclature $\kappa_m^{(i)}$ is used to indicate that the Dirac operator κ_m is applied to the spinor of nucleon i. In this equation we have suppressed the dependence of the operator $\hat{G}$ and the amplitudes ${}^m G$ on the momenta of the interacting nucleons and the density of nuclear matter. With the ansatz of (3.125) one can calculate (see, e.g., Horowitz and Serot 1987) for each set of momenta and density the six independent helicity matrix elements of (3.78) for an antisymmetrized matrix element $\langle \lambda_1' \lambda_2' | \hat{G} | \lambda_1 \lambda_2 \rangle$ and determine the five amplitudes ${}^m G$ in (3.125) to reproduce the matrix elements of the DBHF G matrix in the helicity representation. An analysis along this line has been made, e.g., by Boersma and Malfliet (1994) and Elsenhans et al. (1990) who parameterized the amplitudes ${}^m G$ in terms of effective meson exchange amplitudes.

A simpler parameterization can be obtained from a direct inspection of the self-energy for the nucleon in nuclear matter. From the G matrix elements one calculates the momentum dependence of the single-particle energy of nucleons as a function of momentum by using the BHF approach. The analysis of Boersma and Malfliet (1994) and Elsenhans et al. (1990) shows that the momentum dependence of $\mathcal{M}^s$ and $\mathcal{M}^0$ in (3.112) is rather weak and $\mathcal{M}^v$ is small compared to the other contributions to the self-energy. Taking this into account, Brockmann and Machleidt (1984) determined the values for $\mathcal{M}^s$ and $\mathcal{M}^0$ at various densities from the momentum dependence of the single-particle energies. The results obtained for the scalar part of the self-energy $\mathcal{M}^s$, represented in terms of the effective mass m^* [see (3.114)] are shown in the upper part of Fig. 3.18 in which the OBE potential A of Machleidt (1989) has been considered.

One finds that realistic NN interactions also predict a significant deviation of m^* in nuclear matter from the bare mass of the nucleon in the vacuum. This means that the Dirac spinors of nucleons in nuclear matter contain a small component that is enhanced compared to the Dirac spinor in the vacuum. This feature is more important for larger densities. The effect of this medium-dependence on the calculated energy of nuclear matter is displayed in the lower half of Fig. 3.18. If the medium dependence of the Dirac spinors is ignored, which corresponds to the conventional BHF approximation, the Bonn potential A predicts a saturation point at a Fermi momentum around 1.9 fm^{-1} and a binding energy of -23.5 MeV, which is located on the usual Coester band of Fig. 3.15. Taking the self-consistent determination of the Dirac spinors within the DBHF approximation into account, one finds only a rather small reduction of the energy at small densities up to a Fermi momentum of 1.3 fm^{-1}. A more detailed analysis, discussed, e.g., by Dickhoff and Müther (1992), shows that this small effect is a result of a cancellation since the modification of the Dirac spinors in the medium yields a significant reduction of the calculated attractive potential energy, which is partly compensated for by a reduction in the kinetic energy. At larger densities the loss of attraction in the two-body term overwhelms the loss of repulsion in the one-body term of the kinetic energy and the DBHF energy is less attractive than the energy per nucleon calculated in the BHF approach.

Due to this effect, which strongly depends on density, the saturation point evaluated within the DBHF approach is moved off the Coester band. For the potential A, Brockmann and Machleidt (1984) determine a saturation point very close to the empirical value. The same feature is also observed for other NN interactions considered by various groups of scientists.

Summarizing, one may say that inclusion of the relativistic effects as done within the DBHF approach yields a repulsive contribution that increases strongly as a function of the nuclear density. Therefore the DBHF results obtained by various groups, such as Brockmann and Machleidt (1990), ter Haar and Malfliet (1987), and Anastasio et al.(1983) form a new Coester band, which now, in contrast to non-relativistic calculations meets the empirical saturation point. At present it is hoped that many-body effects beyond the DBHF approximation may shift the calculated results but only along this new Coester band, which would be a copy of the experience made within the framework of non-relativistic many-body theory and has also been discussed in Sect. 3.4. This would mean that the inclusion of relativistic features has solved the problem of the many-body theory of nuclear matter employing realistic NN interactions.

Before such a conclusion can be drawn one would like to test the relativistic effects included in the DBHF approach by inspecting other observables as well. The natural extension of the investigations for the infinite systems, is

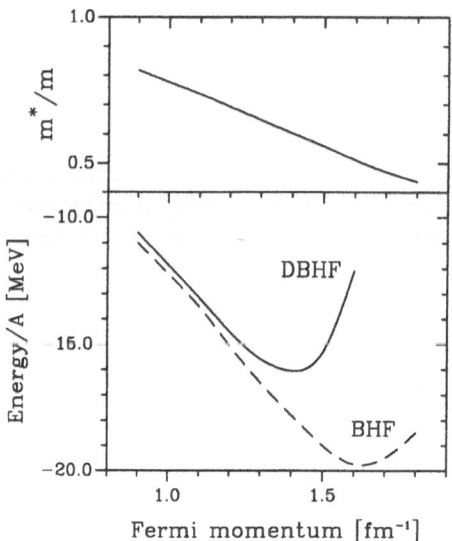

Fig. 3.18. Results for the effective mass m^* and the energy per nucleon of nuclear matter as a function of the Fermi momentum derived from potential A as defined by Machleidt (1989) using the DBHF approach. Also shown is the energy evaluated in the BHF approximation.

of course the study for finite nuclei, as they provide much more experimental data. Such calculations for finite systems, however, are, from a technical point of view, much more complicate because one has to determine the single-particle wave functions within the self-consistent scheme. Various attempts have been made to employ results obtained for the infinite system also for the calculation of finite nuclei in terms of a local density approximation (LDA). One possibility is to use such a LDA in order to account for the density dependence of the correlations. The results of the microscopic DBHF calculations of nuclear matter at density n can be parametrized in terms of an effective Walecka model (see previous subsection) by adjusting the coupling constants g_σ and g_ω in such a way that a Dirac–Hartree–Fock calculation using the Walecka model with these coupling constants reproduces the DBHF result at the density considered. This leads to density-dependent coupling constants and the density dependence reflects the density dependence of the correlations taken into account in the underlying G matrix.

Results for such density-dependent coupling constants are displayed in Fig. 3.19. One finds that the coupling constants decrease with increasing density. This leads to a softening of the equation of state as compared to results derived from a Walecka model with coupling constants not depending on density. The non-linear terms describing the self-interaction of the scalar field, which have been introduced in the Lagrangian (3.108) can be considered as an attempt to simulate this density dependence. It is encouraging to note that the results for the effective coupling constants depend only weakly on the underlying realistic NN interaction.

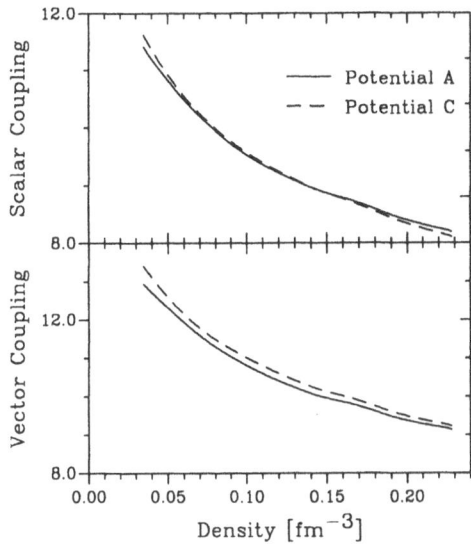

Fig. 3.19. Coupling constants for an effective σ (upper part) and ω (lower part) meson as a function of density, to simulate DBHF calculations with Bonn potentials A and C.

Such a parametrization of the effective interaction has been used by Fritz and Müther (1994) and Boersma and Malfliet (1994) to perform DBHF calculations for finite nuclei. It has been demonstrated by Fritz and Müther (1994) that quite a different LDA, in which the correlation effects are treated directly for the finite system but the Dirac effects are taken into account as derived from nuclear matter, yield almost identical results, which means that both LDA may be considered as reliable approximations to an exact DBHF. The Dirac effects included in DBHF provide a substantial improvement in calculating ground-state properties of finite nuclei. Furthermore Zamick et al. (1992) observe that the relativistic effects lead to an understanding of the spin orbit term in the single-particle spectrum of particle and hole states. Also it is worth noting that the DBHF approach provides a microscopic understanding of the momentum dependence for the optical potential used to describe elastic nucleon–nucleus scattering data (see Kleinmann et al. 1994)

These results and others can be understood as fingerprints for the importance of relativistic effects in the many-body theory of nuclear matter. One must keep in mind, however, that this treatment is a phenomenologic one in the sense that it accounts only for the change of the spinors obtained from the solution of the Dirac equation with positive energies. The corresponding medium dependence of the Dirac sea of solutions with negative energies, the polarization of the Dirac vacuum, is ignored.

Nevertheless, the success of the DBHF approach in describing the saturation properties of nuclear matter has initiated investigations of the equation of state for baryonic matter at higher densities, neutron matter, and asymmetric nuclear matter (see, e.g., Müther et al. 1987, Li et al. 1992, Engvik et al. 1995).

3.6 Subnucleonic Degrees of Freedom

3.6.1 Excitations of the Nucleons

Up to this point we have considered only protons and neutrons as the ingredients of baryonic matter. Protons and neutrons, however, cannot be regarded as elementary constituents of baryonic matter. The interacting nucleons may polarize or excite each other. Already in Sect. 3.3 we have seen that processes like the one displayed in Fig. 3.12d, in which one of the interacting nucleons is excited to the $\Delta(3,3)$ resonance at 1232 MeV, provide a substantial contribution to the medium-range attraction of the NN interaction. In this subsection we will consider excitations of nucleons to the Δ resonance as an example of including the effects of baryon excitations. In discussing this example we want to demonstrate the connections and interplays between the many-body approach considered, the treatment of many-body forces and the mesonic degrees of freedom. This example is used to demonstrate the difficulties in treating these various but connected subjects in a consistent way.

In order to account for the isobar degrees of freedom we have to consider the contributions of the isobar terms to the NN scattering explicitly. For that purpose we split the irreducible NN amplitudes M_{irr} entering the NN scattering equation (3.84) or the corresponding model for the NN interaction V [see (3.88)] into a OBE part and the contributions due to the isobar box diagrams:

$$V(Z) = V_{\mathrm{OBE}} + V_{N\Delta}^{\dagger} \frac{1}{Z - H_0} V_{N\Delta} + V_{\Delta\Delta}^{\dagger} \frac{1}{Z - H_0} V_{\Delta\Delta} . \tag{3.126}$$

As in Sect. 3.3, $V_{N\Delta}$ denotes the transition potential for NN $\to$ NΔ, $V_{\Delta\Delta}$ the corresponding one for NN $\to$ $\Delta\Delta$, Z is the energy of the interacting nucleons and H_0 the unperturbed Hamiltonian for the intermediate states, which also accounts for the NΔ mass difference. The ansatz of (3.126) treats the isobar degrees of freedom in a perturbative way. The coupled channel approach, treating NN, NΔ, and $\Delta\Delta$ channels on the same footing, which has been considered, e.g., by Green (1976) and Wiringa et.al. (1984) yields results which are quite similar. The isobar terms supply a significant contribution to the medium-range attraction of the NN interaction, nuclei would be unbound without these contributions. In a pure OBE model the isobar terms of (3.126) have to be simulated by a strong σ exchange to fit the NN scattering phase shifts. This σ exchange is not modified if one studies the interaction of two nucleons in the medium (ignoring for the moment the relativistic effects discussed in Sect. 3.5). An explicit treatment of the isobar degrees of freedom as indicated in (3.126), however, yields an interaction in the nuclear medium:

$$\tilde{V}(Z) = V_{\text{OBE}} + V_{\text{N}\Delta}^{\dagger} \frac{Q_{\text{Pauli}}}{Z - \tilde{H}_0} V_{\text{N}\Delta} + V_{\Delta\Delta}^{\dagger} \frac{1}{Z - \tilde{H}_0} V_{\Delta\Delta} , \qquad (3.127)$$

which is different from V in the vacuum. The first difference between V and $\tilde{V}$ is due to the Pauli operator Q_{Pauli} in (3.127). The Pauli operator ensures that only those intermediate NΔ states are taken into account for which the nucleon is scattered into a state above the Fermi surface. This constraint on the intermediate NΔ states reduces the attractive NΔ term in the nuclear medium (Pauli quenching). Another reduction of the isobar terms (dispersive quenching) is due to the change of the energy denominators in (3.127) as compared to (3.126). The starting energies Z and the eigenvalues of $\tilde{H}_0$ in (3.127) are defined in terms of single-particle energies for the nucleons and the Δ in the medium. Therefore the absolute values for the energy denominators in $\tilde{V}$ are typically larger than the corresponding ones in V.

All these effects reduce the attractive isobar terms in the effective NN interaction in the nuclear medium and consequently a many-body calculation, which accounts for these quenching mechanisms, will lead to a nuclear binding energy weaker than the corresponding calculation performed without treating these isobar effects explicitly. This phenomenon has been discussed by Green (1976), Anastasio et al. (1978), Manzke and Gari (1978) and other groups within the framework of non-relativistic many-body theories. One observes a considerable decrease in the calculated binding energy accompanied by a lower density for the saturation point. Therefore the inclusion of isobar degrees of freedom on the level discussed so far yields a shift in the calculated saturation point along the Coester band of non-relativistic calculations. Ter Haar and Malfliet (1987) included the effects of isobars in the framework of the DBHF approximation. Also they observe the decrease in binding energy and saturation density, which moves the saturation point in this case along the new Coester band of relativistic many-body calculations. Similar results have been obtained considering the medium dependence of other contributions to the irreducible two-body amplitude M_{irr} beyond the OBE term (see Fig. 3.12).

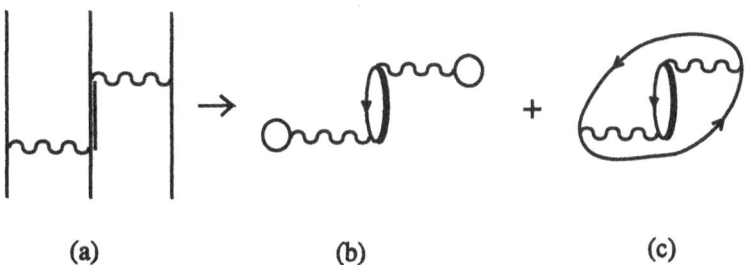

(a)   (b)   (c)

Fig. 3.20. Diagram of second order in the G interaction involving three nucleons and one Δ excitation. The calculation of the expectation value of the effective 3N force displayed in **(a)** leads to the diagrams shown in **(b)** and **(c)**.

After considering isobar degrees of freedom explicitly in the construction of the two-body scattering amplitude, we should also take into account the effects of isobar excitations in processes involving three or more nucleon in nuclear matter. The linked diagram of second order in the baryon–baryon interaction with an intermediate Δ excitation is displayed in Fig. 3.20a. The wavy lines in this figure represent, in analogy to the Brueckner G matrix, the transition potentials corrected for the effect of two-baryon correlations.

In a many-body theory that does not treat isobar excitations explicitly, the three-particle amplitude of Fig. 3.20a would have to be simulated by an effective three-nucleon force. The evaluation of the expectation value of this effective 3N force for the uncorrelated ground state of nuclear matter is represented by two diagrams displayed in Fig. 3.20b,c. Note that all other contractions of the external lines of Fig. 3.20a are also represented by these two diagrams, if we assume the wavy lines to stand for the antisymmetrized matrix elements. The diagram displayed in Fig. 3.20b does not contribute to the energy of the ground state of nuclear matter as one can easily verify: the ground state of nuclear matter has isospin $T_{\mathrm{iso}}=0$. The intermediate state of Fig. 3.20b, however, represents a Δ-hole excitation, which must have isospin $T_{\mathrm{iso}} = 1$ or $T_{\mathrm{iso}} = 2$: the two possibilities to couple the isospin of the Δ (3/2) to the one of the nucleon hole. Since the baryon–baryon interaction conserves isospin, there can be no admixtures to the ground state, which have different isospin. The diagram displayed in Fig. 3.20c represents a lowest-order correction, contained in the Pauli quenching of the isobar terms, which we have just discussed. This discussion demonstrates that the kind of three-nucleon forces to be considered depends on the ingredients of the many-body theory employed. Therefore it is very difficult to include a three-nucleon force in the calculation in a consistent manner, even if it has been deduced, e.g., from πN scattering data.

In order to obtain a non-vanishing contribution to the binding energy from the amplitude in Fig. 3.20a, which has not yet been considered via the medium dependence of the two-body force, one has to consider additional correlations, represented by additional interaction lines between the three nucleons involved. This leads to the ring diagram of third order in G as displayed in Fig. 3.17a with one of the particle lines replaced by a Δ. The effects of this ring diagram plus all others of third order in G involving isobar excitations has been investigated by Dickhoff et al. (1982). The results of these calculations are displayed in Fig. 3.21.

The starting point of the many-body calculations displayed in this figure is the result of a BHF calculation including the quenching of the two-body isobar terms discussed above. As we have already mentioned this yields a saturation point at low density and energy. The effects of the ring diagrams of third and fourth order involving nucleon particle–hole excitations only are rather weak as compared to the results obtained for the Reid soft-core potential (see Fig. 3.14). This is due to the fact that the interaction considered by

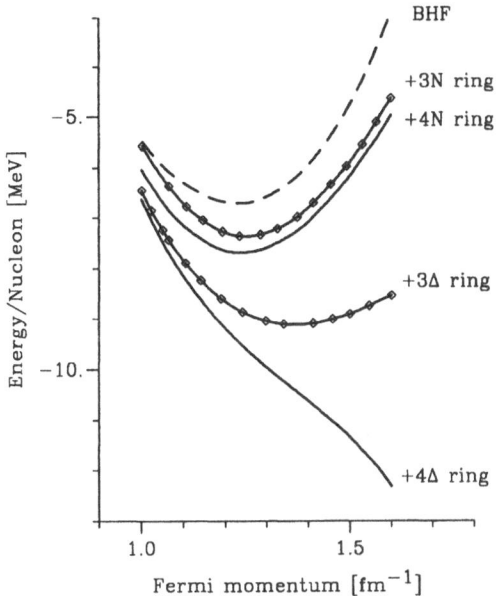

Fig. 3.21. Effects of isobars in nuclear matter. The binding energy as a function of the Fermi momentum is displayed for the BHF approach (including the quenching of the two-body isobar terms), with inclusion of 3N and 4N ring diagrams and with rings including Δ excitations. The results were obtained by Dickhoff et al. (1982).

Dickhoff et al. (1982) has a tensor force that is much weaker than the tensor component of the Reid interaction. In this case the effect of ring diagrams involving Δ excitations is even larger than the one obtained from pure nucleonic configurations. The inclusion of third-order diagrams with Δ shifts the calculated saturation point to higher energies and densities (along the Coester band).

The large contributions of the fourth-order ring diagrams (with Δ), as compared to those of third order, give rise to some doubt about the convergence of the series of ring-diagram contributions. An estimate of the whole series including ring diagrams up to any order has been presented by Müther (1985). One finds for the model of the baryon–baryon interaction considered that the ring-diagram contributions with inclusion of Δ excitations diverge at a density that is about twice the saturation density of nuclear matter. It has been argued that this divergence is related to the phenomenon of pion condensation. To summarize this argument we want to discuss the diagram displayed in Fig. 3.22. It is one example of the ring diagrams we have just discussed. This diagram may also be interpreted as an exchange of two pions between the two nucleons constituting the particle–hole excitations at the very left and very right-hand side of this diagram. The part of this diagram within the box can be regarded as describing the exchange of one of these pions, which is polarizing the nuclear medium. The propagator of such a pion in the medium,

$$P_\pi = \frac{1}{\omega^2 - k^2 - m_\pi^2 - \Pi(k, \omega, n)},\tag{3.128}$$

deviates from the corresponding one in the vacuum in (3.77) due to the self-energy of the pion Π, which depends on the momentum k and the energy ω of the pion as well as the density of the nuclear system. At larger densities this self-energy could become so attractive that the denominator in (3.128) becomes zero for $\omega=0$, which means that pions could be "produced" in the medium without the need for any energy. This phenomenon has been discussed in the literature (see, e.g., Migdal 1978, Brown and Weise 1976) under the name of pion condensation.

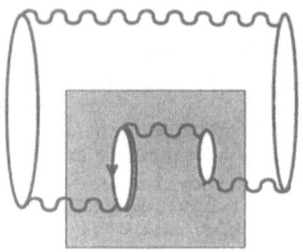

Fig. 3.22. Ring diagram and the exchange of a pion polarizing the medium.

3.6.2 Pion Condensation

The phenomenon called pion condensation may be characterized in simple words by saying that under certain conditions the coupling of a pion to the excitation modes of the nuclear medium, carrying identical quantum numbers, becomes so large that the total energy of this quasiparticle pion is less than zero even with inclusion of the pion rest mass. For asymmetric nuclear matter, which is neutron rich, the possible formation of such quasiparticle pions with negative charge is enhanced because the reaction

$$n \rightarrow p + \pi^-_{qp} \tag{3.129}$$

would be possible when the energy of this quasiparticle pion π^-_{qp} is below the difference of the chemical potentials for protons and neutrons. The question at which critical density a system of nuclear matter or neutron matter may undergo the phase transition from the normal state to the pion condensation has been widely discussed in the literature. There have been essentially two ways to approach this question. One way is to calculate the self-energy of the pion $\Pi(k, \omega, n)$ in (3.128) and investigate at which density the self-energy becomes so attractive that one may observe this phase transition. The alternative way is to perform mean-field calculations for a Lagrangian like the one in (3.108), including a pion field coupled to the nucleons, and study whether there exist regions of density and asymmetry at which the pion field generates a non-vanishing expectation value. In the pion condensed state the medium of baryons will be polarized by the pion field in such a way that one observes alternating layers of nucleons polarized with respect to spin and isospin (see, e.g., Tamiya and Tamagaki 1981).

In order to explore the density at which pion condensation occurs for symmetric nuclear matter, one should calculate the pion self-energy $\Pi(k,\omega,n)$ for pions with momentum k at an energy transfer $\omega=0$. The simplest non-trivial approximation for the self-energy is obtained if one just considers the coupling of the pion to the corresponding uncorrelated particle–hole excitations of the nuclear medium. This yields

$$\Pi_N^0(k) = -4\frac{f_{NN\pi}^2}{m_\pi^2}k^2 U_0(k)\,, \tag{3.130}$$

with $f_{NN\pi}$ denoting the coupling constant for the pion nucleon vertex (3.75). The Lindhard function $U_0(k)$ can be given in analytic form (see, e.g., Fetter and Walecka 1971) as

$$U_0(k) = \frac{m^* k_F}{2\pi^2}\frac{1}{2}\left(1 - \frac{1-\chi^2}{2\chi}\ln\left|\frac{1-\chi}{1+\chi}\right|\right)\,, \tag{3.131}$$

with $\chi = k/2k_F$ if one approximates the relation between single-particle energy and momentum of the nucleon in nuclear matter by

$$\epsilon(k) = \frac{k^2}{2m^*} + C\,. \tag{3.132}$$

In this approximation one can see immediately that the pion self-energy is attractive; its absolute value increases with density (or Fermi momentum k_F) and a maximal effect should be expected for momenta k different from zero. In the next step one may add the absorption of pions by the nucleons leading to Δ excitations, which can also be interpreted as the coupling of the pion to Δ-particle–nucleon-hole excitations of the nuclear medium. This leads to a self-energy

$$\Pi^0(k) = \Pi_N^0(k) + \Pi_\Delta^0(k) = \Pi_N^0(k) - \frac{16}{9}\frac{f_{N\Delta\pi}^2}{m_\pi^2}k^2 U_\Delta(k)\,, \tag{3.133}$$

where $f_{N\Delta\pi}$ is the coupling constant for the $N\Delta\pi$ vertex [see (3.91)] and $U_\Delta(k)$ is the Δ–hole propagator calculated in a way similar to the Lindhard function above. However, with the approximation (3.133) the self-energy becomes so attractive that pion condensation should be observed even at normal nuclear densities. It has been realized that a residual interaction between the particle–hole and Δ–hole states may reduce the absolute value of the self-energy. This means that one tries to calculate a self-energy by including a whole series of diagrams with interacting particle–hole excitations. Two examples for diagrams of second order in the number of particle-hole states are displayed in Fig. 3.23a,b. In order to be acceptable for the evaluation of the pion self-energy these diagrams must be irreducible with respect to a one-pion cut, which means it should be impossible to cut the diagram into two pieces by just cutting a line that represents a pion. This means that the interaction line represented by the wiggly line in the diagram of Fig. 3.23 may contain all contributions to the residual interaction between particle–hole excitations in this channel except of the direct one-pion-exchange contribution. The curled interaction lines in the diagrams of Fig. 3.23b,c, on the other hand, contain the one-pion-exchange part without violating the irreducibility of these diagrams.

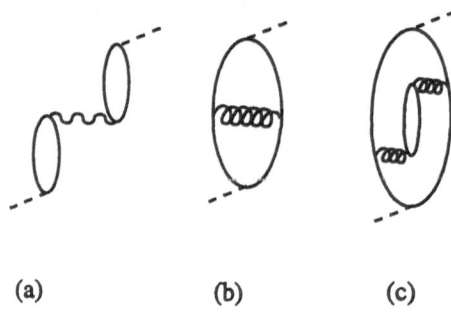

(a)   (b)   (c)

Fig. 3.23. Various contributions to the self-energy of the pion in nuclear matter, as discussed in the text.

If we assume that this irreducible interaction can be represented by one constant γ for the interaction between two nucleon particle–hole states using a normalization in terms of the one-pion-exchange constants $(f_{\mathrm{NN}\pi}^2/m_\pi^2)$ and similar for the interaction particle–hole with Δ–hole $(\gamma_{\mathrm{N}\Delta})$ and the Δ–hole Δ–hole interaction (γ_Δ), the pion self-energy including all these residual interactions can be calculated as

$$\Pi(k) = \Pi_{\mathrm{N}}(k) + \Pi_\Delta(k)\,,\tag{3.134}$$

with

$$\Pi_{\mathrm{N}}(k) = \frac{\Pi_{\mathrm{N}}^0(k)}{D}\left[1 + \frac{\gamma_{\mathrm{N}\Delta} - \gamma_\Delta}{k^2}\Pi_\Delta^0(k)\right],$$

$$\Pi_\Delta(k) = \frac{\Pi_\Delta^0(k)}{D}\left[1 + \frac{\gamma_{\mathrm{N}\Delta} - \gamma}{k^2}\Pi_{\mathrm{N}}^0(k)\right],\tag{3.135}$$

and a denominator

$$D = 1 - \frac{\gamma}{k^2}\Pi_n^0(k) - \frac{\gamma_\Delta}{k^2}\Pi_\Delta^0(k) + \frac{\gamma\gamma_\Delta - \gamma_{\mathrm{N}\Delta}^2}{k^4}\Pi_{\mathrm{N}}^0(k)\Pi_\Delta^0(k)\,.\tag{3.136}$$

The results for the pion self-energy depend rather sensitively on the strength of the various coupling constants γ. In the case of the nucleon particle–hole interaction one can argue that the strength of the residual interaction is related to the Landau–Migdal parameter, G_0', which characterizes the strength of the particle–hole interaction for spin–isospin excitations of the Fermi sea (see, e.g., Brown and Weise 1976, Müther 1985) by

$$\gamma = \frac{m_\pi^2}{f_{\mathrm{NN}\pi}^2}\frac{\pi^2}{2k_{\mathrm{F}}m^*}G_0'.\tag{3.137}$$

This Landau parameter G'_0 at normal nuclear densities can be derived from structure calculation for these excitation modes in finite nuclei (Speth et al. 1977). More sophisticated calculations of Dickhoff et al. (1981) consider a residual interaction derived from the Brueckner G matrix. At normal density these results support the simple estimates obtained with the assumptions made above. In Fig. 3.24 we present the result for the total inverse propagator of the pionic mode $k^2 + m_\pi^2 + \Pi(k)$ calculated for nuclear matter with a Fermi momentum of $k_F = 1.77$, which corresponds to a density about twice as large as the empirical saturation density. If only the nucleon particle–hole excitations are considered, by setting $\Pi_\Delta^0 = 0$ in (3.135) and (3.136), the dashed line is obtained, which shows the effects of a rather attractive pion self-energy. If also the polarization effects due to the isobar excitations are taken into account, one obtains pion condensation ($k^2 + m_\pi^2 + \Pi(k) < 0$) at this density, assuming that the residual interactions are represented by the Brueckner G matrix.

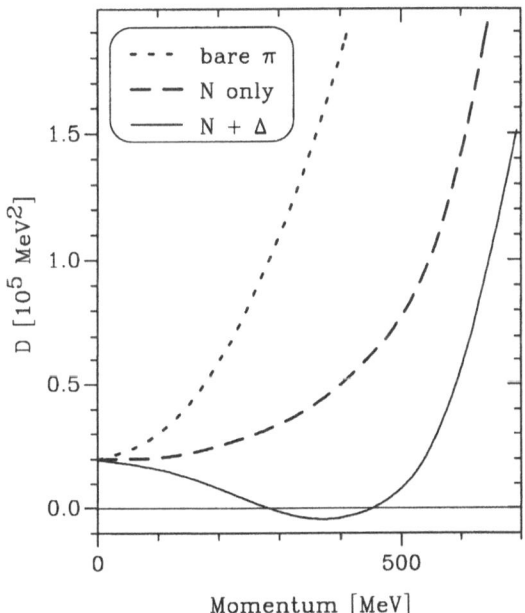

Fig. 3.24. Results for the inverse pion propagator $k^2 + m_\pi^2 + \Pi(k)$ as a function of the momentum k calculated in nuclear matter at about twice the normal density. Results with (solid line) and without (dashed line) inclusion of Δ excitations are compared to the free result ($\Pi = 0$, dotted line).

If, however, the pion propagator is modified in such a dramatic way, one has to take this effect into account in constructing the residual interaction. This modification of the pion-exchange does not affect the "direct" interaction, the wavy interaction line in Fig. 3.23a, as the corresponding contributions to the self-energy would not be irreducible. It does affect, however, the residual "exchange" term of the interaction, represented by the curly line in Fig. 3.23b. Taking this modification of the pion propagator would be presented in terms of diagrams by "bubble-in-bubble" diagrams of the sort displayed in Fig. 3.23c. These contributions lead to the so-called induced interaction, which has been studied by Babu and Brown (1973) for the quantum fluid of liquid ^{3}He. Dickhoff et al. (1983) showed that this induced interaction yields more repulsive residual interactions at densities above the normal density. Whenever there is a tendency to build up pion condensation at these higher densities, the same collective effects yield a more repulsive induced interaction to dampen the attraction in the pion self-energy. This effect has been summarized by Müther (1985) in the statement "Pion condensation prevents pion condensation".

This self-shielding against pion condensation can also be characterized from a different point of view. The strong short-range components of the NN interaction make it impossible to perform nuclear-structure calculations in terms of the bare NN interaction V. It is necessary to solve the NN scattering equation, which means iterating V to all orders in the particle–particle channel. This leads to the G matrix. At higher densities we then observe a strong collectivity in the particle–hole channel with the quantum numbers of the pion, in particular after isobar degrees of freedom are taken into account. This implies that one has to iterate the interaction in the particle–hole channel to all orders as well. The same collective phenomenon is then observed as well in the second independent particle–hole channel, the so-called crossed channel. In the conventional nomenclature of the Mandelstam variables these three channels are identified by total 4-momentum squared s, for the particle–particle channel, t for the particle–hole channel, and u for the crossed channel (see, e.g., Itzykson and Zuber 1980). The fact that collective phenomena show up simultaneously in the t and u channel, make it necessary to consider terms at all orders in both of these channels. This is done by using the induced interaction discussed above. A scheme, that tries to account consistently for strong correlations in all three channels for the two-particle amplitude is the so-called Parquet technique (see, e.g., Jackson et al. 1982).

These investigations show that a phase transition to pion condensation is quite unlikely for the symmetric nuclear matter. Similar rigorous calculations, however, have not yet been performed for asymmetric nuclear matter or plane neutron matter. Therefore one may still speculate that the phase transition may occur for such systems. Such a phase transition would be of interest, first of all, because it would modify the equation of state, making it softer at high densities. It is also of interest, however, because it should modify the cooling rate of a neutron star formed at a very high temperature in the core of a supernova explosion (see, e.g., Shapiro and Teukolsky 1983).

In a first stage at very high temperatures ($T > 10^9$ K) the energy loss of these compact objects is dominated by the energy carried away via neutrinos, which are formed in the so-called URCA reactions

$$n \to p + e^- + \bar{\nu}_e \quad \text{and} \quad e^- + p \to n + \nu_e. \tag{3.138}$$

These processes are highly suppressed at temperatures that are small compared to the Fermi energies of protons, neutrons, and electrons, because of the necessity to conserve energy and momentum (see Shapiro and Teukolsky 1983). In order to enable the neutrino production and thereby a cooling of the system, the URCA processes would need another particle to absorb momentum. This leads to the modified URCA reactions such as

$$n + n \rightarrow n + p + e^- + \bar{\nu}_e \, . \tag{3.139}$$

A much more efficient process of neutrino production would be obtained if a large number of pions were present, as predicted in the scenario of a pion condensate, allowing processes of the form

$$\pi^- + n \rightarrow n + e^- + \bar{\nu}_e \quad \text{and} \quad \pi^- + n \rightarrow n + \mu^- + \bar{\nu}_\mu \, . \tag{3.140}$$

These processes are more efficient than those of the modified URCA in (3.139) because there is no Pauli blocking factor for the pions involved in (3.140) as compared to the second neutron involved in (3.139). Detailed calculations on the thermal evolution of neutron stars with and without inclusion of a phase transition to a pion condensate have recently been performed by Umeda et al. (1993). Comparing their results for the cooling rate with recent measurements of the surface temperature of such objects by the ROSAT experiment (see e.g. Trümper 1995) one finds, that the experimental data are rather well explained within the standard scenario. There is no indication that an enhancement of the cooling rate due to pion condensation is needed. This supports the conclusion about the absence of pion condensation made above.

Beside the speculation about pion condensation in systems of nuclear matter at high density, there exist recent speculations about the possibility of a Kaon (K^-) condensate in the interior of stellar objects such as a neutron star (see e.g. Brown 1994). These ideas are motivated by the assumption that a K^- is composed of one strange quark and one nonstrange $\bar{u}$ antiquark. This $\bar{u}$ antiquark could feel a strong attraction in nucleonic matter at high density due to the exchange of ω mesons, whereas the coupling of the ω meson to the strange quark is weak. The resulting attractive self-energy of the K^- could lower the energies of these Kaons below the chemical potential for the electrons and Kaons could be produced, despite the fact that this requires a process that violates the conservation of strangeness. Also the Kaon condensation should enhance the cooling rate of neutron stars, which presently seems not to be supported by the observations.

Furthermore it has been argued by Glendenning (1985) and others that neutron stars could be like giant hypernuclei and a seizable fraction (20 %) of the baryons may be hyperons like Σ and Λ. This scenario seems rather attractive since the strange baryons are not Pauli blocked by protons and neutrons and the charge neutrality can partly be achieved through the cancellation of charges of massive particles (Σ^+ and Σ^-). The estimates of hyperon contents are based on mean field calculations. The results are sensitive to the interaction considered for the hyperon–nucleon interactions.

3.6.3 Effective Quark Models

At very high densities the baryons should be packed that closely together that they overlap almost completely. This should lead to a system of quarks that are confined to single baryons no longer but move almost freely, interacting weakly by the exchange of gluons. It is evident that it is a real challenge to develop models for a transition from the hadronic phase of matter to the quark–gluon plasma. The most popular model used to obtain an equation of state for the quark–gluon phase is based on the MIT bag model (see, e.g., Close 1979). In this model quarks with small current quark masses for the up (u) and down (d) quarks are confined into a sphere, the bag. This sphere has to be created out of the normal vacuum for the price of energy per volume, which is given by the bag constant B, which has been adjusted to be $B^{0.25} \approx$ 145 MeV. With the masses of the u and d quarks completely ignored and with a coupling constant g_c for the quark–gluon vertex ($g_c^2/16\pi \approx 0.55$), Baym and Chin (1976) evaluate an energy density of

$$\rho(n) = B + \sum_i \left(\frac{3}{4\pi^2} + \frac{g_c^2}{8\pi^4} \right) k_{\mathrm{F}i}^4 , \qquad (3.141)$$

where one has to sum over the flavors i for up and down quarks and the Fermi momenta $k_{\mathrm{F}i}$ are related to the baryon density n by

$$\frac{k_{\mathrm{F}i}^3}{3\pi^2} = f_i n , \qquad (3.142)$$

with f_i the number of quarks with flavor i per baryon. It is obvious that the second term on the right-hand side of (3.141) originates from the relativistic kinetic energy of the Fermi sea of quarks with flavor i, whereas the last term is due to the interaction. As in (3.107) one may calculate the pressure from the energy density, which yields in this case

$$p = \frac{1}{3} (\rho - 4B) . \qquad (3.143)$$

This model for the quark equation of state can easily be generalized to account for the finite masses of the quarks and quarks with other flavors. Various attempts have been made to connect this equation of state for the quark sector with models for the hadronic phase of matter (see, e.g., Kutschera and Kotlorz 1993, and references therein).

As a second quark model for baryon matter we want to discuss the model of Nambu and Jona-Lasinio (NJL). This model allows the study of properties of mesons and nucleons in the vacuum and in a nuclear medium by treating the Dirac sea on the same footing as the Fermi sea. The NJL model is based on a Lagrangian, which was proposed in 1961 by Nambu and Jona-Lasinio

$$\mathcal{L} = \bar{\Psi}(\gamma_\mu p^\mu - M_c)\Psi + \tilde{G}[(\bar{\Psi} \mathbb{1} \Psi)^2 + (\bar{\Psi} \, i\gamma_5 \tau \mathbb{1} \Psi)^2] , \qquad (3.144)$$

where Ψ is a (current) quark field with mass M_c, τ the isospin matrices for the SU(2) flavor and $\mathbb{1}$ the unit operator for the color part [SU(3)] of the quark fields. Originally, this Lagrangian was introduced to describe interacting nucleon fields Ψ and the mesons, as nucleon–antinucleon excitations of the Dirac sea. During the last ten years the interest in the Lagrangian of (3.144) has been revived, but now interpreting the fields Ψ to describe quarks (see e.g. Klevansky 1992 and references cited there). On one hand this revival is based on the possibility to explore the breaking of chiral symmetry within this model (see discussion below). On the other hand, however, the interest in this model has been motivated by the work of Alkofer and Reinhardt (1992) and others, who demonstrated that the Lagrangian of the NJL model can be interpreted as an effective theory representing important features of QCD at low energies.

The coupling constant $\tilde{G}$, which is common for the scalar-isoscalar and the pseudoscalar-isovector parts of the interaction term in (3.144), has dimension mass^{-2}. It is obvious that this Lagrangian does not lead to a renormalizable theory. Therefore all integrals involving states of the Dirac sea should be regularized by introducing a cut-off. The Lagrangian of (3.144) is symmetric under a chiral transformation if we put $M_c = 0$. The self-energy $\mathcal{M}$ for the quarks can be evaluated in the mean-field or Hartree approximation in a way very similar as for the Walecka model discussed in Sect. 3.5:

$$
\begin{aligned}
\mathcal{M} &= \mathbb{1}\mathcal{M}_s + i\gamma_5\,\mathbb{1}\tau\mathcal{M}_{\mathrm{ps}} \\
&= 2\tilde{G}\mathbb{1}\langle\bar{\Psi}\mathbb{1}\Psi\rangle + 2\tilde{G}i\gamma_5\,\mathbb{1}\tau\langle\bar{\Psi}i\gamma_5\tau\mathbb{1}\Psi\rangle\,.
\end{aligned}
\tag{3.145}
$$

Since the vacuum expectation value $\langle\bar{\Psi}i\gamma_5\tau\mathbb{1}\Psi\rangle$ vanishes for systems like the vacuum or symmetric nuclear matter, only the scalar part of the self-energy $\mathcal{M}_s$ survives and leads in a similar way as discussed in (3.113) to an effective mass for the constituent quarks of the form

$$
M^* = M_c - \tilde{G}n_s
\tag{3.146}
$$

$$
n_s = \langle\bar{\Psi}\Psi\rangle = \frac{\gamma}{(2\pi)^3}\int_0^{k_F} d^3k\,\frac{M^*}{E^*(k)} - \frac{\gamma}{(2\pi)^3}\int_0^{\Lambda} d^3k\,\frac{M^*}{E^*(k)}\,,
$$

where, in contrast to Sect. 3.5, the contribution of the Dirac sea to the scalar density is taken into account as well. In order to keep the discussion simple, a three-dimensional cut-off parameter Λ has been used to regularize the vacuum contribution [the last term in (3.146)]. Note that here and in the following we describe the nuclear medium as a Fermi sea of quarks occupying states up to the Fermi momentum k_F.

For the simple Lagrangian of (3.144) it is also very easy to include the Fock exchange terms. For that purpose one should simply evaluate the Fierz transformation of the interaction part of the Lagrangian and add the color-scalar terms of this Fierz transformation to the interaction terms of the original Lagrangian (see, e.g., Henley and Müther 1990). For the scalar interaction this means that by replacing $\tilde{G} \to \frac{13}{12}\tilde{G}$ the Fock terms are taken into account as well.

The parameters of the model (the coupling constant $\tilde{G}$ and the cut-off Λ, the bare mass M_c is fixed to the current quark mass of up and down quarks of $M = 5$ MeV) can be chosen such that in the case of the vacuum ($k_F=0$) the constituent quark mass M^* is close to the empirical value ($M^*=325$ MeV) and also the scalar density $\langle \bar{\Psi}\Psi \rangle$ is close to the condensate value of $(-250$ MeV$)^3$ (see, e.g., Bernard et al. 1987). This means that in this NJL model the constituent quark mass mainly arises due to the polarization of the Dirac vacuum. For the propagation of a quark the vacuum polarization needs also to be propagated, which yields a larger effective mass for the quark quasiparticles.

From (3.146) it is also obvious how the effective mass of the quarks will change as a function of density or Fermi momentum. For large densities, the contribution of the Fermi sea tends to counterbalance the effect of the Dirac sea and the effective mass approaches the bare mass. This can be seen in Fig. 3.25 where the effective quark mass is displayed as a function of density. If we identify the mass of the nucleon with three times the constituent quark mass M^*, we see that the NJL model predicts a similar density dependence of the nucleon mass, as obtained in the DBHF calculations discussed in the previous section. The scale for the densities, e.g., the density at which the nucleon mass is one half of the mass in the vacuum, is determined by the choice of the cut-off parameter Λ.

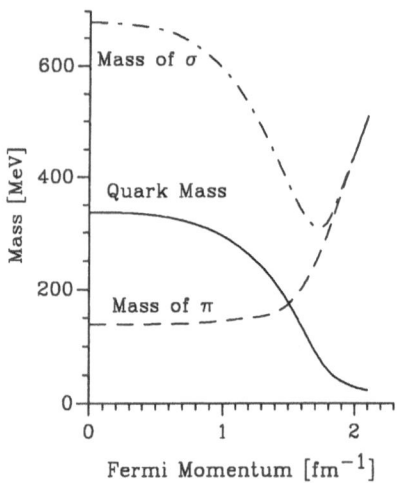

Fig. 3.25. Effective masses for quarks, σ meson, and π meson, calculated in the NJL model as a function of Fermi momentum k_F.

For a vanishing current quark mass M_c the effective mass approaches zero in the limit of large density. The chiral symmetry of the Lagrangian is restored in this limit. At low densities, the chiral symmetry is broken dynamically leading to a scalar density different from zero, and therefore the chiral symmetry is realized in the Nambu–Goldstone mode.

In the next step one may then consider the mesons in the vacuum as the collective modes of the quark excitations from the Dirac sea. For that purpose we calculate as a first step the irreducible "particle–hole" response function for an excitation of the Dirac sea which corresponds to the character of the meson under consideration. For the pion, considered as a pseudo-scalar, iso-vector meson with 4-momentum k this yields

$$\Pi_\pi^0(k) \tag{3.147}$$

$$= 2\mathrm{i}\,\tilde{G}\,n_f n_c \int \frac{d^4p}{(2\pi)^4}\,\mathrm{Tr}\left[\mathrm{i}\,\gamma_5\tau \cdot g_0\left(p+\frac{1}{2}k\right)\mathrm{i}\,\gamma_5\tau\,g_0\left(p-\frac{1}{2}k\right)\right],$$

where g_0 is the Green's function for the quarks and corresponds to (3.70) calculated with the effective quark mass M^* and energy E^* as obtained from the solution of the Hartree equation (3.146) for k_F equal to zero. The constants n_f and n_c refer to the numbers of quark flavors (2) and colors (3) taken into account. The reducible response function is obtained by iterating the Π_π^0 in the particle–hole channel. In the literature this is sometimes referred to as the solution of the Bethe–Salpeter equation in the particle–hole direction or as the RPA approach. For the simple interaction defined by the Lagrangian in (3.144) the reducible response function is obtained by the simple algebraic equation

$$\Pi_\pi(k) = \frac{\Pi_\pi^0(k)}{1 - \Pi_\pi^0(k)}. \tag{3.148}$$

The imaginary part of this response function, which characterizes the excitation modes, is different from zero in the regime of the quark–antiquark excitations ($k_0 > 2M^*$). Due to the strong residual interaction in the pseudo-scalar iso-vector channel the imaginary part of the reducible response function $\Pi_\pi(k)$ shows an additional peak below the quark–antiquark threshold. This peak is referred to as the pion pole. In the limit of a vanishing current quark mass M_c (chiral symmetric case) the pion pole obeys the energy-momentum relation for a massless pion. This pion obtained in the RPA description can be identified as the Goldstone–Boson, which becomes massless if the symmetry is realized in the Nambu–Goldstone mode (see above). For a realistic constituent mass of $M_c=5$ MeV, the parameters of the NJL ($\tilde{G}$ and Λ) can be fixed in such a way that the peak in the imaginary part of the response function can be identified with a pion mass, which is identical to the empirical value.

In a next step one may now consider the response function for densities different from zero. The steps of the calculation are identical to those discussed above, except that now one should use the Green's function calculated at k_F different from zero for the effective mass determined for this density. For the example of the scalar response this yields the following imaginary part for the irreducible response function:

$$\mathrm{Im}\,\Pi_s^0 = \mathrm{Im}\left\{2\mathrm{i}\,\tilde{G}\,n_f n_c \int \frac{d^4p}{(2\pi)^4}\,\mathrm{Tr}\left[g_0\left(p+\frac{1}{2}k\right)g_0\left(p-\frac{1}{2}k\right)\right]\right\}$$

$$= \frac{\tilde{G}n_f n_c}{(2\pi)^2} \int d^3p \left\{ \frac{(EE^{(-)} + \boldsymbol{p} \cdot \boldsymbol{p}^{(-)} - M^{*2})}{EE^{(-)}} \Theta(p - k_F) \right.$$

$$\times \delta(k_0 - E^{(-)} - E)$$

$$+ \frac{E^{(+)}E - \boldsymbol{p}^{(+)} \cdot \boldsymbol{p} + M^{*2}}{E^{(+)}E} \Theta(p^{(+)} - k_F)\Theta(k_F - p)$$

$$\left. \times \left[\delta(k_0 + E - E^{(+)}) + \delta(k_0 - E + E^+) \right] \right\}, \qquad (3.149)$$

with

$$E^{(\pm)} = \sqrt{p^{(\pm)2} + M^{*2}}, \quad \boldsymbol{p}^{(\pm)} = \boldsymbol{p} \pm \boldsymbol{k}, \text{ and } E = \sqrt{p^2 + M^{*2}}.$$

The first term in (3.149) has contributions from $k_0 \geq 4M^{*2}$ and is due to excitations of quarks from the Dirac sea, quenched due to the Pauli principle; i.e., particle–hole excitations with quarks in a state below the Fermi energy are suppressed. The second term represents contributions from normal particle–hole excitations of the Fermi sea. The imaginary part of Π_π^0 essentially yields the same expression with the exception that M^{*2} occurring explicitly in (3.149) has to be replaced by $-M^{*2}$. Typical results for the imaginary part of the irreducible response function (1-loop, dashed line) and for the reducible response function are displayed in Fig. 3.26 as a function of k_0, where a fixed value for $|\boldsymbol{k}|$ is assumed.

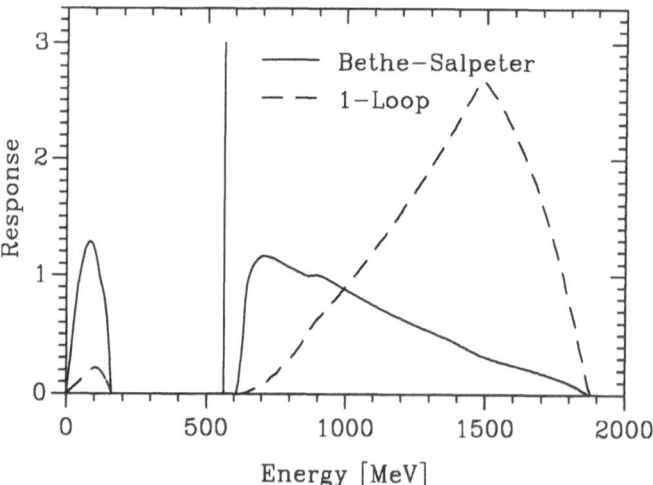

Fig. 3.26. The imaginary part of the self-energy Π_s, calculated for the σ meson in the NJL model. The results for the irreducible and reducible response function are shown as a function of k_0 for $|\boldsymbol{k}| = 200$ MeV and $k_F = 1.3$ fm^{-1}.

Also in the case of the scalar σ meson one finds that the residual interaction yields an enhancement of the response function at low energies and yields an additional peak at energies between those of Fermi excitations and those of Dirac excitations. The energy of this peak (for a given $|k|$) is used to extract a mass for the σ meson.

From this analysis one therefore can determine the density dependence of the masses of π, σ, or other mesons. Such results for this density dependence are displayed in Fig. 3.25. For the mass of the pion one finds that it is almost constant at low densities and suddenly increases at higher densities, where the effective mass of the quarks approaches the current mass value. The constant value at low densities is due to an intriguing balance of two effects. The interplay of the meson mode with the excitations of the Fermi sea tends to reduce the pion mass. This is the same effect as has been discussed in conventional models for pion condensation in the preceding subsection. Within NJL model this reduction of the pion mass is counterbalanced by the Pauli quenching of the Dirac-sea excitations. Therefore the substructure of the pion tends to stabilize nuclear matter against pion condensation, which yields another argument against the pion condensation.

We have seen that the NJL model allows the description of effective quark and meson masses as well as their density dependence. The main shortcomings of the NJL model discussed so far, however, may be summarized as follows (1) The NJL does not contain any mechanism to describe confinement of quarks and therefore a nuclear medium has to be described by a Fermi sea of quarks. (2) The results depend very sensitively on the choice of the cut-off parameter. Therefore one may consider the predictions of the NJL model more as qualitative ones. This means that one does not try to derive from the NJL model at which density the restoration of chiral symmetry occurs, because this depends very much on the choice of the parameters, but one deduces the feature that the masses of the scalar meson and presumably also of most other mesons are reduced with increasing density, whereas the mass of the pion tends to increase because of its special role as the Goldstone boson of chiral symmetry.

Let us assume for a moment that only the mass of the σ meson changes slightly, such that it is reduced by only 3 % at the nuclear-matter saturation density. Such reductions of the mass of the σ meson have been predicted not only within the NJL model, but also by models that describe the σ in terms of irreducible 2π exchange terms (see, e.g., Chanfray et al. 1991). Taking such a medium dependence into account, even if it is as weak as just described, BHF or DBHF calculations predict dramatic changes in the calculated binding energy, which move the predictions far away from the empirical data.

Guided by the NJL model, one may assume that the density dependence of all meson masses is similar to the density dependence of the nucleon masses (see Fig. 3.25, and Brown et al. 1990). This assumption asks for some kind of conspiracy in the medium dependence of all hadrons involved. In this case, the predictions of the BHF or DBHF calculations are not affected as drastically as in the model which allows for a medium dependence of the σ only. The success of these descriptions is still spoiled and one has to speculate about different mechanisms (higher order loop corrections), which generate the saturation properties (see Brown et al. 1990). The success of the OBE model without assuming any medium dependence of meson properties gives rise to the questions: Why don't we observe any subnucleonic degrees of freedom or quark effects at normal density of nuclear matter? Does this imply that dense matter is described very well in terms of a pure hadronic model even at densities much larger than the saturation density?

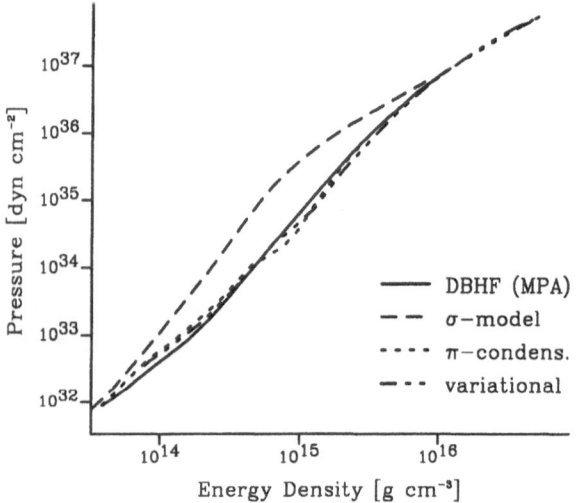

Fig. 3.27. Equation of state for neutron matter at temperature $T = 0$. Results are presented for various approaches as discussed in the text.

At the end of this chapter we would like to present a few examples for results on the equation of state for neutron matter at zero temperature. The resulting pressures as a function of energy density are displayed in Fig. 3.27. Note that this figure covers only a small part of the equation of state shown in Fig. 3.2. It is restricted to energy densities above 10^{13} g cm^{-3}. The solid line in this figure referes to the equation of state MPA, evaluated by Müther et al. (1987) using the DBHF approximation as discussed in Sect. 3.5.2. The OBE potential A defined by Machleidt (1989) has been used for the NN interaction. This potential is a realistic one in the sense that it has been fitted to reproduce the NN scattering data up to 300 MeV for the scattered nucleon. DBHF calculations for nuclear matter with this interaction reproduce the empirical saturation properties as displayed in Fig. 3.18. Therefore the MPA equation of state stands for a prediction of the properties of neutron matter at high densities derived from a many-body calculation that yields the properties of nuclear matter at saturation density without adjusting any free parameter.

Also the equation of state calculated by Pandharipande (1971) has been derived from a realistic NN interaction (here the Reid soft-core potential) without any adjustment. It is represented by the dashed-dotted line, labeled "variational" in Fig. 3.27. In this case a lowest-order variational calculation has been performed ignoring the relativistic effects that are considered in the DBHF approximation. Therefore the differences in these two equation of states (MPA and "variational") are due to the NN interaction, the approximation used for the many-body calculation, and the neglect of relativistic effects in the variational approach. Because of these differences the variational calculation fails to reproduce the saturation point of nuclear matter. The differences between these two microscopic approaches for the equation of state are hardly visible on the logarithmic scales used in Fig. 3.27. Therefore we present these results in Table 3.3 as well.

Table 3.3. Pressure p as a function of the energy density ρ for various microscopic calculations of the equation of state: MPA (Müther et al. 1987), the variational calculation by Pandharipande (1971), and the equation of state in the presence of pion condensation evaluated by Weise and Brown (1975).

ρ [g cm^{-3}]	p [dyn cm^{-2}]		
	MPA	variational	π-condens.
3.767×10^{13}	8.774×10^{31}	6.293×10^{31}	8.774×10^{31}
5.641×10^{13}	1.573×10^{32}	1.993×10^{32}	2.045×10^{32}
9.768×10^{13}	3.874×10^{32}	5.139×10^{32}	5.731×10^{32}
1.554×10^{14}	8.205×10^{32}	1.051×10^{33}	1.203×10^{33}
2.324×10^{14}	1.845×10^{33}	2.122×10^{33}	2.515×10^{33}
3.325×10^{14}	4.174×10^{33}	4.608×10^{33}	5.225×10^{33}
4.586×10^{14}	9.352×10^{33}	9.496×10^{33}	1.127×10^{34}
6.153×10^{14}	1.987×10^{34}	1.987×10^{34}	1.595×10^{34}
8.083×10^{14}	3.951×10^{34}	3.230×10^{34}	2.366×10^{34}
1.044×10^{15}	7.408×10^{34}	5.275×10^{34}	4.616×10^{34}
1.332×10^{15}	1.315×10^{35}	8.771×10^{34}	1.001×10^{35}
1.682×10^{15}	2.361×10^{35}	1.669×10^{35}	1.978×10^{35}
2.125×10^{15}	4.121×10^{35}	3.080×10^{35}	3.801×10^{35}
2.683×10^{15}	6.926×10^{35}	5.350×10^{35}	6.672×10^{35}
3.388×10^{15}	1.123×10^{36}	9.054×10^{35}	1.106×10^{36}
4.281×10^{15}	1.758×10^{36}	1.507×10^{36}	1.763×10^{36}
5.407×10^{15}	2.631×10^{36}	2.418×10^{36}	2.693×10^{36}
6.829×10^{15}	3.991×10^{36}	3.727×10^{36}	3.968×10^{36}
8.626×10^{15}	5.754×10^{36}	5.534×10^{36}	5.687×10^{36}
1.129×10^{16}	8.438×10^{36}	8.449×10^{36}	8.348×10^{36}

Weise and Brown (1975) developed a simple model to account for the effects of a possible pion condensation on the equation of state for neutron matter. This calculation is based on the variational calculation of Pandharipande (1971) that we have just discussed. Also these results are given in Fig. 3.27 and Table 3.3. The pion condensation leads to a softening of the equation of state at energy densities around 8×10^{14} g cm^{-3}.

All these microscopic calculations yield predictions for the equation of state which are similar if they are inspected on a level of accuracy as displayed in Fig. 3.27. Quite different results can be obtained from phenomenological models, like the Walecka model discussed in Sect. 3.5.1 with inclusion of the non-linear terms for the scalar field σ in the Lagrangian (3.108). Waldhauser et al. (1988) performed mean field calculations for this Lagrangian, fixing the parameter of the Lagrangian to reproduce the saturation properties of nuclear matter. Results for the equation of state for neutron matter are displayed in Fig. 3.27 by the dashed line (σ model). The discrepancy between the results obtained in the various microscopic calculations demonstrates the limitations of such phenomenological approaches.

4. Neutron Stars: Spherically Symmetric and Rotating Models

It is obvious that the extreme densities beyond nuclear matter density, described in Chap. 3, are of outstanding importance for the structure of neutron stars. Early after the discocery of the neutron, theoreticians speculated about the existence of stars consisting "of extremely closely packed neutrons", as cited from the first paper by Baade and Zwicky (1934) in which the term *neutron star* was mentioned. Somewhat later, it was recognized that general relativity is needed to describe the structure of such compact objects. The first quantitative calculation was made by Oppenheimer and Volkoff (1939) who computed the gravitational equilibrium by using the equation of state of a cold Fermi gas of neutrons. Although their limit of the maximum mass ($\sim 0.7 M_\odot$) was, due to their simple equation of state, not very accurate, the qualitative result of the existence of a maximum mass was fully correct.

To convince the reader that general relativity is an indispensable ingredient for the description of neutron stars, we compare in Figs. 4.1 and 4.2 the results for masses and radii calculated relativistically (with general relativity) with the corresponding ones calculated non-relativistically (with Newton's gravitational theory). From these figures it is obvious that a general relativistic treatment is necessary.

Before we turn to rapidly rotating neutron stars, we want to describe spherically symmetric stars in the next section.

4.1 Spherically Symmetric Neutron Stars

The structure equations for non-rotating neutron stars are, compared to the equations for normal stars as presented in Chap. 2, on the one hand more complicated since general relativity must be employed, and on the other hand simpler since temperature effects can be neglected.

After the derivation of these equations in the next section, we will discuss their solution and questions of stability.

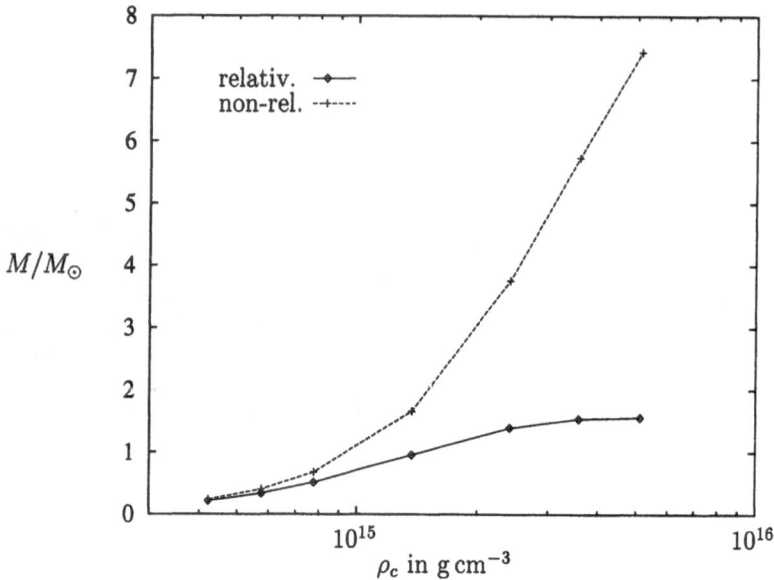

Fig. 4.1. Mass as function of central density (with EOS MPA) for non-rotating neutron stars, calculated relativistically and non-relativistically.

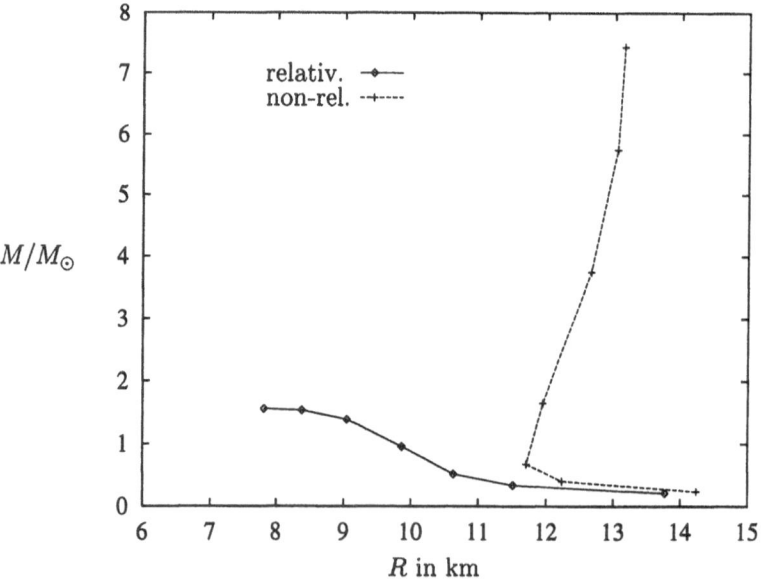

Fig. 4.2. Mass as function of radius (with EOS MPA), for non-rotating neutron stars, calculated relativistically and non-relativistically.

4.1.1 Relativistic Structure Equations

Let us start with the metric for static, spherically symmetric problems. If we use the coordinates t, r, θ, φ for time and space, where θ and φ represent spherical angular coordinates, the metric can be written as

$$ds^2 = -A(r)\,dt^2 + B(r)\,dr^2 + C(r)\left(dr^2 + r^2\,d\theta^2 + r^2\sin^2\theta\,d\varphi^2\right) . \quad (4.1)$$

In this form, the term $\sim dt\,dr$ has been removed by a transformation of the time coordinate, see, e.g., Weinberg (1972).

In (4.1), the three metric functions $A(r)$, $B(r)$, and $C(r)$ can still be reduced by a transformation of the radial coordinate: $\hat{r} = \hat{r}(r)$. There are different possibilities for such a transformation. For instance, if one wants the so-called isotropic form, one has to set $d\hat{r}/\hat{r} = (B(r)/C(r) + 1)^{1/2}\,dr/r$; this leads to

$$ds^2 = -D(\hat{r})\,dt^2 + E(\hat{r})\left(d\hat{r}^2 + \hat{r}^2\,d\theta^2 + \sin^2\theta\,d\varphi^2\right) . \quad (4.2)$$

Another radial coordinate is introduced by setting $\hat{r}^2 = r^2 C(r)$ which yields the metric

$$ds^2 = -F(\hat{r})\,dt^2 + G(\hat{r})\,dr^2 + \hat{r}^2\left(d\theta^2 + \sin^2\theta\,d\varphi^2\right) . \quad (4.3)$$

In the following we will use this metric for non-rotating stars, writing it in the standard form

$$ds^2 = -e^{2\nu(r)}\,dt^2 + e^{2\lambda(r)}\,dr^2 + r^2\left(d\theta^2 + \sin^2\theta\,d\varphi^2\right) , \quad (4.4)$$

where we have introduced exponential functions for the positive coefficients $F(\hat{r})$ and $G(\hat{r})$ and dropped the 'hat' on the radial coordinate. Thus, the two metric functions $\nu(r)$ and $\lambda(r)$ are the unknowns, which have to be determined from the gravitational field equations.

Considering Einstein's field equations

$$R_{\mu\nu} - \frac{1}{2}g_{\mu\nu}R = 8\pi G\,T_{\mu\nu} , \quad (4.5)$$

one recognizes that one needs the Ricci tensor $R_{\mu\nu}$ and the curvature scalar $R = g^{\mu\nu}R_{\mu\nu}$ in order to calculate the left-hand side of (4.5). These quantities contain the Christoffel symbols and their derivatives. Therefore, one has to compute the Christoffel symbols via their defining formula,

$$\Gamma^\lambda_{\mu\nu} = \frac{1}{2}g^{\lambda\kappa}\left(g_{\kappa\mu,\nu} + g_{\kappa\nu,\mu} - g_{\mu\nu,\kappa}\right) . \quad (4.6)$$

A straightforward calculation with the help of the metric (4.4) yields the following nonvanishing Christoffel symbols:

$$
\begin{aligned}
\Gamma^t_{tr} &= \Gamma^t_{rt} = \nu' , \\
\Gamma^r_{tt} &= e^{2\nu - 2\lambda}\,\nu' , & \Gamma^r_{rr} &= \lambda' , \\
\Gamma^r_{\theta\theta} &= -r\,e^{-2\lambda} , & \Gamma^r_{\varphi\varphi} &= -r\,e^{-2\lambda}\sin^2\theta ,
\end{aligned}
\quad (4.7)
$$

$$\Gamma^\theta_{r\theta} = \Gamma^\theta_{\theta r} = 1/r, \quad \Gamma^\theta_{\varphi\varphi} = -\sin\theta\,\cos\theta\,,$$

$$\Gamma^\varphi_{r\varphi} = \Gamma^\varphi_{\varphi r} = 1/r, \quad \Gamma^\varphi_{\theta\varphi} = \Gamma^\varphi_{\varphi\theta} = \cot\theta\,.$$

Derivatives with respect to r are denoted by a prime. From this the Riemann curvature tensor

$$R^\lambda_{\mu\kappa\nu} = \Gamma^\lambda_{\mu\nu,\kappa} - \Gamma^\lambda_{\mu\kappa,\nu} + \Gamma^\lambda_{\sigma\kappa}\Gamma^\sigma_{\mu\nu} - \Gamma^\lambda_{\sigma\nu}\Gamma^\sigma_{\mu\kappa} \tag{4.8}$$

and the Ricci tensor $R_{\mu\nu} = R^\lambda_{\mu\lambda\nu}$ can be derived. Here one obtains the components

$$R_{tt} = e^{2\nu-2\lambda}\left(\nu'' + \nu'^2 - \nu'\lambda' + \frac{2}{r}\nu'\right),$$

$$R_{rr} = -\nu'' - \nu'^2 + \nu'\lambda' + \frac{2}{r}\lambda'\,, \tag{4.9}$$

$$R_{\theta\theta} = e^{-2\lambda}\left(r\lambda' - r\nu' - 1\right) + 1\,,$$

$$R_{\varphi\varphi} = \sin^2\theta\,R_{\theta\theta}\,.$$

For symmetry reasons all the off-diagonal components vanish.

The Einstein equations (4.5) contain the energy-momentum tensor $T_{\mu\nu}$ of the matter on the right-hand side. For stationary neutron stars it can be assumed that the matter is a perfect fluid, i.e., one can write

$$T^{\mu\nu} = (\rho + p)\,u^\mu u^\nu + p\,g^{\mu\nu}\,, \tag{4.10}$$

where ρ is the energy density (or mass density) and p is the pressure, both measured in the rest frame of matter. For a static star, the four-velocity (u^μ) has only a time component, i.e., $u^r = u^\theta = u^\varphi = 0$; the normalization condition $g_{\mu\nu}u^\mu u^\nu = -1$ then leads to $g_{tt}(u^t)^2 = -1$, from which one gets $u^t = e^{-\nu}$.

It is convenient to write the Einstein equations (4.5) in the form

$$R_{\mu\nu} = 8\pi G\left(T_{\mu\nu} - \frac{1}{2}g_{\mu\nu}T\right), \tag{4.11}$$

energy-momentum tensor with $T = T^\lambda{}_\lambda$.

In order to evaluate the right-hand side of (4.11), one first calculates from (4.10)

$$T_{tt} = \rho\,e^{2\nu}, \quad T_{rr} = p\,e^{2\lambda}, \quad T_{\theta\theta} = p\,r^2, \quad T_{\varphi\varphi} = p\,r^2\sin^2\theta\,. \tag{4.12}$$

Then, with $T = -\rho + 3p$, one gets from the tt, rr, and $\theta\theta$ components of (4.11) the equations

$$e^{-2\lambda}\left(\nu'' + \nu'^2 - \nu'\lambda' + \frac{2}{r}\nu'\right) = 4\pi G\,(\rho + 3p)\,, \tag{4.13}$$

$$e^{-2\lambda}\left(-\nu'' - \nu'^2 + \nu'\lambda' + \frac{2}{r}\lambda'\right) = 4\pi G\,(\rho - p)\,, \tag{4.14}$$

$$e^{-2\lambda}\left(\frac{\lambda'}{r} - \frac{\nu'}{r} - \frac{1}{r^2}\right) + \frac{1}{r^2} = 4\pi G\,(\rho - p)\,. \tag{4.15}$$

It is now easy to eliminate the metric function ν by adding (4.13), (4.14), and two times (4.15); this leads to

$$e^{-2\lambda}\left(\frac{2}{r}\lambda' - \frac{1}{r^2}\right) + \frac{1}{r^2} = 8\pi G\rho \qquad (4.16)$$

or the integrable form

$$\left(r\,e^{-2\lambda}\right)' = 1 - 8\pi G\,\rho\,r^2 \, . \qquad (4.17)$$

Using the regularity at the center ($r = 0$), one obtains from (4.17)

$$e^{-2\lambda} = 1 - \frac{2Gm(r)}{r} \qquad (4.18)$$

with the mass function [compare with the non-relativistic formula (2.2)]

$$m(r) = 4\pi \int_0^r \rho(r')\,r'^2\,\mathrm{d}r' \, . \qquad (4.19)$$

Obviously, outside the star the mass density ρ vanishes and $m(r)$ is identical with the total mass M. Thus, we have there

$$g_{rr} \equiv e^{2\lambda} = \left(1 - \frac{2GM}{r}\right)^{-1} \, , \qquad (4.20)$$

which is well-known from the vacuum Schwarzschild solution.

Evaluating $\lambda'(r)$ from (4.18) and inserting this into (4.15), one can solve for the derivative of ν,

$$\nu' = \frac{G(m(r) + 4\pi r^3 p)}{r^2(1 - 2Gm(r)/r)} \, . \qquad (4.21)$$

Outside the star, with pressure $p = 0$, this leads to $\nu' = GM/(r^2 - 2GMr)$, which can be immediately integrated to

$$g_{tt} \equiv e^{2\nu} = \mathrm{const}\left(1 - \frac{2GM}{r}\right) \, . \qquad (4.22)$$

This result is again, after an appropriate normalization of time, in agreement with the Schwarzschild metric.

In order to close the relativistic structure equations, it is convenient to take the energy-momentum conservation explicitly into account (although one could consider the field equations alone), i.e.,

$$T^{\mu\nu}_{\ \ ;\nu} \equiv T^{\mu\nu}_{\ \ ,\nu} + \Gamma^\mu_{\lambda\nu}T^{\lambda\nu} + \Gamma^\nu_{\lambda\nu}T^{\mu\lambda} = 0 \, . \qquad (4.23)$$

Here, inserting (4.12) and (4.7), one sees that only the r component of (4.23) gives a nontrivial relation, namely

$$p' = -(\rho + p)\,\nu' \, . \qquad (4.24)$$

This equation is nothing but the relativistic formulation of the pressure equilibrium in r direction.

Combining (4.24) with (4.21) we get the Tolman–Oppenheimer–Volkoff (TOV) equation

$$\frac{dp}{dr} = -\frac{G(m + 4\pi r^3 p)(\rho + p)}{r^2(1 - 2Gm/r)}. \tag{4.25}$$

Therein, the relativistic effects can be discussed very easily, if one compares it with the non-relativistic formula (2.1). The following replacements can be recognized:

- $\rho \to \rho + p$,
  i.e., the pressure strengthens the passive gravitation,
- $m \to m + 4\pi r^3 p$,
  i.e., the pressure strengthens the active gravitation,
- $r^2 \to r^2(1 - 2Gm/r)$,
  i.e., the curved space causes this effect.

In summary, (4.25) and (4.19) combined with an equation of state $p = p(\rho)$ or $\rho = \rho(p)$ determine the structure of spherically symmetric neutron stars.

4.1.2 Solution Method and Results

Before we discuss the determination of the structure of realistic neutron stars, we want to consider an example in which the TOV equations can be solved analytically, namely the interior Schwarzschild solution (cf. Kramer et al. 1980). In this case one assumes that the mass density of the matter inside the star is constant, i.e. $\rho(r) = \rho_0 = $ const. This is an unrealistic equation of state, since (the square of) the sound velocity $dp/d\rho$ would be infinite. Nevertheless, this is an instructive case.

The mass function (4.19) is here given by $m(r) = \frac{4\pi}{3} r^3 \rho_0$. Inserting this into (4.25), we can see that a separation of variables is possible, namely

$$\frac{dp}{(\rho_0 + 3p)(\rho_0 + p)} = -\frac{4\pi G}{3}\frac{r dr}{1 - \frac{8\pi G}{3}\rho_0 r^2}. \tag{4.26}$$

With the abbreviation $r_0 = \left(\frac{8\pi G}{3}\rho_0\right)^{-1/2}$ one obtains

$$\frac{\rho_0 + 3p}{\rho_0 + p} = \text{const}\left(1 - \frac{r^2}{r_0^2}\right)^{1/2}, \tag{4.27}$$

where the integration constant will be determined by the condition $p = 0$ at $r = R$, with R being the stellar radius. Solving for p, one gets

$$p(r) = \rho_0 \frac{(1 - r^2/r_0^2)^{\frac{1}{2}} - (1 - R^2/r_0^2)^{\frac{1}{2}}}{3(1 - R^2/r_0^2)^{\frac{1}{2}} - (1 - r^2/r_0^2)^{\frac{1}{2}}}. \tag{4.28}$$

Considering this pressure function, one recognizes that there is a nonsingular solution only if the pressure $p_c = p(0)$ at the center is positive. This condition reduces to $R^2/r_0^2 < 8/9$. Since R^2/r_0^2 is identical to the ratio between the Schwarzschild radius and the actual radius, $2GM/R$, where the mass $M = \frac{4\pi}{3}R^3\rho_0$ has been introduced, an interior Schwarzschild solution exists only for $2GM/R < 8/9$. (In the limit $2GM/R = 8/9$ one would get $p_c = \infty$.) It is also not difficult to calculate the metric functions. One obtains

$$e^\lambda = \left(1 - \frac{r^2}{r_0^2}\right)^{-1/2}, \quad e^\nu = \frac{1}{2}\left[3\left(1 - \frac{R^2}{r_0^2}\right)^{1/2} - \left(1 - \frac{r^2}{r_0^2}\right)^{1/2}\right]. \quad (4.29)$$

For realistic equations of state, the maximum compactness $2GM/R$ of a star turns out to be not as high as the maximum value for the case in which $\rho = $ const. Additionally, it is obvious that the structure must be determined numerically, the more so since the equation of state itself is usually given not analytically but in tabular form.

The general numerical procedure is as follows. One integrates the equation $dm/dr = 4\pi r^2 \rho$, the differential form of (4.19), together with the TOV equation (4.25) starting at $r = 0$ with the initial conditions $m(0) = 0$ and $p(0) = p_c$, where the central pressure p_c has been given (equivalently the central density ρ_c can be prescribed). The integration stops at the point $r = R$ where the monotonically decreasing pressure vanishes, $p(R) = 0$ (assuming vacuum outside the star). Every standard ordinary differential equation solver is suited for this task. Thus the radius R of the neutron star and the total mass M via $M = m(R)$ are determined. By variation of the central density the mass-radius relation for the considered equation of state is obtained, as shown in Fig. 4.2 for the EOS MPA. Typical results for other EOSs can be found, for instance, in Shapiro and Teukolsky (1983).

In Fig. 4.3 we present the density profiles (mass density versus radial coordinate) of three neutron stars with different masses, calculated with the EOS MPA. One sees that the outer crust region, with $\rho \lesssim 4 \times 10^{11} \mathrm{g\,cm}^{-3}$ where there are no free neutrons (density below the neutron drip point, the clearly visible kink in the profile of the lightest star), is thicker if the star is lighter.

The heaviest star shown (with $M \approx 1.56 M_\odot$) has approximately the maximum mass allowed for a stable configuration (see next section). Of course, this mass limit depends on the equation of state and is still under active discussion.

4.1.3 Stability and Maximum Mass

In this section we concentrate on the problem of radial perturbations of non-rotating neutron stars with the stress on questions of stability. Non-radial oscillations will be treated in Chap. 5.1, and especially for neutron stars in Sect. 5.5.

It is remarkable that the stability of static models with respect to radial perturbations can be decided just by considering the mass-radius relation without any actual calculation of the frequencies of oscillations [see, e.g., Harrison et al. (1965), Weinberg (1972), Shapiro and Teukolsky (1983)].

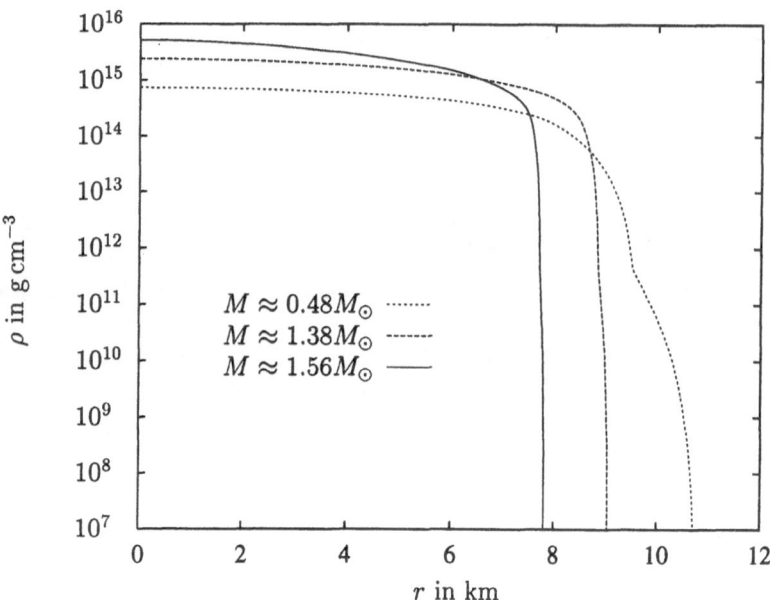

Fig. 4.3. The density profiles of three spherically symmetric neutron stars with different masses. The equation of state MPA has been used here.

First, it turns out that the TOV equilibrium equation can be obtained by an extremum principle, in which one looks for an extremum of the mass, given by $M = 4\pi \int_0^\infty \rho(r)\, r^2\, dr$, with a fixed total baryon number N. The number of baryons N is defined by the integral

$$N = 4\pi \int_0^\infty n(r)\, e^{\lambda(r)} r^2\, dr \ , \tag{4.30}$$

where $n(r)$ is the radial profile of the number density of baryons inside the star. That N as given by (4.30) is a conserved quantity follows from the baryon-number conservation law

$$(n\, u^\mu)_{;\mu} = \frac{1}{\sqrt{-g}} \frac{\partial}{\partial x^\mu} \left(\sqrt{-g}\, n\, u^\mu \right) = 0 \ , \tag{4.31}$$

which yields as a conserved integral

$$N = \int n\, u^t \sqrt{-g}\, d^3 x \ . \tag{4.32}$$

For spherical symmetry this is identical with (4.30).

By performing the variation, using the Lagrange multiplier method, i.e.,

$$\delta M - \mu_0 \delta N = 0 \ , \tag{4.33}$$

where μ_0 is a constant (this is the Tolman chemical potential, $\mu_0 = \mu(r) e^{\lambda(r)}$, with μ being the normal chemical potential), and expressing the variation of n by the variation of ρ via (cf. (3.107))

$$\delta n = \frac{n}{\rho + p} \delta \rho \,, \tag{4.34}$$

one arrives at the TOV equation (4.25). Thus, considering, e.g., the dependence of mass M and baryon number N on the central density ρ_c, one gets the relation

$$\frac{\mathrm{d}M}{\mathrm{d}\rho_c} = \mu_0 \frac{\mathrm{d}N}{\mathrm{d}\rho_c} \,, \tag{4.35}$$

which means that the extrema of M and N occur at the same points in the one-parameter equilibrium sequence.

The above-mentioned static-stability criterion is based on the observation that a radial normal mode, i.e., an oscillation with fixed frequency ω_n, changes its character from stable to unstable only at the extrema of M (or N). The reason for this is that such a change occurs when $\omega_n^2 = 0$, which means that another equilibrium solution $\rho(r) + \delta\rho(r)$ exists in the infinitesimal neighborhood, where $\delta M = 0$ and $\delta N = 0$ due to energy and baryon-number conservation, but $\delta\rho_c \neq 0$. It is not essential to use ρ_c as parameter, another possibility is the radius as in the mass-radius relation $M = M(R)$. A more accurate discussion of the sign of $\mathrm{d}R/\mathrm{d}\rho_c$ at the extrema of the curve in Fig. 4.4, where a wide range of the mass-radius relation for the EOS MPA shows which branches are stable and which unstable. It turns out that for neutron stars there is one stable branch, starting at a minimum mass (point B) of approximately $0.1 M_\odot$ ($R \approx 200 \,\mathrm{km}$) and going up to the maximum mass (point A), which sensitively depends on the equation of state because the central density there is far beyond the standard nuclear density. This uncertainty also extends to the radii of massive neutron stars, but the rough estimate $R \approx 10\,\mathrm{km}$ is not extremely bad. (Point C in Fig. 4.4 corresponds to the maximum mass of white dwarfs, the Chandrasekhar limit).

4.2 Rapidly Rotating Neutron Stars

The first observation of millisecond pulsars in 1982 (Backer et al. 1982) has stimulated the research on rapidly rotating neutron stars, particularly as in the meantime the number of observed sources in this category has steadily increased.

General relativistic calculations of *rotating* neutron stars are a lot more complicated than those of non-rotating stars. The case of slow rotation has been treated for the first time by Hartle and Thorne [see Hartle (1967), Hartle and Thorne (1968)]. In this case the angular velocity Ω is assumed to be small and effects up to the order Ω^2 are taken into account in the determination of the structure and the gravitational fields of the star.

If we want to give up the approximation of slow rotation, i.e., if we want to construct realistic models for *rapidly* rotating neutron stars, we have to employ different methods. A first approach in this direction was the work

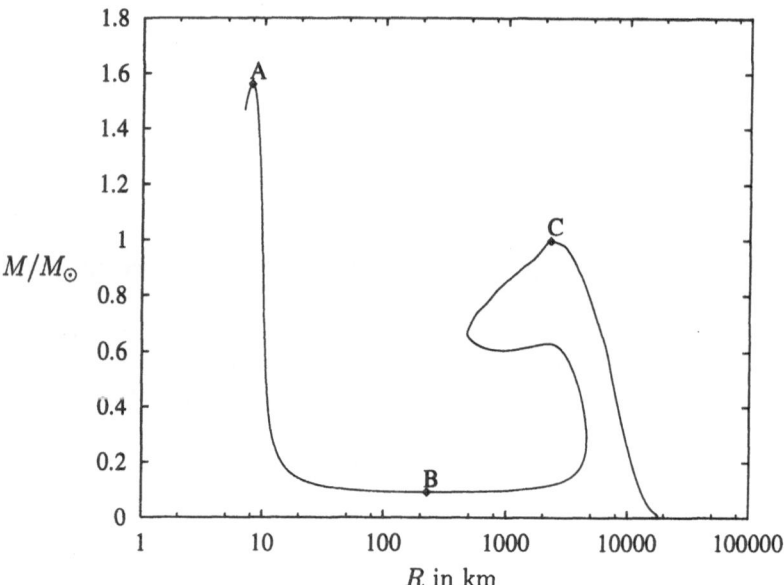

Fig. 4.4. The mass-radius relation for the equation of state MPA.

by Butterworth and Ipser (1976), who treated homogeneous bodies of constant density. Based on this method, Friedman et al. (1986) calculated solutions for rotating neutron stars with realistic equations of states [see also Friedman et al. (1989)]. The numerical method employed by these authors is based on the idea of directly discretizing Einstein's field equations for a chosen parametrization of the metric – to be more precise, they deal with the linearized equations obtained by Newton's method. Recently, spectral methods instead of difference methods have been used to solve the partial differential equations, which describe rotating bodies in general relativity (see Bonazzola et al. (1993) and further references to other approaches therein).

Here, we will show that the minimal surface formalism, as described by Neugebauer and Herold (1992), is a very convenient basis for the development of numerical methods for the calculation of the gravitational fields and the structure of rapidly rotating neutron stars. Especially the minimal surface variational principle in particular facilitates the procedure of discretizing very much [see Herold and Neugebauer (1992)], since it is not necessary to consider the field equations explicitly, which are rather complicated indeed. For this purpose we first have to develop the basic equations used for the description of the structure of rapidly rotating neutron stars and their gravitational fields.

4.2.1 Basic Formulation

The space-time generated by an isolated rotating star admits two commuting Killing vectors: the (at least asymptotically) time-like vector $(\xi^\mu) = \partial/\partial t$, which expresses the stationarity of the problem, and the space-like (azimuthal) vector $(\eta^\mu) = \partial/\partial\varphi$, which exists due to the axial symmetry. The Killing property means that the metric coefficients do not depend on the coordinates $x^0 = t$ and $x^3 = \varphi$,

$$\frac{\partial}{\partial t}g_{\mu\nu} = 0, \quad \frac{\partial}{\partial\varphi}g_{\mu\nu} = 0. \tag{4.36}$$

The other spatial coordinates x^1, x^2 (in the following denoted by x^a with $a = 1, 2$) will not be specified at this stage; they may be, e.g., cylindrical coordinates ϱ, z or spherical coordinates r, θ.

Although there is no time-reversal ($t \to -t$) invariance of the problem due to the rotation, we have the combined symmetry $t \to -t$, $\varphi \to -\varphi$. Therefore, it follows that $g_{t1} = g_{t2} = 0$ and $g_{\varphi 1} = g_{\varphi 2} = 0$. Thus, the metric can be written as ($a, b = 1, 2$)

$$ds^2 = g_{tt}\,dt^2 + 2g_{t\varphi}\,dt\,d\varphi + g_{\varphi\varphi}\,d\varphi^2 + g_{ab}\,dx^a\,dx^b, \tag{4.37}$$

where the metric coefficients are functions of x^1, x^2. There are different possibilities for parametrizing this metric. In Neugebauer and Herold (1992) [see also Neugebauer and Herlt (1984)] the following form is used:

$$ds^2 = -e^{2U}(dt + A\,d\varphi)^2 + e^{-2U}\left(\gamma_{ab}\,dx^a\,dx^b + W^2 d\varphi^2\right). \tag{4.38}$$

This means that the metric coefficients are given by

$$g_{tt} = -e^{2U}, \quad g_{t\varphi} = -e^{2U}A \tag{4.39}$$

$$g_{\varphi\varphi} = e^{-2U}W^2 - e^{2U}A^2, \quad g_{ab} = e^{-2U}\gamma_{ab}. \tag{4.40}$$

It should be noted that g_{tt}, $g_{t\varphi}$, and $g_{\varphi\varphi}$ and therefore also U, A, and W are invariant quantities, since they can be expressed in terms of the two Killing vectors, namely

$$g_{tt} = \boldsymbol{\xi} \cdot \boldsymbol{\xi}, \quad g_{t\varphi} = \boldsymbol{\xi} \cdot \boldsymbol{\eta}, \quad g_{\varphi\varphi} = \boldsymbol{\eta} \cdot \boldsymbol{\eta}. \tag{4.41}$$

In the following, we will use another variant of the parametrization of the metric, which has some advantages, namely the form

$$ds^2 = -e^{2\nu}dt^2 + e^{-2\nu}\left[\gamma_{ab}\,dx^a\,dx^b + W^2\left(d\varphi - \omega dt\right)^2\right]. \tag{4.42}$$

The metric coefficients are then given by

$$g_{tt} = -e^{2\nu} + e^{-2\nu}W^2\omega^2, \quad g_{t\varphi} = -e^{-2\nu}W^2\omega \tag{4.43}$$

$$g_{\varphi\varphi} = e^{-2\nu}W^2, \quad g_{ab} = e^{-2\nu}\gamma_{ab}, \tag{4.44}$$

and the old potentials U and A are algebraically related to the new ones ν and ω by

$$e^{2U} = e^{2\nu}\left(1 - \omega^2 W^2 e^{-4\nu}\right) \tag{4.45}$$

$$A = \omega W^2 e^{-4\nu}\left(1 - \omega^2 W^2 e^{-4\nu}\right). \tag{4.46}$$

The version of the metric with the potentials ν and ω instead of U and A is preferable because then the sign of g_{tt} is allowed to change, i.e., the possible appearance of ergoregions (in which the frame dragging does not allow any static observer) is not forbidden, whereas from (4.39) one sees that the exponential ansatz for g_{tt} is too restrictive. The exponential in (4.44) for $g_{\varphi\varphi}$, however, is no problem, since $g_{\varphi\varphi} > 0$ is always true because of the azimuthal character of the φ coordinate.

The metric forms (4.37), (4.38), or (4.42) still permit transformations of the (x^1, x^2) coordinates, $\bar{x}^1 = f(x^1, x^2)$, $\bar{x}^2 = g(x^1, x^2)$. Thus, within the (x^1, x^2) surface, which is orthogonal to the Killing trajectories, for instance isotropic coordinates may be introduced, i.e.,

$$\gamma_{11} = \gamma_{22}, \quad \gamma_{12} = 0, \quad \text{or} \quad \gamma_{ab} = e^{2k} \delta_{ab}, \tag{4.47}$$

with the (non-invariant) potential $k = k(x^1, x^2)$. Thus (4.42) can be written as

$$ds^2 = -e^{2\nu} dt^2 + e^{-2\nu} \left[e^{2k} \left(d\varrho^2 + dz^2 \right) + W^2 \left(d\varphi - \omega dt \right)^2 \right], \tag{4.48}$$

where the notation $x^1 = \varrho$, $x^2 = z$ for cylindrical coordinates has been used.

Having discussed the ansatzes for the metric of stationary, axially symmetric spacetimes, we now want to consider the matter variables. As for non-rotating neutron stars, the matter is assumed to be a perfect fluid with the energy-momentum tensor (4.10). Additionally, it is assumed that this fluid rotates rigidly with the angular velocity Ω. Thus, we have for the components of the four-velocity

$$u^t = u^t(x^1, x^2), \quad u^\varphi = u^\varphi(x^1, x^2), \quad u^1 = u^2 = 0, \tag{4.49}$$

and due to $\Omega = d\varphi/dt = u^\varphi/u^t$ the relation $u^\varphi = \Omega u^t$. This means that the four-velocity is given by

$$(u^\mu) = e^{-V} \left[\frac{\partial}{\partial t} + \Omega \frac{\partial}{\partial \varphi} \right] = e^{-V} \left[(\xi^\mu) + \Omega \, (\eta^\mu) \right], \tag{4.50}$$

where the factor e^{-V} is needed for the normalization, $u \cdot u = -1$. Inserting here the metric (4.42) together with (4.41), one obtains

$$e^{2V} = e^{2\nu} \left[1 - (\Omega - \omega)^2 W^2 e^{-4\nu} \right], \tag{4.51}$$

or explicitly

$$V = \nu + \frac{1}{2} \ln \left[1 - (\Omega - \omega)^2 W^2 e^{-4\nu} \right]. \tag{4.52}$$

In the non-relativistic limit, where $\nu \ll 1$, $W = \varrho$, $\omega = 0$, and $\Omega \varrho \ll 1$, the expression (4.52) reduces to $V = \nu - \frac{1}{2} \Omega^2 \varrho^2$, i.e., the sum of the Newtonian and the centrifugal potential.

The energy-momentum conservation (4.23) can be evaluated for the four-velocity discussed above, and leads to a generalization of the pressure equation

(4.24) of the spherically symmetric case. A straightforward calculation yields the two pressure equilibrium relations $(a = 1, 2)$

$$\frac{\partial p}{\partial x^a} = -(\rho + p)\frac{\partial V}{\partial x^a} \,, \tag{4.53}$$

from which one immediately concludes that the pressure p and the mass density ρ depend only on the potential V, i.e., $p = p(V)$ (and $\rho = \rho(V)$). This function has to satisfy the equation

$$\rho(p) + p = -\frac{\mathrm{d}p}{\mathrm{d}V} \,. \tag{4.54}$$

For a given equation of state $\rho = \rho(p)$, the differential equation (4.54) can be integrated to obtain the function $p = p(V)$. The zero point of V is fixed by the prescription of vanishing pressure on the surface (due to the vacuum outside),

$$p(V_0) = 0 \,. \tag{4.55}$$

Then the surface $V = V_0$ describes the surface of the star and has to be determined from the self-consistent solution of the full set of equations, which will be presented later.

As an illustration we consider matter of constant density, $\rho = \rho_0 = \mathrm{const.}$ Here, the differential equation (4.54) can be solved analytically, and we get

$$p(V) = \begin{cases} \rho_0 \left(e^{V_0 - V} - 1\right) & \text{for } V \leq V_0 \text{ (inside the star)} \,, \\ 0 & \text{for } V > V_0 \text{ (outside the star)} \,. \end{cases} \tag{4.56}$$

The pressure function for this equation of state is shown in Fig. 4.5.

For realistic equations of state, which are usually given in tabular form, the pressure equation (4.54) must be integrated numerically to determine $p(V)$. Typical results for several equations of state are compared in Fig. 4.6, apart from the EOS MPA, the EOS BPS (cf. Baym et al. 1971), the EOS G [cf. Canuto and Chitre (1974); notation as in Arnett and Bowers (1977)], and the EOS π with pion condensation (cf. Weise and Brown 1975).

The variational principle that we are going to formulate originates from the action integrals for matter and gravitation. We consider first the matter part. It is well known [see ,e.g., Weinberg (1972)] that the variation of the action I_M of a general material system with respect to the metric $g_{\mu\nu}$ yields the energy-momentum tensor of the matter,

$$\delta I_M = \frac{1}{2}\int \mathrm{d}^4x\sqrt{-g}\,T^{\mu\nu}\delta g_{\mu\nu} \,. \tag{4.57}$$

Formulating this for a perfect fluid in rigid rotation, one finds that one can employ the action integral

$$I_M = \int \mathrm{d}^4x\sqrt{-g}\,p(V) \tag{4.58}$$

with the pressure function discussed above. This can be made plausible by calculating the variation

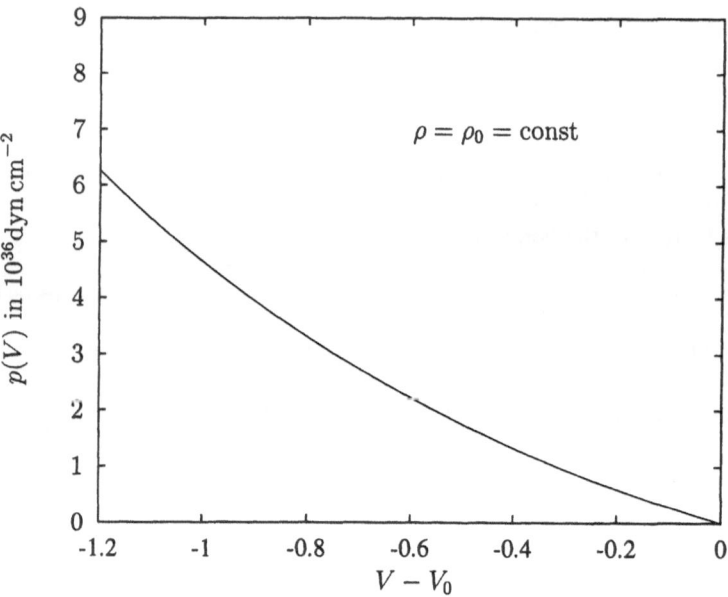

Fig. 4.5. Pressure function for the (unrealistic) equation of state $\rho = $ const.

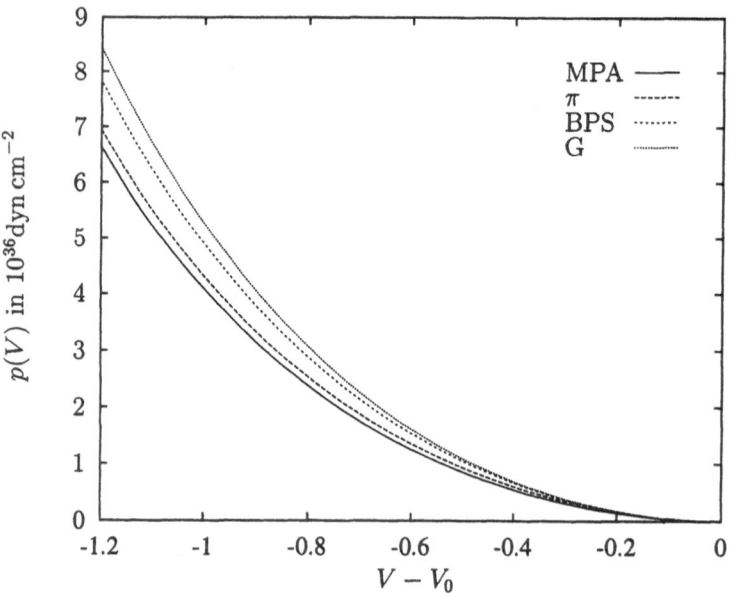

Fig. 4.6. Pressure functions for several realistic equations of state.

$$\delta I_{\mathrm{M}} = \int \mathrm{d}^4 x \sqrt{-g} \left[\frac{1}{2} \delta g_{\mu\nu} g^{\mu\nu} p(V) + \frac{\mathrm{d}p}{\mathrm{d}V} \delta V \right] \tag{4.59}$$

and determining δV from the relation $\mathrm{e}^{2V} = -g_{\mu\nu}(\xi^\mu + \Omega\eta^\mu)(\xi^\nu + \Omega\eta^\nu)$ as $\delta V = -\frac{1}{2} u^\mu u^\nu \delta g_{\mu\nu}$. The last result, inserted into (4.59), leads with the help of (4.54) to

$$\delta I_{\mathrm{M}} = \frac{1}{2} \int \mathrm{d}^4 x \sqrt{-g} \, \delta g_{\mu\nu} \left[p \, g^{\mu\nu} + (\rho + p) u^\mu u^\nu \right] , \tag{4.60}$$

which is the correct formula for the energy-momentum tensor of a perfect fluid.

Apart from the matter action (4.58) we need the action for the gravitational field, i.e., the Einstein–Hilbert action integral

$$I_{\mathrm{G}} = \frac{1}{16\pi G} \int \mathrm{d}^4 x \sqrt{-g} \, R . \tag{4.61}$$

The integrand of (4.61) can be split into a term quadratic in the Christoffel symbols and a complete divergence,

$$\sqrt{-g} \, R = \sqrt{-g} \, \tilde{R} + \left(\sqrt{-g} \, S^\lambda \right)_{,\lambda} , \tag{4.62}$$

where the abbreviations

$$\tilde{R} = g^{\mu\nu} \left(\Gamma^\kappa_{\mu\lambda} \Gamma^\lambda_{\nu\kappa} - \Gamma^\lambda_{\mu\nu} \Gamma^\kappa_{\lambda\kappa} \right) \tag{4.63}$$

$$S^\lambda = g^{\mu\nu} \Gamma^\lambda_{\mu\nu} - g^{\mu\lambda} \Gamma^\kappa_{\mu\kappa} \tag{4.64}$$

have been introduced. Inserting the metric (4.42) into these expressions, we obtain by a straightforward calculation, using for instance $\sqrt{-g} = W\mathrm{e}^{-2\nu}\sqrt{\gamma}$ with $\gamma = \det(\gamma_{ab})$,

$$\sqrt{-g} \, \tilde{R} = \sqrt{\gamma} \Big[-2W\nu_{,a}\nu^{,a} + \frac{1}{2} W^3 \mathrm{e}^{-4\nu} \omega_{,a}\omega^{,a}$$
$$+ W_{,a} \left(\gamma^{ab} \hat{\Gamma}^c_{bc} - \gamma^{bc} \hat{\Gamma}^a_{bc} \right) + W\gamma^{ab} \left(\hat{\Gamma}^c_{ad} \hat{\Gamma}^d_{bc} - \hat{\Gamma}^c_{ab} \hat{\Gamma}^d_{cd} \right) \Big] \tag{4.65}$$

and

$$\sqrt{-g} S^a = \sqrt{\gamma} \Big[-2W^{,a} + 2W\nu^{,a} + W \left(\gamma^{bc} \hat{\Gamma}^a_{bc} - \gamma^{ab} \hat{\Gamma}^c_{bc} \right) \Big] . \tag{4.66}$$

In these expressions indices are lifted with the (inverse) 2-metric γ^{ab}, and $\hat{\Gamma}^a_{bc}$ are the Christoffel symbols of the 2-metric γ_{ab}.

The Ricci scalar (4.62) can also be written in a somewhat different form by introducing the Ricci scalar $^{(2)}R$ of the 2-metric γ_{ab}, namely

$$\sqrt{-g} \, R = \sqrt{\gamma} \left(-2W\nu_{,a}\nu^{,a} + \frac{1}{2} W^3 \mathrm{e}^{-4\nu} \omega_{,a}\omega^{,a} \right)$$
$$+ \left[\sqrt{\gamma} \left(-2W^{,a} + 2W\nu^{,a} \right) \right]_{,a} + W\sqrt{\gamma} \, {}^{(2)}R . \tag{4.67}$$

For the evaluation of $^{(2)}R$ it is convenient to employ the conformally flat form of the 2-metric γ_{ab}, which has been mentioned in connection with (4.47), i.e.

$$^{(2)}ds^2 \equiv \gamma_{ab}\, dx^a dx^b = e^{2k}\left(d\varrho^2 + dz^2\right) . \tag{4.68}$$

Then we get a simple expression for $^{(2)}R$,

$$^{(2)}R = -2\,e^{-2k}\, k_{,aa} . \tag{4.69}$$

A disadvantage of k compared with the other metric potentials ν, ω, W is that it is no invariant quantity; therefore we introduce the scalar quantity ζ via the definition

$$e^{-2\zeta} := \gamma^{ab}\, W_{,a}\, W_{,b} = e^{-2\nu}\, W_{,\mu}\, W^{,\mu} \tag{4.70}$$

(cf. Neugebauer and Herlt 1984), from which the relation

$$e^{2k} = e^{2\zeta}\,(\nabla W)^2 \tag{4.71}$$

follows, where the the gradient operator is used here in the normal Euclidean meaning, i.e.,

$$(\nabla W)^2 = W_{,\varrho}^2 + W_{,z}^2 . \tag{4.72}$$

The transformation from the special isotropic coordinates back to arbitrary (x^1, x^2) coordinates yields

$$^{(2)}ds^2 = e^{2\zeta}\left(W_{,c}W_{,d}h^{cd}\right) h_{ab}\, dx^a dx^b \tag{4.73}$$

with an arbitrary 2-metric h_{ab}. It should be noted that h_{ab} is conformally equivalent to γ_{ab}, i.e., $\gamma_{ab} = F\, h_{ab}$, and the conformal factor F is not relevant in (4.73).

The elimination of k from (4.69) via (4.71) leads to

$$\sqrt{\gamma}\,^{(2)}R = -2\left[\sqrt{\gamma}\left(\zeta^{,a} + e^{2\zeta}W^{,b}{}_{|b}W^{,a}\right)\right]_{,a} , \tag{4.74}$$

where the vertical bar denotes the covariant derivative with respect to the 2-metric γ_{ab}.

Combining all this, we arrive at a relatively compact expression for the integrand of the gravitational action integral, namely

$$\sqrt{-g}\,R = \sqrt{\gamma}\left(-2W\nu_{,a}\nu^{,a} + \frac{1}{2}W^3 e^{-4\nu}\omega_{,a}\omega^{,a} + 2W_{,a}\zeta^{,a}\right) + \left(2\sqrt{\gamma}Q^a\right)_{,a} \tag{4.75}$$

where

$$Q^a = W\left(U^{,a} - \zeta^{,a} - e^{2\zeta}W^{,b}{}_{|b}W^{,a}\right) . \tag{4.76}$$

As one can see immediately, in (4.75) the metric γ_{ab} may be replaced by h_{ab} everywhere.

In summary, writing the total action integral as

$$I_G + I_M = \int_{t_1}^{t_2} (\mathcal{L}_G + \mathcal{L}_M)\, dt = \int_{t_1}^{t_2} \mathcal{L}\, dt , \tag{4.77}$$

we obtain for the gravitational part (omitting the total divergence term)

$$\mathcal{L}_{G} = \frac{1}{8G} \int dx^1 dx^2 \sqrt{h} h^{ab} \left(-2W\nu_{,a}\nu_{,b} + \frac{1}{2} W^3 e^{-4\nu} \omega_{,a}\omega_{,b} + 2W_{,a}\zeta_{,b} \right) (4.78)$$

and for the matter part

$$\mathcal{L}_{M} = 2\pi \int dx^1 dx^2 \sqrt{h} h^{ab} W e^{-2\nu} e^{2\zeta} W_{,a} W_{,b} \, p(V) \, . \tag{4.79}$$

An essential part of this formulation is that now the invariant metric potentials ν, ω, W, ζ may be considered as independent quantities for the variation, which is not difficult to prove.

It can be shown (Neugebauer and Herlt 1984) that, for the situation described the formulation given so far can be interpreted as a minimal surface problem in an abstract Riemannian potential space (whose coordinates are ν, ω, W, ζ) with a well-defined indefinite metric. Einstein's field equations are then equivalent to the solution of such a problem. [For details of this minimal surface interpretation, see Neugebauer and Herold (1992)].

Here, it is important that the Lagrangian $\mathcal{L} = \mathcal{L}_{G} + \mathcal{L}_{M}$ yields a variational principle, which is the basis of the numerical solution. Thus, the Lagrangian, whose variation must vanish, $\delta\mathcal{L} = 0$, is given by

$$\mathcal{L} = \frac{1}{2G} \int_0^{\frac{\pi}{2}} d\theta \int_0^\infty dr \, r \left[-W(\nabla\nu)^2 + \frac{1}{4} W^3 e^{-4\nu}(\nabla\omega)^2 + \nabla\zeta \cdot \nabla W \right.$$
$$\left. + 8\pi G \, W e^{2\zeta - 2\nu} p(V)(\nabla W)^2 \right] \tag{4.80}$$

and should be considered as a functional of the four potentials $\nu(r,\theta)$, $W(r,\theta)$, $\zeta(r,\theta)$, $\omega(r,\theta)$. Therein, we have specialized the, in principle arbitrary, "meridional" coordinates x^1, x^2 to spherical coordinates r, θ, which are related to the quasi-Euclidean cylindrical coordinates ϱ, z by $\varrho = r \sin\theta$, $z = r \cos\theta$. (Note that in the 2-space orthogonal to t, φ we use isotropic coordinates. Thus the coordinate r is *different* from the Schwarzschild-like coordinate r of Sect. 4.1.) Here, we have additionally assumed symmetry with respect to the equatorial plane ($z = 0$ or $\theta = \pi/2$). The metric in these coordinates obviously reads

$$ds^2 = -e^{2\nu} dt^2 + e^{-2\nu} \left[e^{2\zeta}(\nabla W)^2 (dr^2 + r^2 d\theta^2) + W^2 (d\varphi - \omega dt)^2 \right] \, . \tag{4.81}$$

The admissible functions ν, W, ζ, ω have to fulfill the boundary conditions

– on the rotation axis ($\varrho = r \sin\theta = 0$):

$$W = 0 \, , \quad \zeta = 0 \, , \tag{4.82}$$

– at infinity ($r \to \infty$):

$$\nu = 0 \, , \quad \zeta = 0 \, , \quad \omega = 0 \, , \quad W - r \sin\theta = 0 \, . \tag{4.83}$$

The potentials ν and ω must be regular on the rotation axis. Furthermore, the unconstrained variation of (4.80) yields the natural boundary conditions on the equatorial plane

$$\frac{\partial \nu}{\partial \theta} = 0 \,, \quad \frac{\partial W}{\partial \theta} = 0 \,, \quad \frac{\partial \zeta}{\partial \theta} = 0 \,, \quad \frac{\partial \omega}{\partial \theta} = 0 \,, \tag{4.84}$$

which express the reflection symmetry.

One could write up the fields equations explicitly by performing the functional variation of the Lagrangian (4.80) with respect to ν, W, ζ and ω, leading to a system of four coupled elliptic partial differential equations, but in our procedure this is not necessary at all.

Note that it is an essential feature of our formulation that there is no distinction between the interior and the exterior of the star; both regions are treated simultaneously. The outside is just characterized by $p(V) = 0$, i.e. the last term in (4.80) vanishes. The position and shape of the surface come out automatically from a self-consistent solution.

In summary, the parameters which characterize a rigidly rotating neutron star are the angular velocity Ω and the surface gravity V_0, which describes the compactness of the system.

4.2.2 Numerical Solution Method

In order to calculate the structure of rapidly rotating stars, a convenient procedure is to start from a non-rotating star and to increase the angular velocity gradually. Therefore, one has to connect our formulation to the usual one for non-rotating neutron stars. The method for doing this has been described by Herold and Neugebauer (1992) and can be transferred easily to the choice of the potentials we use here. For this purpose it is appropriate to transform the TOV equations for non-rotating stars to the coordinates that appear in our general metric (4.81) for rotating bodies. Specializing (4.81) to the case $\Omega = 0$ (no rotation) is equivalent to setting $\omega = 0$. Thus, the comparison with the "Schwarzschild coordinate" form (4.4) – remember that the r coordinate used now is different from that in (4.4), therefore we denote the Schwarzschild-like coordinate in (4.4) now by $\tilde{r}$ – reveals that a radial coordinate transformation is sufficient, while the angle coordinates θ and φ remain unchanged. The explicit transformation formulae read:

$$\nu = \nu \,, \tag{4.85}$$

$$W = e^{\nu} \tilde{r} \sin \theta =: W_0(r) \sin \theta \,, \tag{4.86}$$

$$\frac{d\tilde{r}}{dr} = \frac{\tilde{r}}{r} e^{-\lambda} \,, \tag{4.87}$$

$$e^{-2\zeta} = 1 + \left(\frac{r^2 W_0'^2}{W_0^2} - 1 \right) \sin^2 \theta \,, \tag{4.88}$$

so that in the non-rotating case the potentials ν and $W_0 = W/\sin \theta$ depend only on r, and for ζ we obtain the relation $\exp(-2\zeta) = 1 + f(U) W^2$ with a well-defined function $f(U)$. [In special cases, e.g., for the inner and outer Schwarzschild solution this function has a simple analytic form, see Neugebauer and Herlt (1984).]

Thus, combining (4.19), (4.25), and (4.85–4.88) yields the following ordinary differential equations

$$\frac{d\tilde{r}}{dr} = \frac{1}{r}[\tilde{r}(\tilde{r} - 2Gm)]^{\frac{1}{2}} , \tag{4.89}$$

$$\frac{dm}{dr} = 4\pi\frac{\tilde{r}^2}{r}[\tilde{r}(\tilde{r} - 2Gm)]^{\frac{1}{2}} \rho(p) , \tag{4.90}$$

$$\frac{dU}{dr} = \frac{G(m + 4\pi\tilde{r}^3 p)}{r[\tilde{r}(\tilde{r} - 2Gm)]^{\frac{1}{2}}} . \tag{4.91}$$

Since for $\Omega = 0$ the quantity V is identical to the potential ν, the pressure function $p = p(V)$, which characterizes the matter, is actually a function of ν, and $\rho(p)$ in (4.90) must be determined from $\rho = -p - dp/dV$.

The integration of (4.89–4.91) is performed as usual. Starting at $r = 0$ with appropriate initial conditions (the essential parameter is here the difference $\nu_c - \nu_0$ between the potential ν at the center and the potential ν on the surface) one arrives at the surface when $p = 0$. There, matching to the outer Schwarzschild solution, which can be written in our coordinates as

$$e^{2\nu} = \frac{r - \frac{1}{2}GM}{r + \frac{1}{2}GM} , \tag{4.92}$$

$$W = (r - (GM)^2/4r) \sin\theta , \tag{4.93}$$

$$e^{-2\zeta} = 1 + \frac{(GM)^2}{(r - (GM)^2/4r)^2} \sin^2\theta , \tag{4.94}$$

where the Schwarzschild radial coordinate $\tilde{r}$ is

$$\tilde{r} = \frac{(r + \frac{1}{2}GM)^2}{r} , \tag{4.95}$$

yields the mass and the radius of the non-rotating star.

For rapidly rotating stars, we must solve the variational principle $\delta\mathcal{L} = 0$, i.e., we have to determine those metric potentials ν, W, ζ, ω for which the variation of the integral (4.80) vanishes.

The domain $0 \leq r < \infty$, $0 \leq \theta \leq \pi/2$ of the coordinates (r, θ) in (4.80) is unbounded and thus not very suitable for the numerical treatment. Therefore, we transform the coordinate r to a new coordinate $\bar{r}$, which has a finite domain, e.g., $0 \leq \bar{r} \leq 1$, by the definition $r = S(\bar{r})$ with a monotonic function S, which should satisfy $S(0) = 0$ and $S(1) = \infty$. There are various possibilities, but a choice that was flexible enough in our calculations is

$$r = S(\bar{r}) = c_0\frac{\bar{r}}{1 - \bar{r}} \quad \text{or} \quad \bar{r} = \frac{r}{r + c_0} , \tag{4.96}$$

where the constant c_0 can be adapted approximately to the radius of the star (this means that the surface of the star is in the middle of the $\bar{r}$ domain).

It turned out during the numerical calculations that it is the best procedure to implement the boundary condition (4.83) for W in the following form. Since $W - r\sin\theta = O(1/r)$ for $r \to \infty$, a new function $\widetilde{W}$ is introduced by

$$W = r\sin\theta + (1 - \bar{r})\,\widetilde{W}\ . \tag{4.97}$$

Then, because of (4.96) the modified potential $\widetilde{W}$ takes finite values at infinity, which in general are not zero. As the potential ζ is strongly coupled to W in the functional (4.80), we also have to employ instead of ζ a new function $\widetilde{\zeta}$ defined by

$$\zeta = (1 - \bar{r})^2\,\widetilde{\zeta}\ . \tag{4.98}$$

Using the behavior $\zeta = O(1/r^2)$, which can be deduced from the field equations, it follows that also $\widetilde{\zeta}$ takes non-vanishing values at infinity.

In a similar way, the asymptotic behavior of the potentials ν and ω is taken into account by introducing modified potentials $\widetilde{\nu}$ and $\widetilde{\omega}$,

$$\nu = (1 - \bar{r})\,\widetilde{\nu}\ , \quad \omega = (1 - \bar{r})^3\,\widetilde{\omega} \tag{4.99}$$

which are non-vanishing at infinity.

In summary, the actual functional used in the numerical calculations can be written in the form

$$\mathcal{L} = \int_0^1 d\bar{r} \int_0^{\frac{\pi}{2}} d\theta\ I\big(\bar{r},\theta,\widetilde{\nu},\widetilde{W},\widetilde{\zeta},\widetilde{\omega},\widetilde{\nu}_{,\bar{r}},\widetilde{\nu}_{,\theta},\widetilde{W}_{,\bar{r}},\widetilde{W}_{,\theta},\widetilde{\zeta}_{,\bar{r}},\widetilde{\zeta}_{,\theta},\widetilde{\omega}_{,\bar{r}},\widetilde{\omega}_{,\theta}\big)\ . \tag{4.100}$$

To determine the four functions $\widetilde{\nu} = \widetilde{\nu}(\bar{r},\theta)$, $\widetilde{W} = \widetilde{W}(\bar{r},\theta)$, $\widetilde{\zeta} = \widetilde{\zeta}(\bar{r},\theta)$, and $\widetilde{\omega} = \widetilde{\omega}(\bar{r},\theta)$, one has to discretize the integral (4.100). A natural way is to apply the *finite element* approach [see, e.g., Zienkiewicz (1977)]. The domain is divided into N_e elements with a number of node points (usually at the corners and edges of the elements) that simultaneously belong to neighboring elements. On each element an unknown function is approximated by a low-order polynomial interpolation through its values at the node points. Thus, e.g., the potential $\widetilde{\nu}$ is given by

$$\widetilde{\nu}(\bar{r},\theta) = \sum_n \widetilde{\nu}_n f_n(\bar{r},\theta)\ , \tag{4.101}$$

where $\widetilde{\nu}_n$ are the node point values of $\widetilde{\nu}$ and $f_n(\bar{r},\theta)$ are the polynomial shape functions. In total, one has N_n node points and approximately $4N_n$ unknown function values. (The number of unknowns is somewhat smaller than $4N_n$ after the boundary conditions have been taken into account.) Since the functional (4.80) is non-linear in the potentials, in the integration over each element one has to employ a numerical integration procedure, usually a Gauss integration formula.

Since here the domain is rectangular, we have used a very simple finite element discretization, namely rectangular 4-node bilinear finite elements with Gauss–Legendre integration formula. As the results with these ansatz were satisfactory, it was not necessary to turn to more complicated finite elements.

If we denote the set of unknown potential values by X_i ($i = 1, ..., N$), then the discretized Lagrangian (4.100) is a non-linear function of these variables,

$$\mathcal{L} = \mathcal{L}(X_1, \ldots, X_N)\ . \tag{4.102}$$

The discretized field equations are equivalent to

$$F_i(X) \equiv \frac{\partial \mathcal{L}}{\partial X_i}(X) = 0 \qquad (i = 1, \ldots N) . \tag{4.103}$$

This non-linear system of algebraic equations is then solved by the Newton-Raphson method. The Jacobi matrix,

$$\frac{\partial F_i}{\partial X_j} = \frac{\partial^2 \mathcal{L}}{\partial X_i \partial X_j} , \tag{4.104}$$

needed for this is a symmetrical matrix and can be calculated analytically. (Unfortunately it is not positive definite, otherwise the solution would be easier.) At each Newton step we use a direct sparse matrix solver as the linear equation solver.

After convergence we obtain a solution represented by the node point values of the four potentials ν, W, ζ, and ω. The actual procedure that turns out to be most efficient is to fix the surface gravity parameter V_0 (this determines the mass, at least to a great extent) and to increase the angular velocity Ω starting from $\Omega = 0$.

4.2.3 Results

As has already been described in Sect. 4.1, an essential ingredient for the calculation of *realistic* neutron star models is the equation of state (EOS), i.e., the relation $p = p(\rho)$ between pressure p and energy (or mass) density ρ. A lot of different EOSs for neutron stars exist in the literature [for a review see Arnett and Bowers (1977); see also Glendenning et al. (1992), e.g.]. In producing the results given in the next sections, we have used the EOS MPA [see Wu et al. (1991)].

Global Properties. In this section we will discuss some global properties of typical solutions. First, there is the gravitational mass (total mass) M that characterizes a star. The baryon mass M_0 is always greater than the total mass M because of the gravitational binding energy. From the angular momentum J the moment of inertia I, defined by the (Newtonian) relation $I = J/\Omega$, can be calculated. In Table 4.1 typical results are presented. Additionally to the total mass M, the baryon mass M_0 and the moment of inertia I, the central density ρ_c and the equatorial radius R (measured by the circumference) are given.

In Figs. 4.7 and 4.8 it is shown how the central density ρ_c and the equatorial radius R depend on the rotation for fixed values of V_0. A large increase in radius can be observed mainly for the fastest stars, which are near their stability limit against mass shedding.

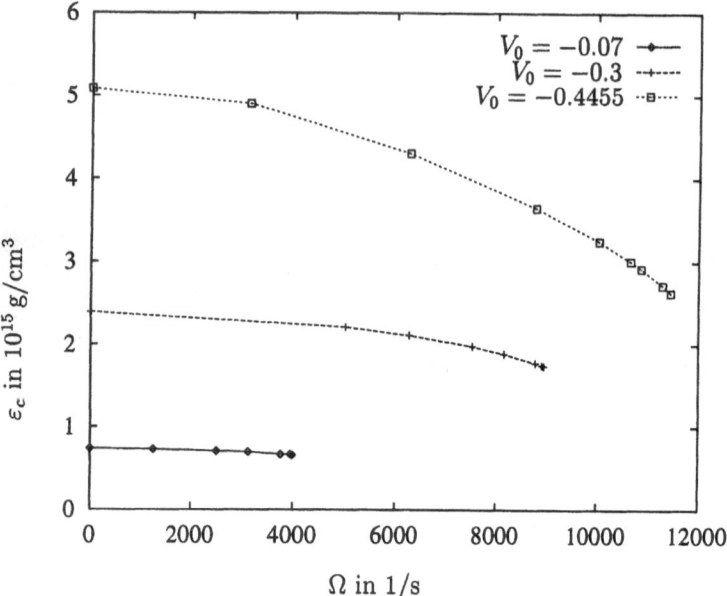

Fig. 4.7. Central density as function of Ω for different values of V_0.

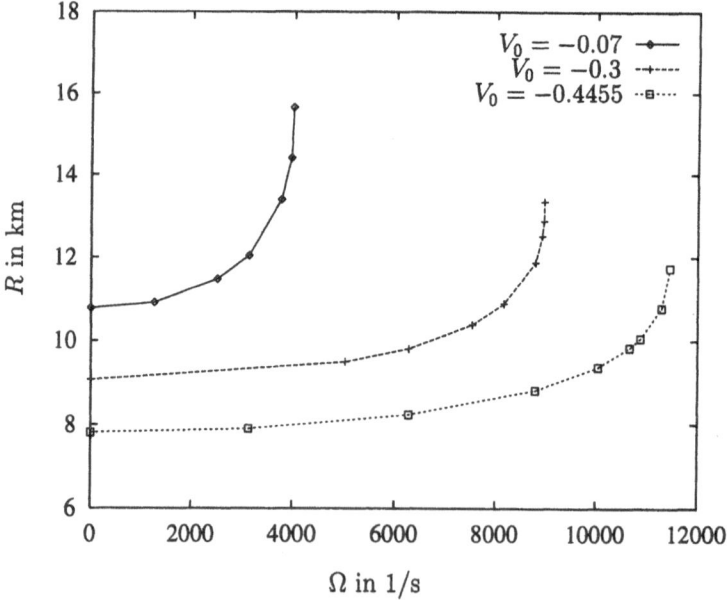

Fig. 4.8. Equatorial radius as function of Ω for different values of V_0.

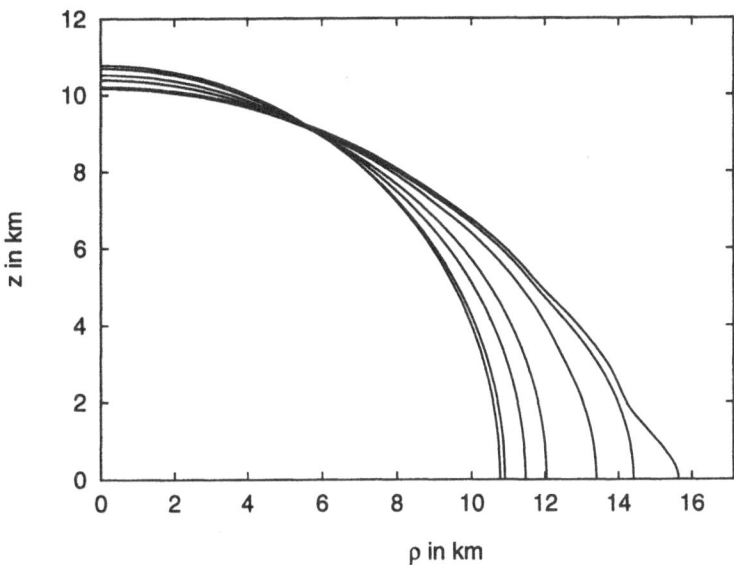

Fig. 4.9. Embedding diagrams of the surfaces of stars with $V_0 = -0.07$. The angular velocity takes the values indicated by markers in Figs. 4.7 and 4.8. As Ω increases, the deformation increases.

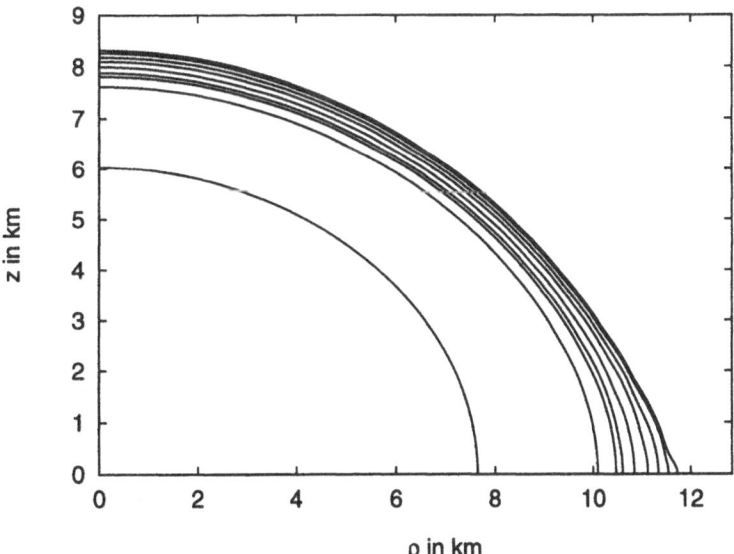

Fig. 4.10. Embedding diagrams of some internal constant-density surfaces of the fastest star of Table 4.1, which rotates just at the mass shedding limit. The outermost curve represents the surface of the star itself, while the other ones belong to the density values 10^7, 10^9, 10^{10}, 10^{11}, 10^{12}, 10^{13}, 10^{14}, 10^{15} g/cm^3.

Visualization by Embedding and 4D Ray-Tracing. To get a coordinate-independent impression of the structure of rotating neutron stars, we have calculated embedding diagrams that visualize the intrinsic geometry of the surface of the neutron star and of internal surfaces of constant pressure (or density). Details of such an embedding procedure may be found in Nollert (1996).

Figure 4.9 shows the embedded surfaces of relatively light neutron star models ($M \approx 0.5 M_\odot$). Additionally, in Fig. 4.10 the internal structure of a fast-rotating heavy neutron star is depicted. It can be recognized that the mass shedding at the equator begins with a bump there caused by just the outermost layers.

As another visualization method of the space-time structure of rapidly rotating neutron stars we show some pictures of "what a rotating neutron star looks like". Here the idea is to assume that from the surface of the considered neutron star photons are emitted which are moving through the curved space-time and eventually reach an observer located far away (near infinity, in the asymptotically flat region). Obviously, this visualization method may be considered to be complementary to the embedding pictures of the interior of the star, as the photons propagate in the region outside the star. Practically, we use a ray-tracing (or ray-casting) approach: from the observer's position the paths of photons are followed back in different directions by integrating the null geodesic equations, using the Christoffel symbols of the numerically determined metric (4.81), until each photon hits (or does not hit) the surface of the rotating star. We call this procedure *4D ray-tracing*, since in addition to three-dimensional space, the time also plays a role (the time when a photon hits the star determines the position on the surface). This approach is described in detail by Nollert (1996).

In Figs. 4.11, 4.12, and 4.13 three examples are presented. All three are models with $V_0 = -0.4455$; the first one is a relatively slow neutron star with angular velocity $\Omega \approx 0.63 \times 10^4 \, \mathrm{s}^{-1}$, the second one has $\Omega \approx 1.07 \times 10^4 \, \mathrm{s}^{-1}$, the third one is the fastest one with $\Omega \approx 1.15 \times 10^4 \, \mathrm{s}^{-1}$ (near the mass-shedding

Table 4.1. Results for the EOS MPA for $V_0 = -0.4455$. The angular velocity Ω is given in s^{-1}, the central density ρ_c in $10^{15} \, \mathrm{g/cm}^3$, the total mass M and the baryon mass M_0 in units of the solar mass $M_\odot$, the equatorial radius R in km, the moment of inertia I in $10^{45} \, \mathrm{g \, cm}^2$

Ω	ρ_c	M	M_0	R	I
0	5.09	1.559	1.834	7.81	0.847
3137	4.90	1.565	1.839	7.90	0.863
6274	4.30	1.590	1.862	8.24	0.933
8783	3.65	1.624	1.890	8.82	1.048
10038	3.25	1.650	1.910	9.37	1.152
10665	3.01	1.670	1.926	9.84	1.236
10872	2.92	1.677	1.932	10.06	1.271
11293	2.72	1.699	1.950	10.78	1.373
11456	2.63	1.711	1.961	11.74	1.434

limit: the Kepler frequency Ω_K, i.e., the angular frequency of a particle in circular orbit at the equator, is here $\Omega_K \approx 1.19 \times 10^4\,\mathrm{s}^{-1}$) and is identical to the star whose internal structure is depicted in Fig. 4.10. For the sake of visibility the surfaces of the stars are painted in a checkerboard pattern (with 30° by 30° patches in the angles θ and φ).

Note that these are not three pictures of the same star with different rotation rates: the number of baryons differs; the first one has $M_0 \approx 1.86 M_\odot$, the second one $M_0 \approx 1.925 M_\odot$, and the third one $M_0 \approx 1.96 M_\odot$. For this reason also the optical images have different sizes; the lightest one ($M \approx 1.59 M_\odot$) in Fig. 4.11 appears as smallest, whereas the other ones ($M \approx 1.67 M_\odot$ in Fig. 4.12 and $M \approx 1.71 M_\odot$ in Fig. 4.13) are bigger due to the different light deflection.

Relativistic light deflection is also responsible for the fact that one can see both poles simultaneously and more than the front hemisphere in equatorial regions (for non-rotating stars, see also Nollert et al. (1989)). The effects of rotation are only weakly visible in Fig. 4.11 (a slight bending of meridional lines), while in Fig. 4.12 they can clearly be seen. More spectacular is Fig. 4.13. Generally, one can say that the rotation leads to the asymmetric appearance which is caused by time-of-flight effects and Lense–Thirring frame dragging of photon paths in combination.

Ergoregions for Solutions with ρ = const. The results presented in the previous sections have been calculated with a realistic equation of state. If we want to investigate solutions for rotating bodies with even stronger gravity – the redshift z_0 at the poles of the star, which can be calculated from $1 + z_0 = e^{-V_0}$, is a reasonable measure for this – we have to consider for example the equations of state ρ = const, which means the relativistic generalization of Maclaurin ellipsoids or the interior Schwarzschild solutions set into rotation. For this case one can obtain much more compact rotating solutions with redshift values greater than 5. Then an effect appears (cf. Butterworth and Ipser (1976)) which is well known from rotating black holes, namely a change of sign of the metric component g_{tt}. This is shown in Fig. 4.14 where the region with $g_{tt} > 0$ is a toroidal ergoregion in which the frame dragging does not allow any static observer. In the case of Fig. 4.14 the ergoregion is partially inside and partially outside the body.

Fig. 4.11. 4D ray-tracing picture of a slowly rotating neutron star
($V_0 = -0.4455$; $\Omega \approx 0.63 \times 10^4\,\mathrm{s}^{-1}$).

Fig. 4.12. 4D ray-tracing picture of a faster rotating neutron star
($V_0 = -0.4455$; $\Omega \approx 1.07 \times 10^4\,\mathrm{s}^{-1}$).

Fig. 4.13. 4D ray-tracing picture of a very fast rotating neutron star ($V_0 = -0.4455$; $\Omega \approx 1.15 \times 10^4\,\text{s}^{-1}$) near the mass shedding limit.

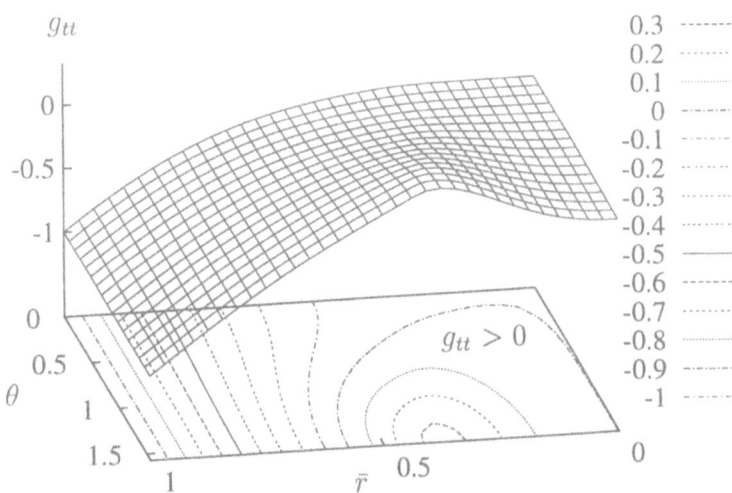

Fig. 4.14. The metric coefficient $g_{tt} = g_{tt}(\bar{r}, \theta)$ for the $\rho = \text{const}$ solution with $V_0 = -1.813$ and maximal rotation.

5. Asteroseismology

Asteroseismology gives us the opportunity to probe the interior of planets, ordinary stars, white dwarfs, and neutron stars with a resolution depending on the number of modes available. With this information, we have the potential to determine the internal structure, i.e., the mass distribution, the chemical composition, the rotation period, the temperature profile, etc. The principal method consists of the observation of oscillation frequencies, the calculation of eigenmodes from appropriate physical models of the oscillating objects, and the subsequent fit of the model parameters in order to match the observed and the calculated frequencies. The measurement of a large number of eigenmodes for the Earth, the Sun, and a few white dwarfs has, within the last decades, tremendously increased our knowledge of the interior properties of those objects. This subject has been recently reviewed in an article by Brown and Gilliland (1994) and is discussed on a more comprehensive level in the book by Hansen and Kawaler (1994).

5.1 Oscillations of Spheres

We have already studied the structure of spherically symmetric distributions of matter in Sect. 1, and oscillations are considered to be small time-dependent perturbations of this static structure. Including dynamic effects will on the one hand modify the present hydrostatic equation (2.1), and on the other hand increase the overall number of equations. It turns out that the resulting system of equations is non-linear, but it can be linearized by assuming that the amplitudes of the oscillations are small. Thus, any given dynamical variable $A(r,t)$ will be decomposed as

$$A(r,t) = A_0(r) + \delta A(r,t) , \tag{5.1}$$

where the perturbation $\delta A(r,t)$ is a small quantity

$$|\delta A(r,t)| \ll |A_0(r)| . \tag{5.2}$$

Inserting the above ansatz into all equations and ignoring non-linear contributions from small quantities leads to equations that are linear in the perturbations, which have to be solved together with appropriately linearized boundary conditions. In this chapter we will present the perturbation equations for oscillating solid bodies and gas spheres. For free oscillations these equations lead to an eigenvalue problem, and the resulting eigenfrequencies and eigenfunctions can be identified by three integer indices n, l, and m. The eigenvalue problem corresponds to an analysis in terms of standing waves (or modes). A different point of view is a travelling-wave representation of small oscillations which is usually performed as a ray-theory, applicable in the short-wavelength limit. The standard example of such a theory is the investigation of seismic waves from earthquakes. Travelling waves suffer from various modifications as they propagate, such as attenuation, reflection, refraction, etc. It is the detailed wave analysis of earthquake data within decades of geophysical research that lead to our current knowledge of the Earth's internal structure and composition. We will, however, not persue this aspect of small oscillations but rather focus on the theory of modes. Only the non-relativistic limit is considered, and the results are therefore applicable to free oscillations of planets like the Earth, ordinary stars, and white dwarfs. Oscillating neutron stars have to be treated in the framework of general relativity, and will be considered in Chap. 3.

We start our investigation by rewriting the hydrostatic equation (2.1) in a slightly different form

$$\nabla \cdot \mathsf{T} = \rho \nabla \Psi \ , \tag{5.3}$$

with the stress tensor T, and the gravitational potential Ψ follows from Poisson's equation

$$\Delta \Psi = 4\pi G \rho \ . \tag{5.4}$$

In the static case T is diagonal and depends only on the pressure

$$T_{jk} = -p\,\delta_{jk} \ , \tag{5.5}$$

where δ_{jk} is the Kronecker symbol and the indices j and k assume the values (1,2,3). These two relations (5.3) and (5.4) are equivalent to (2.1) and (2.2) in case of spherical symmetry where $\nabla \cdot \mathsf{T} = -\nabla p = -\partial p/\partial r\,e_r$. If dynamical effects are taken into account, forces will lead to an acceleration of matter

$$\frac{d\boldsymbol{v}}{dt} = \frac{\partial \boldsymbol{v}}{\partial t} + (\boldsymbol{v}\cdot\nabla)\boldsymbol{v} \ , \tag{5.6}$$

where $\boldsymbol{v}$ denotes the velocity field. The derivative on the left-hand side corresponds to the "Lagrangian" point of view, i.e., changes are considered in a frame locally moving with the matter. It is common practice to introduce a displacement vector $\boldsymbol{\xi}$ in place of the velocity

$$\boldsymbol{v} = \frac{\partial \boldsymbol{\xi}}{\partial t} \ , \tag{5.7}$$

which is the usual variable in the theory of elasticity. Since $\boldsymbol{\xi}$ describes the amplitude of oscillations it is assumed to be a small quantity, thus it can be utilized for fluids or gases only if $\boldsymbol{v}$ represents the oscillating part of the velocity field. Large-scale motions such as rotation have to be subtracted before investigating oscillations and would therefore lead to additional terms in the unperturbed structure equations, and we will ignore such terms in our present considerations. In addition to the acceleration, oscillations will cause a modification of the internal state of matter, i.e., the stress tensor must be complemented accordingly. For isotropic matter the additional internal stresses can be decomposed into a compressional part which is diagonal and can be written similar to (5.5), and a contribution from shear motions that is trace-free. Thus, we add a small correction δT to the unperturbed stress tensor $T^{(0)}$

$$T_{jk} = T_{jk}^{(0)} + \delta T_{jk} = -p_0\,\delta_{jk} - \delta p\,\delta_{jk} + S_{jk} , \tag{5.8}$$

with $\mathrm{Tr}(S) = 0$. Here p_0 indicates the pressure in the absence of oscillations, and the small perturbation δp is connected to the volume compression. In order to obtain this latter relation we have to make an assumption about the thermodynamics of the perturbations, and one widely used approch is to consider adiabatic changes of the internal state of matter. For the case of stellar oscillations the adiabatic approximation leads to a considerable simplification since it eliminates the perturbation equations for the energy generation and transport. Non-adiabatic oscillations are usually investigated in the theory of radial stellar pulsations [see Kippenhahn and Weigert (1990)]. The approximation to treat adiabatic oscillations is justified by the fact that the corresponding temporal changes are too fast to allow for any significant exchange of heat with the environment. This statement holds of course only in the local rest frame of matter, therefore

$$dp = \left(\frac{dp}{d\rho}\right)_{\mathrm{ad}} d\rho , \tag{5.9}$$

where $d\rho$ denotes the Lagrangian density perturbation associated with dp. Since $d\rho/\rho$ is the relative volume change it can be expressed by the divergence of the displacement vector (Landau and Lifshitz 1975)

$$\frac{d\rho}{\rho} = -\nabla\cdot\boldsymbol{\xi} . \tag{5.10}$$

The adiabatic derivative of the pressure with respect to density is usually written in terms of appropriate material coefficients

$$\rho\left(\frac{dp}{d\rho}\right)_{\mathrm{ad}} = K \quad \text{(solid body)} , \qquad \frac{\rho}{p}\left(\frac{dp}{d\rho}\right)_{\mathrm{ad}} = \gamma_{\mathrm{ad}} \quad \text{(gas)} , \tag{5.11}$$

with the bulk modulus K and the adiabatic exponent γ_{ad}. These coefficients can be calculated from the EOS. For an ideal monoatomic gas γ_{ad} is constant ($\gamma_{\mathrm{ad}} = 5/3$), and for a mixture of an ideal gas and radiation it is a function of the ratio of the gas pressure to the total pressure.

In the following we will linearize all equations with respect to small perturbations, that is only first-order terms in the pressure, density, and gravitational potential and the displacement vector are retained. The bulk modulus

K is assumed to be a zero-order quantity, thus K is independent of the material strain, and the pressure change dp depends linearly on $\nabla \cdot \boldsymbol{\xi}$, which is an expression of Hooke's law. Finally, we make a transition from the Lagrangian pressure perturbation dp to the value at a fixed position

$$dp = \delta p + \boldsymbol{\xi} \cdot \nabla p_0 \ . \tag{5.12}$$

and insert the above relations (5.9), (5.10), (5.11), and (5.12) into the stress tensor (5.8)

$$
\begin{aligned}
T_{jk} &= -(p_0 - \boldsymbol{\xi} \cdot \nabla p_0)\,\delta_{jk} + (\gamma_{\mathrm{ad}}\, p_0\, \nabla \cdot \boldsymbol{\xi})\,\delta_{jk} \quad \text{(gas)} \\
T_{jk} &= -(p_0 - \boldsymbol{\xi} \cdot \nabla p_0)\,\delta_{jk} + (K \nabla \cdot \boldsymbol{\xi})\,\delta_{jk} + S_{jk} \quad \text{(solid body)} \ .
\end{aligned}
\tag{5.13}
$$

The first term in both equations includes a contribution from self-gravity because the pressure gradient is proportinal to the gravitational force; the expression $p_0(\boldsymbol{r}) - \boldsymbol{\xi} \cdot \nabla p_0(\boldsymbol{r}) \approx p_0(\boldsymbol{r} - \boldsymbol{\xi})$ can be understood as the pressure at the position $(\boldsymbol{r} - \boldsymbol{\xi})$ from which the mass element was shifted to its actual location $\boldsymbol{r}$. For pure shear motions, the displacement vector has no radial components. The matter then moves along equipotential surfaces and gravity will not influence this type of strain. Within the approximation of Hooke's law, the trace-free tensor S is linear in the strain and can be expressed as (Landau and Lifshitz 1975)

$$S_{jk} = 2\mu \left(u_{jk} - \frac{1}{3} u_{nn}\delta_{jk} \right) = \mu \left(\frac{\partial \xi_j}{\partial x_k} + \frac{\partial \xi_k}{\partial x_j} - \frac{2}{3} \frac{\partial \xi_n}{\partial x_n}\delta_{jk} \right) , \tag{5.14}$$

with the shear modulus μ, and the strain tensor u_{jk}. Here we have written S in terms of its cartesian components, and any index that appears twice is understood to be summed over. More general forms of the stress tensor for self-gravitating elastic bodies including rotation are discussed by Dahlen (1972). As mentioned above, the bulk modulus K can be calculated from the EOS. If the pressure is estimated from the free energy F (see Chap. 2), K follows from a second derivative of this thermodynamic potential [see Bina and Helffrich (1992)]. This procedure can be extended to the calculation of the shear modulus as well, provided the strain tensor and the temperature are used as independent thermodynamic variables [see for example Leibfried and Ludwig (1961)]

$$T_{jk} = \rho \frac{\partial F}{\partial u_{jk}} \ ,$$

and, in general, a set of elastic coefficients $C_{jk,lm}$ is obtained from

$$C_{jk,lm} = \frac{\partial T_{jk}}{\partial u_{lm}} = \rho \frac{\partial^2 F}{\partial u_{jk} \partial u_{lm}} - \rho \frac{\partial F}{\partial u_{jk}} \delta_{lm} \ ,$$

where the derivatives can be taken isothermally or adiabatically depending on the problem under consideration (Davis 1973, 1974). For symmetry reasons the $C_{jk,lm}$ can be expressed through 36 moduli $c_{\alpha\beta}$ $(\alpha, \beta = 1, \ldots, 6)$ [see Born and Huang (1954)], but for isotropic matter there are only two independent quantities, and we get $\mu = c_{44}$ and $K = c_{12} + \frac{2}{3} c_{44}$. The practical calculation of the free energy has to be performed by employing the methods of solid state physics discussed in Chap. 2.

Returning to the perturbation equations, we obtain the linearized relations

$$
\begin{aligned}
\rho_0 \ddot{\boldsymbol{\xi}} &= -\rho_0 \nabla(\delta \Psi) - \delta\rho \nabla \Psi_0 + \nabla \cdot \delta\mathsf{T} , \\
\Delta(\delta\Psi) &= 4\pi G \, \delta\rho , \\
\delta\rho &= -\rho_0 \nabla \cdot \boldsymbol{\xi} - \boldsymbol{\xi} \cdot \nabla\rho_0 ,
\end{aligned}
\tag{5.15}
$$

where dots over a variable indicate time derivatives. The last relation follows from (5.10) when expressed in terms of local changes, and corresponds to the continuity equation in hydrodynamics. For a solid body, the divergence of $\delta\mathsf{T}$ can be expressed as

$$
\begin{aligned}
\nabla \cdot \delta\mathsf{T} &= \left(K + \tfrac{4}{3}\mu\right) \nabla(\nabla \cdot \boldsymbol{\xi}) - \mu \nabla \times (\nabla \times \boldsymbol{\xi}) - \boldsymbol{\xi} \Delta\mu \\
&\quad + (\nabla \cdot \boldsymbol{\xi}) \nabla \left(K + \tfrac{1}{3}\mu\right) + \nabla \times (\boldsymbol{\xi} \times \nabla\mu) + \nabla(\boldsymbol{\xi} \cdot \nabla\mu) \\
&\quad + \nabla(\boldsymbol{\xi} \cdot \nabla p_0) ,
\end{aligned}
\tag{5.16}
$$

whereas for gas spheres

$$
\nabla \cdot \delta\mathsf{T} = \nabla(\boldsymbol{\xi} \cdot \nabla p_0) + \nabla(\gamma_{\mathrm{ad}} \, p_0 \, \nabla \cdot \boldsymbol{\xi}) .
\tag{5.17}
$$

K is sometimes expressed in terms of the Lamé coefficient λ

$$
K = \lambda + \tfrac{2}{3}\mu .
\tag{5.18}
$$

Note that zero-order functions depend only on the radial coordinate r, therefore

$$
\nabla\mu = \mu' \, \boldsymbol{n} , \qquad \nabla K = K' \, \boldsymbol{n} , \qquad \text{etc.}
\tag{5.19}
$$

where the prime denotes derivatives with respect to r, and $\boldsymbol{n} = \boldsymbol{r}/r$ is a radial unit vector.

The dynamic equations have to be solved together with suitable boundary conditions. In the center of the spherical mass distribution all functions must be regular, and on the outer surface we assume a free boundary, i.e., there are no external forces acting on this surface. Let the boundary of the unperturbed sphere be given by $r = R$, then for the perturbed surface $r = R + \boldsymbol{\xi}$, and the free surface boundary conditions read

$$
T_{jk} s_k = 0 \qquad (r = R + \boldsymbol{\xi}) ,
\tag{5.20}
$$

where $\boldsymbol{s}$ is the unit normal vector of the boundary surface. The gravitational potential and its derivative normal to the boundary must be continuous

$$
\Psi = \Psi^{(\mathrm{out})} , \qquad \boldsymbol{s} \cdot \nabla\Psi = \boldsymbol{s} \cdot \nabla\Psi^{(\mathrm{out})} \qquad (r = R + \boldsymbol{\xi}) ,
\tag{5.21}
$$

and the potential $\Psi^{(\mathrm{out})}$ outside the material body follows from the Laplace equation

$$\Delta \Psi^{(\text{out})} = 0 \qquad (r > R + \xi) \tag{5.22}$$

together with a regularity condition at infinity. We now expand the boundary conditions to first order. From (5.13) and (5.14) we obtain for spherical solid bodies and gas spheres at $r = R$, respectively

$$\left(K + \tfrac{1}{3}\mu\right)(\nabla \cdot \xi)\, n + \mu \left[\nabla \times (\xi \times n) + \nabla(\xi \cdot n) - \tfrac{2}{r}\xi\right] = 0$$
$$\nabla \cdot \xi = 0, \tag{5.23}$$

where we have used the boundary condition of the unperturbed state $p_0(R) = 0$. Next, we expand the conditions (5.21) for the gravitational potential taking into account the continuity of the undisturbed functions Ψ_0 and $(n \cdot \nabla \Psi_0)$ at $r = R$

$$\delta \Psi = \delta \Psi^{(\text{out})},$$
$$n \cdot \nabla(\delta \Psi) + (n \cdot \xi)\, \Psi_0'' = n \cdot \nabla(\delta \Psi^{(\text{out})}) + (n \cdot \xi)\, \Psi_0''^{(\text{out})}. \tag{5.24}$$

Using the Poisson equation (5.4) for Ψ_0 and the Laplace equation (5.22) for $\Psi_0^{(\text{out})}$ we obtain

$$n \cdot \nabla(\delta \Psi) + 4\pi G \rho_0 (n \cdot \xi) = n \cdot \nabla(\delta \Psi^{(\text{out})}) \qquad (r = R). \tag{5.25}$$

In order to gain some insight into the nature of the perturbation equations for solid bodies, we first consider a simplified situation assuming constant values for K, μ, ρ_0, and ignoring the effects of self-gravity. Then

$$\rho_0 \ddot{\xi} = (K + \tfrac{4}{3}\mu)\nabla(\nabla \cdot \xi) - \mu \nabla \times (\nabla \times \xi). \tag{5.26}$$

Decomposing the displacement vector into a source-free (transverse) and a curl-free (longitudinal) part

$$\xi = \xi_T + \xi_L \qquad \text{with} \qquad \begin{aligned} \nabla \cdot \xi_T &= 0 \\ \nabla \times \xi_L &= 0 \end{aligned} \tag{5.27}$$

leads to two decoupled wave equations for ξ_L and ξ_T with different phase velocities

$$\ddot{\xi}_T = c_T^2 \Delta \xi_T, \qquad c_T = \sqrt{\tfrac{\mu}{\rho_0}},$$
$$\ddot{\xi}_L = c_L^2 \Delta \xi_L, \qquad c_L = \sqrt{\tfrac{3K+4\mu}{3\rho_0}}. \tag{5.28}$$

Considering plane-wave solutions with wave vector k, the relations (5.27) read

$$k \cdot \xi_T = 0, \qquad k \times \xi_L = 0. \tag{5.29}$$

The transverse wave is called the S wave in seismology; it is entirely due to shear motions and does not exist in a gaseous medium. The longitudinal wave, called the P wave, is connected to the compressibility of matter.

The decomposition of the oscillation behavior into two independent wave types, one of which is a pure shear mode, can be obtained for the full set of equations (5.15) as well. Since the static (unperturbed) structure is spherically symmetric, it is useful to expand the angular dependence of all quantities in terms of spherical harmonics Y_{lm} of degree l and order m

$$Y_{lm}(\theta, \phi) = (-1)^m \sqrt{\frac{2l+1}{4\pi} \frac{(l-m)!}{(l+m)!}} \, P_l^m(\cos\theta) \, e^{im\phi} \,, \tag{5.30}$$

where $P_l^m(\cos\theta)$ are the associated Legendre functions, l is an integer ($l = 0, 1, 2, \ldots$), and $|m| = 0, 1, 2, \ldots, l$. The Y_{lm} are eigenfunctions of the square of the angular momentum operator $\boldsymbol{L} = -i\boldsymbol{r} \times \nabla$

$$\boldsymbol{L}^2 Y_{lm} = l(l+1)Y_{lm} = \left(-r^2\Delta + \frac{\partial}{\partial r}r^2\frac{\partial}{\partial r}\right) Y_{lm} = -r^2\Delta Y_{lm} \,. \tag{5.31}$$

In order to represent $\boldsymbol{\xi}$ in terms of Y_{lm} we use the following vector spherical harmonics

$$
\begin{aligned}
\boldsymbol{B}_{lm} &= iLY_{lm} = \boldsymbol{r} \times \nabla Y_{lm} = \boldsymbol{n} \times \boldsymbol{E}_{lm} \,, \\
\boldsymbol{E}_{lm} &= r\nabla Y_{lm} = -\boldsymbol{n} \times \boldsymbol{B}_{lm} \,, \\
\boldsymbol{Q}_{lm} &= \boldsymbol{n} \, Y_{lm} \,,
\end{aligned}
\tag{5.32}
$$

which are mutually orthogonal

$$\int_{4\pi} \boldsymbol{F}_{lm}^* \cdot \boldsymbol{G}_{l'm'} \, d\Omega = a_l \, \delta_{FG} \, \delta_{ll'} \, \delta_{mm'} \,. \tag{5.33}$$

Here $\boldsymbol{F}_{lm}$ and $\boldsymbol{G}_{l'm'}$ represent any of the functions $\boldsymbol{B}_{lm}$, $\boldsymbol{E}_{lm}$, $\boldsymbol{Q}_{lm}$, and $\boldsymbol{F}_{lm}^*$ stands for the complex conjugate of $\boldsymbol{F}_{lm}$. Here $\boldsymbol{B}_{lm}$ and $\boldsymbol{E}_{lm}$ are transverse vectors, and $\boldsymbol{Q}_{lm}$ points in the radial direction; $\boldsymbol{E}_{lm}$ and $\boldsymbol{Q}_{lm}$ have the same parity $(-1)^l$ whereas $\boldsymbol{B}_{lm}$ has parity $(-1)^{l+1}$. Because of the orthogonality relation (5.33) the three vector fields (5.32) can serve as a basis for the decomposition of an arbitrary vector field into vector spherical harmonics. The choice of the transverse components $\boldsymbol{B}_{lm}$ and $\boldsymbol{E}_{lm}$ corresponds to the multipole expansion of wave fields in electrodynamics, thus these functions are sometimes called "magnetic" and "electric" components, respectively. From the above definition the divergence and the curl of the three vector fields can be easily calculated

$$
\begin{aligned}
\nabla \cdot \boldsymbol{B}_{lm} &= 0 \,, & \nabla \times \boldsymbol{B}_{lm} &= -\tfrac{1}{r}\left(\boldsymbol{E}_{lm} + l(l+1)\,\boldsymbol{Q}_{lm}\right) \,, \\
\nabla \cdot \boldsymbol{E}_{lm} &= -\tfrac{l(l+1)}{r}Y_{lm} \,, & \nabla \times \boldsymbol{E}_{lm} &= \tfrac{1}{r}\boldsymbol{B}_{lm} \,, \\
\nabla \cdot \boldsymbol{Q}_{lm} &= \tfrac{2}{r}Y_{lm} \,, & \nabla \times \boldsymbol{Q}_{lm} &= -\tfrac{1}{r}\boldsymbol{B}_{lm} \,.
\end{aligned}
\tag{5.34}
$$

We now make an ansatz for the various perturbations by expanding all functions in terms of appropriate spherical harmonics. Since we are interested in free oscillations of the spherical mass distributions, we assume periodic time dependencies with frequency ω. Considering terms for fixed values of l and m we have

$$\begin{aligned}
\xi &= [W(r)B_{lm} + V(r)E_{lm} + U(r)Q_{lm}]\, e^{i\omega t}\,, \\
\delta\rho &= \hat{\rho}(r)Y_{lm}\, e^{i\omega t}\,, \\
\delta p &= \hat{p}(r)Y_{lm}\, e^{i\omega t}\,, \\
\delta\Psi &= \hat{\Psi}(r)Y_{lm}\, e^{i\omega t}\,.
\end{aligned}$$

$$(5.35)$$

Inserting this into the continuity equation and Poisson's equation using (5.31), (5.34), and dropping the common factor $Y_{lm}\exp{(i\omega t)}$ leads to

$$\hat{\rho} + \rho_0\left[U' + \frac{2}{r}U - \frac{l(l+1)}{r}V\right] + \rho_0' U \;=\; 0$$

$$\hat{\Psi}'' + \frac{2}{r}\hat{\Psi}' - \frac{l(l+1)}{r^2}\hat{\Psi} \;=\; 4\pi G\hat{\rho}\,. \qquad (5.36)$$

Note that for a fixed radius the quantity

$$\frac{\sqrt{l(l+1)}}{r} = k_\perp \qquad (5.37)$$

can be considered as the wave number in the horizontal direction, i.e., perpendicular to the radial direction; therefore, large l correspond to small horizontal wavelengths. For the exterior gravitational potential $\Psi^{(\mathrm{out})}$ we write

$$\Psi^{(\mathrm{out})} = \frac{c_l}{r^{l+1}}Y_{lm}\, e^{i\omega t}\,, \qquad (5.38)$$

which fulfills the Laplace equation (5.22). From that, the two boundary conditions (5.24) and (5.25) reduce to a single relation

$$\hat{\Psi}' + \frac{l(l+1)}{r}\hat{\Psi} + 4\pi G\rho_0\, U = 0 \qquad (r = R)\,. \qquad (5.39)$$

We will now focus our attention on oscillations of solid bodies. The various terms in the divergence of the stress tensor (5.16) can be easily calculated with the aid of the relations (5.34). As a result, the equation of motion is decomposed into two independent parts, one of which is in the direction of B_{lm} and contains only the coefficient $W(r)$

$$\left(r^2\mu W'\right)' - [r\mu' + l(l+1)\mu]\,W + r^2\rho_0\,\omega^2 W = 0\,, \qquad (5.40)$$

and the other part consists of two coupled equations for $U(r)$ and $V(r)$

$$\begin{aligned}
0 = \;& \left(r^2\mu V'\right)' - \left[r\mu' + l(l+1)\left(K + \tfrac{4}{3}\mu\right)\right]V + r^2\rho_0\,\omega^2 V \\
&+ \left(K + \tfrac{1}{3}\mu\right)rU' + \left[r\mu' + 2\left(K + \tfrac{4}{3}\mu\right)\right]U \\
&- r\rho_0\left(\hat{\Psi} + r\Psi_0'\right)U\,,
\end{aligned} \qquad (5.41)$$

$$0 = \left[r^2 \left(K + \tfrac{4}{3}\mu\right) U'\right]' + r^2 \rho_0 \, \omega^2 U$$
$$+ \left[2r \left(K - \tfrac{2}{3}\mu\right)' - l(l+1)\mu - 2\left(K + \tfrac{4}{3}\mu\right)\right] U$$
$$-l(l+1)\left[\left(K + \tfrac{1}{3}\mu\right) rV' - \left(K + \tfrac{7}{3}\mu\right) V + \left(K - \tfrac{2}{3}\mu\right)' rV\right] \tag{5.42}$$
$$-r\rho_0 \left[r\hat{\Psi}' + r\Psi_0'' U - \Psi_0' \left(2U - l(l+1)V\right)\right] \ .$$

From (5.23) we have for the boundary conditions at $r = R$

$$rW' - W = 0 \ ,$$

$$rV' - V + U = 0 \ , \tag{5.43}$$

$$\left(K + \tfrac{4}{3}\mu\right) rU' + 2\left(K - \tfrac{2}{3}\mu\right) U - l(l+1)\left(K - \tfrac{2}{3}\mu\right) V = 0 \ .$$

This set of differential equations together with the homogeneous boundary conditions is self-adjoint and corresponds to the well known Sturm–Liouville type (Courant and Hilbert 1968). As a result, a solution exists only for special values of ω, the eigenvalues of the system. They form an infinite set of real and distinct frequencies ω_{nl} where the index n labels the frequencies for fixed values of l which appears as a parameter in the equations. Eigenfunctions belonging to two different eigenvalues are real and orthogonal to one another. Note that none of the equations or boundary conditions contains the order m of the spherical harmonics, thus the problem is degenerate with respect to this parameter. This is a consequence of the underlying spherical symmetry that will be broken for example for rotating configurations. Since the equation for W is decoupled from the remaining equations, there exist two types of oscillations which have different parity. The first one has a purely transverse displacement vector proportional to $\boldsymbol{B}_{lm}$ and is called the toroidal mode $_nT_l^m$. Its radial dependence $W(r)$ follows from equation (5.40) which is independent of the bulk modulus K. Thus, toroidal oscillations are pure shear waves (S waves). The second type is called the spheroidal mode $_nS_l^m$ and the displacement vector consists of a transverse part ($\propto \boldsymbol{E}_{lm}$) and a logitudinal part ($\propto \boldsymbol{Q}_{lm}$), i.e., S waves and P waves are coupled in this type of oscillation. In general, the full set of equations must be solved numerically. Analytic solutions, however, can be obtained for homogeneous spheres, i.e., for constant values of K, μ, and ρ_0 (Lapwood and Usami 1981). This is particulary simple for the toroidal oscillations and from equation (5.40) we get

$$r^2 W'' + 2rW' + \left[k^2 r^2 - l(l+1)\right] W = 0 \ ,$$
$$rW' - W = 0 \qquad (r = R) \ , \tag{5.44}$$

where $k^2 = \omega^2 \rho_0 / \mu = \omega^2 / c_{\mathrm{T}}^2$ with the phase velocity c_{T} defined in (5.28). This is the differential equation for spherical Bessel functions (Abramowitz and Stegun 1970) and the only solution that is regular at $r = 0$ is given by

$$j_l(kr) = \sqrt{\frac{\pi}{2kr}} J_{l+1/2}(kr) \ . \tag{5.45}$$

Note that the case $l = 0$ can be excluded since $Y_{00} = 1/\sqrt{4\pi}$ and from (5.32) $B_{00} = 0$ which is just the trivial solution. Using a recurrence relation for the j_l, the boundary condition can be expressed as

$$(l+2)j_l - kR\,j_{l-1} = 0 \qquad (r = R)\,, \tag{5.46}$$

and the roots of this equation determine the eigenvalues ω_{nl} of the toroidal oscillations. The eigenfrequencies are usually arranged in an ascending sequence with $n = 0$ denoting the smallest eigenvalue called the fundamental mode. In the case of toroidal oscillations n also labels the number of radial nodes of the corresponding eigenfunction, i.e., the eigenfunction of the fundamental mode has no zero. For $l = 1$, (5.46) reads

$$\tan(kR) = \frac{3\,kR}{3 - k^2 R^2}\,, \tag{5.47}$$

and the first non-zero root is $(kR) \approx 5.67$ whereas the eigenfunction j_1 has its first zero at about 4.493. Thus, this solution corresponds to $n = 1$ and the fundamental mode is missing for $l = 1$. Eigenfrequencies for various l and n values are shown in Fig. 5.1.

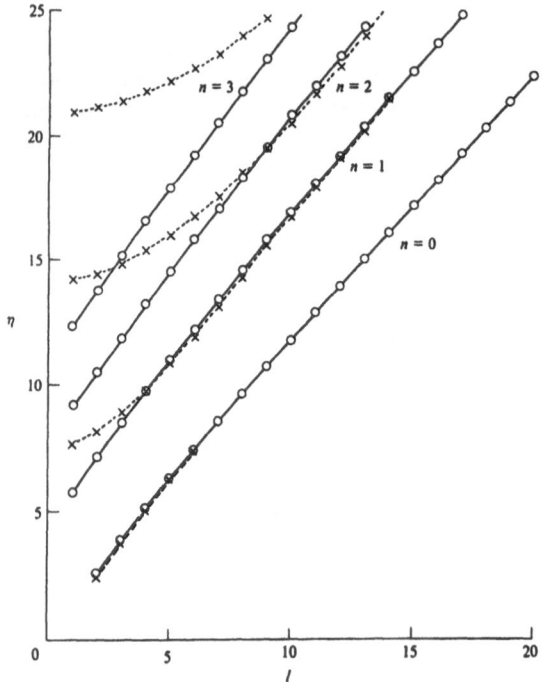

Fig. 5.1. Frequencies of toroidal oscillations ($\eta = \omega R\sqrt{\rho/\mu}$) as a function of n and l. The various curves connect the frequencies for fixed n. Solid lines correspond to a uniform sphere of radius R, dashed curves are for a uniform shell with a ratio of its inner to outer radius of 0.545 [from Lapwood and Usami (1981)].

The equations for the spheroidal oscillations are directly coupled to the gravitational potential, and, in contrast to the toroidal modes, there is a distortion of the shape of the body with respect to the unperturbed case. For a homogeneous elastic sphere one can again obtain analytic solutions and express the radial dependence of all functions in terms of spherical Bessel functions (Pekeris and Jarosch 1958). In general, the connection between the order n of a mode and the number of nodes of the corresponding eigenfunction is more complicated than for the toroidal oscillations. Figure 5.2 shows the eigenfrequencies for the homogeneous model as a function of l and n.

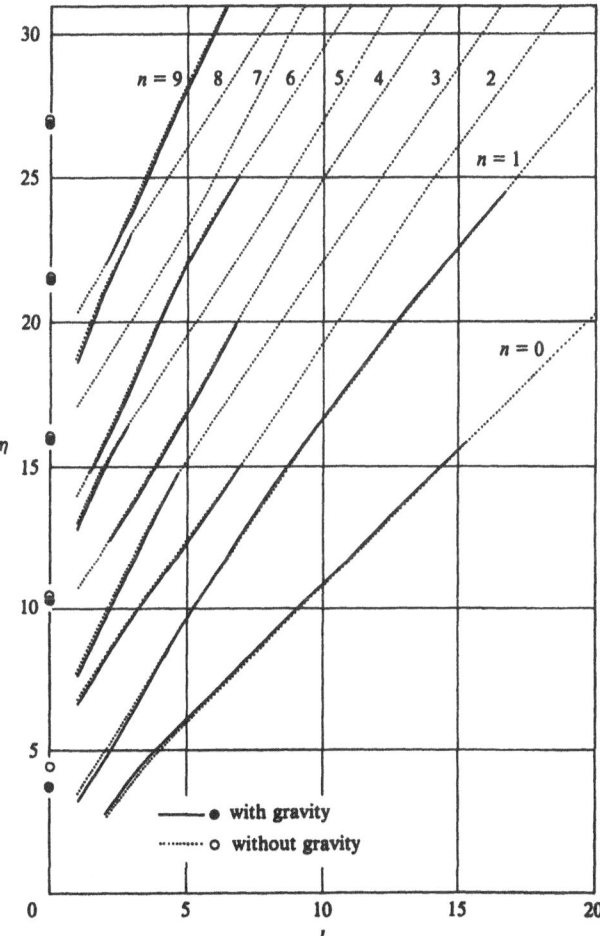

Fig. 5.2. Frequencies of spheroidal oscillations ($\eta = \omega R \sqrt{\rho/\mu}$) as a function of n and l for a uniform sphere of radius R (frequencies for fixed n are connected by lines). The dashed curves correspond to a model without gravity. Model parameters have been chosen to agree with a radially averaged Earth: $R = 6370$ km, $\rho_0 = 5.52$ g cm^{-3}, $c_L = 6.667$ km s^{-1}, $K = 5/3\,\mu$ [from Lapwood and Usami (1981)].

The case $l = 0$ corresponds to a "breathing mode" where the displacement is purely radial ($E_{00} = 0$) and the motion consists of a compression and dilation of the entire body. The fundamental mode ($n = 0$) is again missing for $l = 1$. In the limit of large l this mode is equivalent to a Rayleigh wave, i.e., a surface wave that is neither purely transverse nor longitudinal. Asymptotic investigations of the distribution of eigenfrequencies show that for large n ($\omega R/c_T \gg l$) the longitudinal part of the spheroidal modes decouples from the transverse part, i.e., the S waves and the the P waves propagate independent of one another [for details see Lapwood and Usami (1981)].

The analysis of oscillations of gas spheres has to start from the stress tensor (5.17). Again we insert the ansatz (5.35) into the perturbation equations. This leads to

$$
\begin{aligned}
\omega^2 W &= 0, \\
\rho_0 \omega^2 V - \frac{1}{r}\hat{p} - \frac{1}{r}\rho_0\hat{\Psi} &= 0, \\
\rho_0 \omega^2 U - \hat{p}' - \rho_0\hat{\Psi}' - \hat{\rho}\Psi_0' &= 0,
\end{aligned}
\tag{5.48}
$$

with the boundary condition

$$
rU' + 2U - l(l+1)V = 0 \qquad (r = R). \tag{5.49}
$$

For the pressure perturbation we have

$$
\hat{p} = -\gamma_{\mathrm{ad}}\, p_0 \left(U' + \frac{2}{r}U - \frac{l(l+1)}{r}V \right) - p_0' U, \tag{5.50}
$$

and the equations for $\hat{\rho}$ and $\hat{\Psi}$ are again given by (5.36). When we ignore the trivial case $\omega = 0$, the first of the equations (5.48) yields the solution $W = 0$, i.e., there is no toroidal mode since a gas cannot provide a restoring force for shear motions. The remaining relations form a system of differential equations of fourth order. It can be shown that it is also self-adjoint but has a different structure than the standard Sturm–Liouville eigenvalue problem. Neither the differential equations nor the boundary conditions depend on the parameter m and the solutions are therefore again degenerate with respect to the order of the spherical harmonics.

The complete set of equations can be simplified considerably by an approximation introduced by Cowling (1941) where the contribution due to the perturbation of the gravitational potential is neglected, i.e., the function $\hat{\Psi}$ is ignored in (5.48). That decouples the Poisson equation (5.36) from the remaining relations and reduces the system to second order which, after some algebraic manipulations, can be represented as [see also Ledoux and Walraven (1958)]

$$
\frac{\mathrm{d}}{\mathrm{d}r}\left[\frac{\omega^2 c_{\mathrm{s}}^2}{l(l+1)c_{\mathrm{s}}^2 - r^2\omega^2}\, \rho_0\, p_0^{-2/\gamma_{\mathrm{ad}}}\, \frac{\mathrm{d}X}{\mathrm{d}r} \right] -
$$

$$
\frac{1}{r^2}\left(\omega^2 + gA \right) \rho_0\, p_0^{-2/\gamma_{\mathrm{ad}}}\, X = 0, \tag{5.51}
$$

where

$$X = r^2 p_0^{1/\gamma_{\text{ad}}} U \ ,$$

and

$$A = \frac{\rho_0'}{\rho_0} - \frac{p_0'}{\gamma_{\text{ad}} \, p_0} \ . \tag{5.52}$$

Note that a constant value for γ_{ad} has been assumed in the derivation of (5.51). Here c_{s} is the adiabatic sound speed which follows from (5.11), and g denotes the local gravity

$$c_{\text{s}}^2 = \left(\frac{\mathrm{d}p}{\mathrm{d}\rho}\right)_{\text{ad}} = \gamma_{\text{ad}}\frac{p_0}{\rho_0} \ , \qquad g = -\frac{p_0'}{\rho_0} = \frac{Gm(r)}{r^2} \ . \tag{5.53}$$

The boundary condition (5.49) at the outer radius $r = R$ can also be expressed in terms of the variable X

$$\frac{\omega^2 c_{\text{s}}^2}{l(l+1)c_{\text{s}}^2 - r^2\omega^2}\frac{\mathrm{d}X}{\mathrm{d}r} + \frac{g}{r^2}X = 0 \qquad (r = R) \ . \tag{5.54}$$

The analysis of the self-adjoint equation (5.51) is difficult because of its non-linear dependence on the parameter ω^2. There are, however, two asymptotic regimes where the equation is reduced to the standard Sturm–Liouville form. If ω^2 is large an eigenvalue problem is obtained with a spectrum of increasing eigenvalues ω_{nl}. The corresponding oscillations have been called p modes by Cowling (1941) because the restoring force is mainly due to the gas pressure. These modes are usually labelled with the order of the eigenfrequencies n, i.e., p_1, p_2, p_3, etc. For stellar models with a density profile that is not too concentrated towards the center, the order of the mode agrees with the number of nodes of the eigenfunction. For small values of ω an eigenvalue problem is obtained where $1/\omega^2$ corresponds to the parameter of the Sturm–Liouville equation. Thus, for each l a sequence of frequencies ω_{nl} is obtained that decrease with increasing order n. However, this holds only if the function A of (5.52) is negative, otherwise the eigenfrequencies are purely imaginary and the perturbations grow exponentially with time. The quantity

$$\omega_{\text{V}} = \sqrt{-gA} \tag{5.55}$$

is the Brunt–Väisälä frequency that describes the oscillation of a gas element inside a gravitating sphere if it is displaced adiabatically in the radial direction. Since the restoring force of these modes is dominated by gravity they are called g modes and they can also be labelled according to the number of nodes of the corresponding eigenfunctions, i.e., g_1, g_2, etc. Note that the condition $\omega_{\text{V}}^2 > 0$ for the existence of g mode oscillations is identical to the Schwarzschild criterion (2.40) for convectional stability, thus in a stellar convection zone the g modes are all unstable.

A sequence of eigenmodes for $l = 2$ from a stellar model with a polytropic EOS (2.10) is shown in Fig. 5.3 where also numbers and positions of the

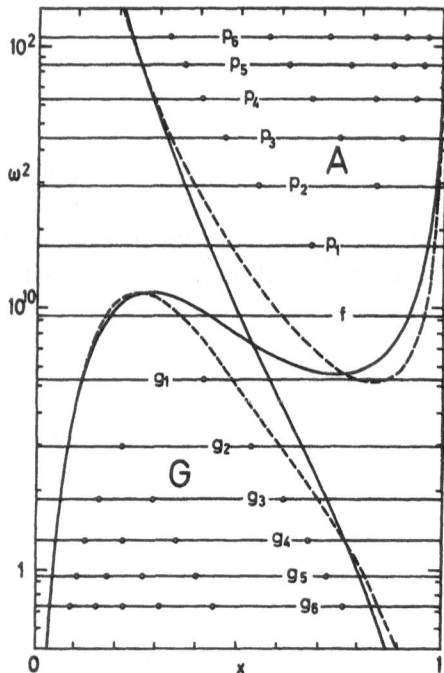

Fig. 5.3. Eigenfrequencies for a polytropic stellar model as a function of the (normalized) radius $x = r/R$. The circles indicate the zeros of the eigenfunctions of the various modes. Solid lines correspond to the Brunt–Väisälä frequency and to the Lamb frequency S_l. Dashed curves denote acoustical cutoff frequencies for plane waves [from Scuflaire (1974)].

corresponding nodes can be seen. If the polytropic exponent is not too small there is a unique relation between the order n and the number of zeros of the eigenfunctions. and $n > 0$ denotes the p modes whereas $n < 0$ holds for g modes. For realistic stellar models or polytropes with smaller exponents this relation is more complicated [for details see Scuflaire (1974) or Unno et al. (1989)]. The modes with $n = 0$ are again called fundamental or f modes; for large l they are restricted to the outer layers of the star and can be identified as surface waves. For realistic stellar models (5.51) must be solved numerically and the eigenfrequencies for a solar model are illustrated in Fig. 5.4.

We close this chapter with a few comments on the short wavelength limit of the oscillations. Using an appropriate transformation, we can express the equation of motion in the Cowling approximation (5.51) as a Helmholtz equation with radial wave number k_r

$$\frac{d^2 Z}{dr^2} + k_r^2 Z = 0 , \tag{5.56}$$

where

$$Z = \left| l(l+1)c_s^2 - r^2\omega^2 \right|^{-1/2} r^2 \omega c_s \sqrt{\rho_0}\, U .$$

For wavelenghts much smaller than the stellar radius a local approximation yields

$$k_r^2 c_s^2 \approx \omega^2 - \omega_c^2 + k_\perp^2 c_s^2 \left(\frac{\omega_V^2}{\omega^2} - 1 \right) . \tag{5.57}$$

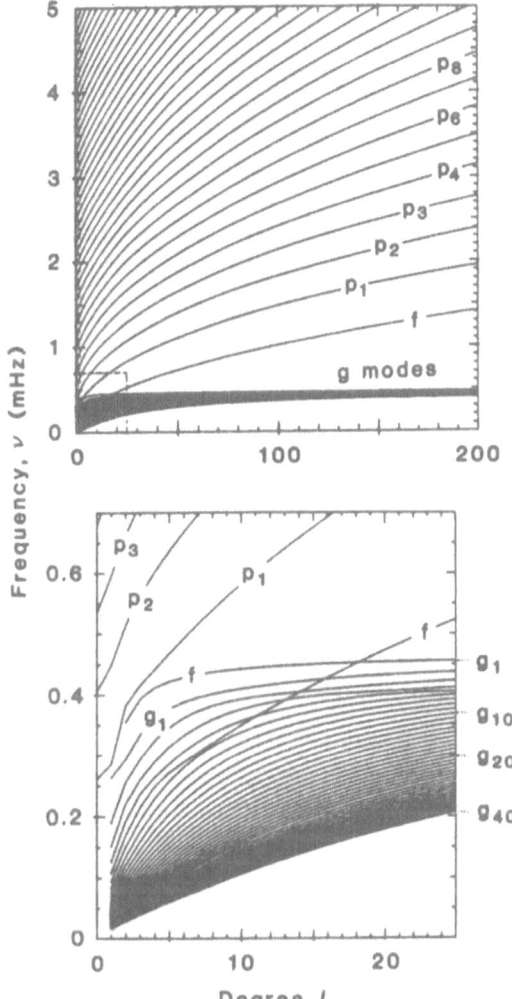

Fig. 5.4. Eigenfrequencies for a solar model as a function of n and l (frequencies for fixed n are connected by lines). The p and g modes are labelled according to their order n. The lower figure shows an enlarged part of the upper one in order to obtain a better resolution for the g modes. Note that the various lines never cross [from Christensen-Dalsgaard et al. (1985)].

The quantity $k_\perp c_{\mathrm{s}} = S_l$ is sometimes called the Lamb frequency, and the acoustic cutoff frequency ω_{c} is defined by

$$\omega_{\mathrm{c}}^2 = \frac{c_{\mathrm{s}}^2}{4H^2}\left(1 - 2H'\right) , \qquad (5.58)$$

with the density scale height $H = \rho/\rho'$. The expression (5.57) has been derived in the limit of large l ignoring derivatives of the gravity g. It describes the dispersion properties of waves propagating in a self-gravitating medium. When k_{r} is a slowly varying function of r the wave equation (5.56) can be solved approximately by the WKB method [see for example Morse and Feshbach (1953)]

$$Z \approx \frac{\text{const}}{\sqrt{k_r}} \exp\left(\pm i \int^r k_r \, dr\right) ,$$

and a wave-like solution is obtained only if $k_r^2 > 0$. The condition $k_r^2 = 0$ determines the cutoff frequencies

$$\omega_\pm^2 = \tfrac{1}{2}(\omega_c^2 + S_l^2) \pm \tfrac{1}{2}\sqrt{(\omega_c^2 + S_l^2)^2 - 4S_l^2\omega_V^2} , \qquad (5.59)$$

and in an inhomogeneous medium waves will be reflected once their frequency equals a cutoff frequency at some point along their path. Thus, wave propagation will be possible if

$$0 < \omega^2 < \omega_-^2 , \quad \text{or} \quad \omega^2 > \omega_+^2 > 0 , \qquad (5.60)$$

where the two cases correspond to g and p modes, respectively. For illustrative purposes, the dispersion relation (5.57) can be simplified for various special cases. Ignoring gravity and density gradients, i.e., $\omega_V = \omega_c = 0$, leads to the well-known relation for acoustic waves

$$\omega^2 = \left(k_r^2 + k_\perp^2\right) c_s^2 ,$$

whereas the dispersion relation

$$\omega^2 = \left(k_r^2 + k_\perp^2\right) c_s^2 + \omega_c^2$$

is obtained in a medium with density gradients but without gravity. If we consider an exponential density structure with a constant scale height H, i.e., $\omega_c^2 = c_s^2/(4H^2)$, wave reflection will occur if the acoustic wave length $\lambda_a = 2\pi c_s/\omega$ is of the order of H. A detailed discussion of the theory of waves in fluids including gravity and rotation can be found for example in the review article by Tolstoy (1963).

5.2 Free Oscillations of the Earth

Obviously, our Earth is the best studied cosmic object. This is also true for the mechanical properties and its global oscillations. Probably the beginning of terrestial seismology may be dated in the second half of the nineteenth century by Lord Kelvin's remark of "the grand idea of learning the physical condition of the interior from phenomena of rotary motion presented by the surface". In contrast to the common opinion of his time he claimed that the observed precession and nutation of the Earth's axis imply that the spinning Earth behaves dynamically more like a rigid body than a fluid ball. In 1863 Kelvin compared the response of a sphere of perfect liquid to a tide-raising potential to the response of a perfectly elastic sphere of given rigidity. Tidal measurements showed that the observed fortnightly tide, which due to its long period can be treated in a static approximation, was in fact about $\tfrac{2}{3}$ of the theoretical tide, and thus Kelvin concluded that the Earth responds to long-period tidal forces like an elastic sphere as rigid as steel. Using the measured speed of shear waves in steel he found a travel time of 68 min for a shear wave passing through a steel sphere of the size of the Earth, thus obtaining for the first time an estimate of the lowest eigenfrequency for a free oscillation of the whole Earth.

In 1882 Lamb solved completely the problem of oscillations of a perfectly elastic, uniform, non-gravitating, non-rotating sphere. A modern presentation of this problem has been given in some detail in Sect. 5.1. The hundred years since he wrote his beautiful paper have seen the laborous efforts to model the Earth more and more realistically by taking into account step-by-step self-gravitation, non-uniformity, rotation, ellipticity and anelasticity. One step which should be mentioned is the chapter "Vibrations of a gravitating compressible planet" in the monograph of Love in 1911, where he calculated that the period of the slowest vibration for a homogeneous Earth-sized steel sphere is almost exactly 60 min. The full problem, however, is only feasible numerically with the help of modern computers.

Parallel to these theoretical investigations, also dramatic improvements in building sensitive seismographs have been achieved, culminating in the construction of a strain meter (a quartz rod 25 m long and anchored to the Earth at one end) by Benioff, which was capable of recording free oscillation for as long as one hour. The evidence in the seismogram of this instrument of an oscillation with a period of 57 min after the Kamchatkan earthquake of 1952 strongly stimulated both theoreticans and experimentalists. A further progress was caused by the task of monitoring underground atomic bomb explosions.

Today there exist more than a thousand stations scattered over the Earth's surface where seismograms are continuously recorded with high sensitvity. These data can be analyzed to deliver arrival times and amplitudes of body and surface waves as well as power spectra of free oscillations excited by strong earthquakes. As an example, Fig. 5.5 shows the power spectra of free oscillations of the Earth after the Chilean earthquake of 1960 and the Alaskan earthquake of 1964. These oscillations were detectable for more than 11 days. The identification of the observed frequencies as spheroidal and toroidal eigenmodes requires a realistic model of the Earth.

The internal structure of the Earth, in particular its shell-like structure including the solid inner and liquid outer core, the lower and upper mantle and the crust, was obtained from the analysis of seismic waves since the beginning of this century. The basic method consists in a detailed measurement of the travel times of various seismic waves, from which the distribution of the phase velocities of P and S waves (5.28) with depths are derived. The density structure is more difficult to obtain, some information can be gained from the multipole moments of the mass distribution. It should be noted that a calculation of the internal structure from basic physical principles as outlined in Chap. 2 is currently impossible since neither the chemical composition nor

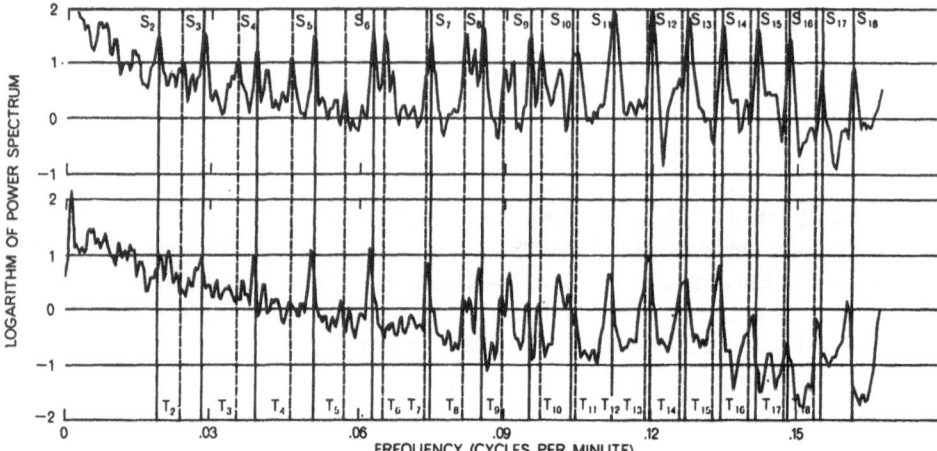

Fig. 5.5. Power spectra of free oscillations of the Earth after the Chilean earthquake of 1960 (top) and the Alaskan earthquake of 1964 (bottom). Spectral peaks corresponding to spheroidal (S) and toroidal (T) oscillations are labeled with l. (From Press 1965)

the EOS are known with sufficient reliability. Fig. 5.6 summarizes the level of our knowledge about the Earth's interior.

The available set of observational data consists of the above-mentioned seismic travel times and the frequencies of the eigenmodes. Extracting from this information a global model of the Earth is a typical inverse problem which is extremely hard to solve. Therefore a direct method is usually applied where eigenfrequencies and travel times of seismic waves are calculated numerically for a trial model of the Earth. Starting with a reasonable initial guess for the radial profiles of the density, the bulk and the shear modulus (see Sect. 5.1), these functions are iteratively refined until the calculated frequencies and travel times agree with the observed data.

As a result the best models today are the spherically symmetric Earth models 1066 A and 1066 B calculated by Gilbert and Dziewonski (1975) and the "Preliminary Reference Earth Model" (PREM) by Dziewonski and Anderson (1981). More than a thousand eigenfrequencies have been included with periods ranging from about one minute to 54 min and, in addition, a data base of about two million arrival times for P and S wave was used. In Fig. 5.7 and Fig. 5.8 observed and calculated eigenfrequencies of toroidal and spheroidal modes depending on l and n are shown. The precision of the measurements varies between 4×10^{-3} for some toroidal modes to 4×10^{-6} for the fundamental P mode. The agreement between the calculated and the identified observed modes is quite impressive.

For the n and l values shown in Figs. 5.7 and 5.8 a typical spatial resolution of a few hundred kilometers is obtainable. More detailed structures are resolved by the analysis of travel-time data; however, the fitting procedure

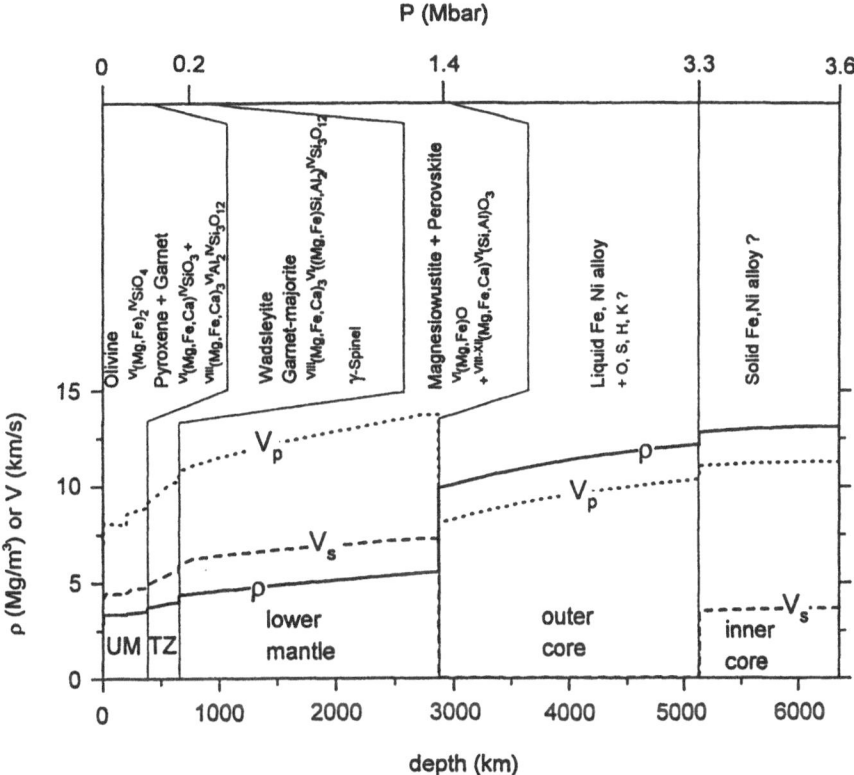

Fig. 5.6. Internal structure of the Earth, i.e., the density, velocities of seismic waves, pressure, and chemical composition as functions of depth [from Bukowinski (1994)]

is far more complex. A good agreement can only be achieved by taking into account anisotropic material properties in the upper mantle, isotropic dissipation of the shear and compressional wave energies, and frequency-dependent elastic moduli. The analysis of earthquake data in the framework of sophisticated models of the entire Earth, therefore, also serve as a tool to investigate the properties of matter at high pressures.

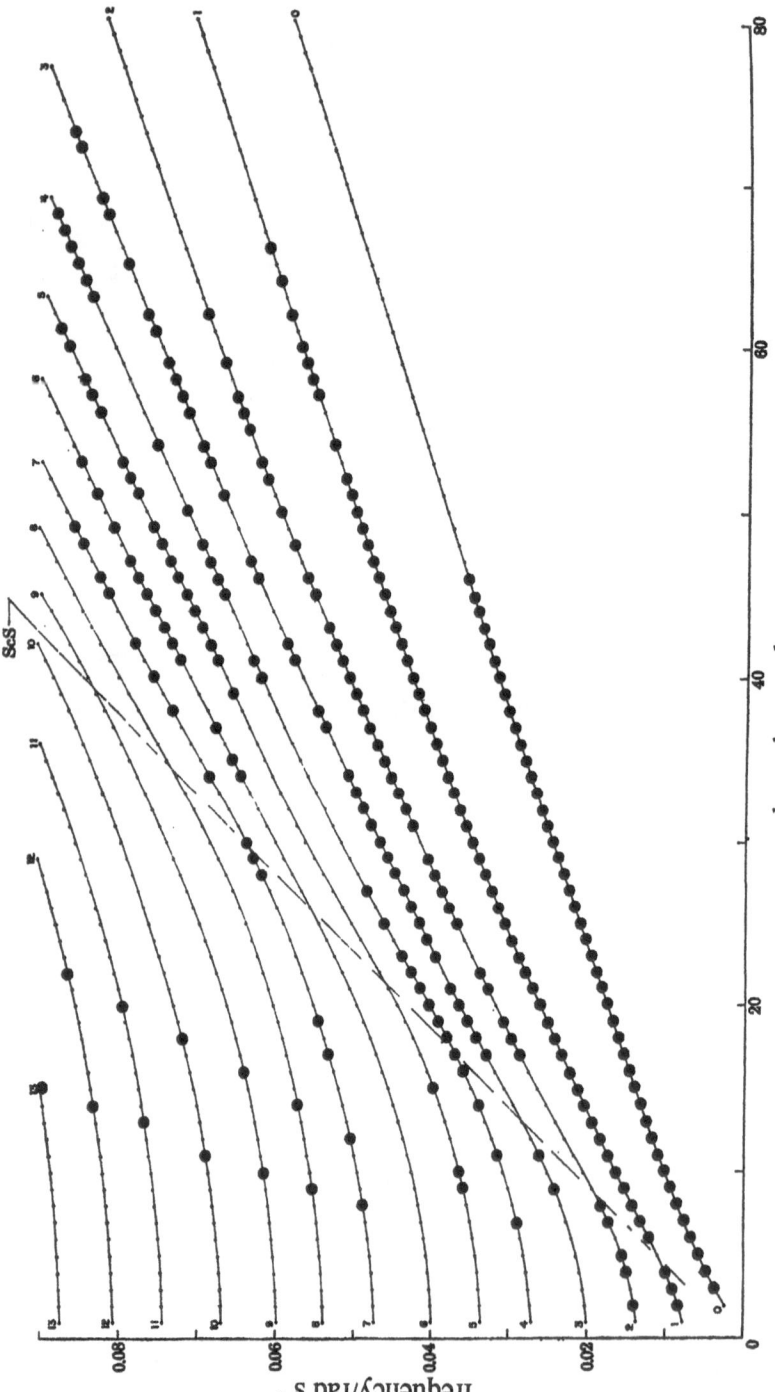

Fig. 5.7. Toroidal eigenmodes in the (ω, l) plane. Small dots represent calculated frequencies and large dots are observed values. Frequencies with the same n are connected by lines, the lowest branch corresponds to the fundamental mode. (From Gilbert and Dziewonski 1975)

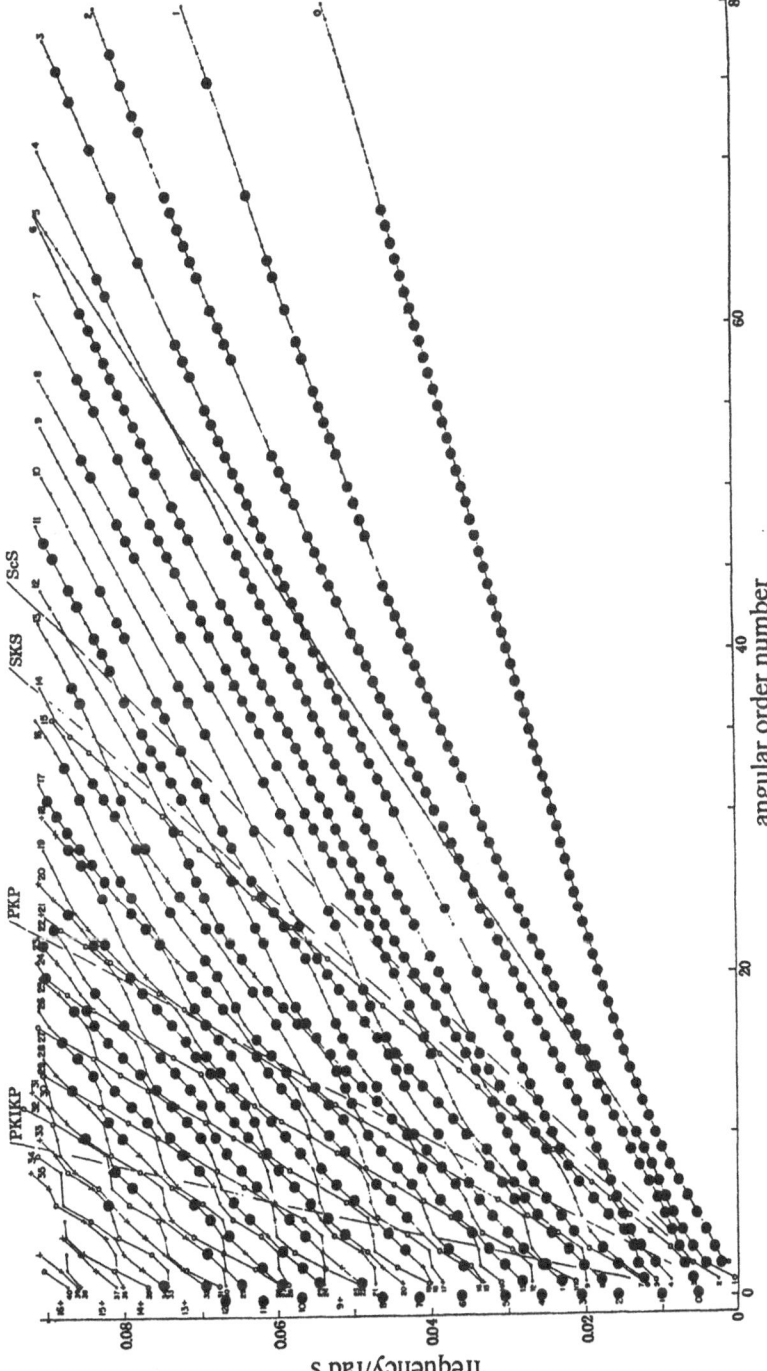

Fig. 5.8. Spheroidal eigenmodes in the (ω, l) plane. Small dots represent calculated frequencies and large dots are observed values. Frequencies with the same n are connected by lines, the lowest branch corresponds to the fundamental mode. (From Gilbert and Dziewonski 1975)

5.3 Helioseismology

In this section we will discuss the properties of solar oscillations and the information that can be gained about the internal structure of the sun. Temporal brightness fluctuations have been well known for a long time for many stars such as the RR Lyrae stars or the classical Cepheids, which typically change their luminosity by a factor of two within a few days. There is actually a region in the Hertzsprung–Russel diagram, the instability strip, that is populated by vibrationally unstable stars. These stellar pulsations are excited by the κ mechanism (see Sect. 5.4), and only low-degree modes are observed in those cases [for a review on stellar pulsations see Gautschi and Saio (1995) and the references therein]. Oscillations of the sun are, however, quite different since there is a huge number of high-order and high-degree modes excited at any time with very small amplitudes. The subject of solar oscillations, or helioseismology as it is frequently called, has been recently reviewed in number of papers including: Deubner and Gough (1984), Christensen-Dalsgaard et al. (1985), Libbrecht (1988a), Gough and Toomre (1991), Harvey (1995) [see also Stix (1991)].

Oscillations on the surface of the sun have been first reported by Leighton et al. (1962) who discovered Doppler-shift variations of a spectral line with periods of about five minutes and coherence times of approximately 20 minutes. The observations showed that the gas motion is mainly vertical with velocities of the order of $0.5 \ldots 1.0 \text{ km s}^{-1}$. The oscillation patterns are not stationary but rather change continuously, and at any given time they are visible on about one third of the solar surface. It was first believed that these oscillations were just a local phenomenon excited by the turbulent motion of the convection zone underneath the surface. However, analyzing the spatial and temporal oscillation structures, Ulrich (1970) and Leibacher and Stein (1971) suggested that the oscillations are p modes trapped in a resonant cavity inside the sun. A dispersion relation for those trapped waves was calculated by Ulrich and later by Ando and Osaki (1976), and shortly afterwards these predictions could be observationally confirmed by Deubner (1975). It turned out that there are about 10^7 trapped acoustic waves excited inside the sun, which propagate and interfere with each other to form standing wave patterns that correspond to the various p, f, and g modes discussed in Sect. 5.1. It is now generally believed that the modes are excited by acoustic noise generated in the solar convection zone, an idea that has been first proposed by Goldreich and Keeley (1977) and worked out in detail by Goldreich et al. (1994).

Solar oscillations have been measured in two different ways. First, there are temperature fluctuations associated with the modes that result in small brightness changes (a few times 10^{-6} per mode) that can be observed from spacecraft or with ground-based instruments (Harvey et al. 1982). The most important observation technique, however, is based on measurements of Doppler shifts of spectral lines, which are caused by the up-and-down movements of

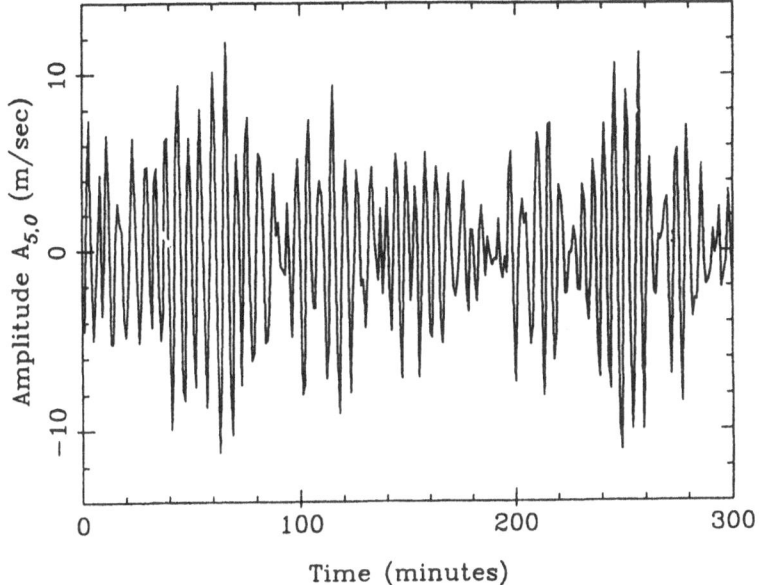

Fig. 5.9. Time dependence of the spherical harmonic fit coefficient $A_{lm}(t)$ for $l = 5$, $m = 0$ obtained from a time series of solar Doppler images. The beats that are visible in the figure are due to the fact that $A_{lm}(t)$ is a superposition of modes with different orders n. Note that the velocity scale has to be reduced by a factor of $4\pi^{1/2}$ to obtain the mode velocity [from Libbrecht and Zirin (1986)].

the oscillating matter. Full-disk solar images are produced in the blue and red wing of a particular line, for example the 643.9 nm Ca line which has a width of about 0.015 nm. This is done by using small-bandwidth filters that can be chopped by about ±0.01 nm around the line center. Subtracting the "blue" from the "red" image then leads to a Doppler image of the sun, and a few hundred of such images are averaged in order to increase the signal-to-noise ratio [for details see Libbrecht and Zirin (1986) and Libbrecht (1988b)]. The resulting time series of Doppler images is further processed by projecting each image onto a set of spherical harmonics Y_{lm} by employing a least-square fitting procedure. From that, time-dependent fitting coefficients $A_{lm}(t)$ are obtained which represent the superposition of radial velocity components for all modes with fixed values of l and m (Libbrecht and Zirin 1986). Some mixing of modes with different l and m is, however, unavoidable since data are only availiable for half the solar surface and the Y_{lm} are not orthogonal to one another on the half-sphere. Figure 5.9 shows an example of the function $A_{lm}(t)$ for $l = 5$, $m = 0$; note that the amplitude of the oscillation velocity is of the order of 1 m s^{-1}. Next the power spectra of the $A_{lm}(t)$ are calculated from a Fourier transform, and the mode frequencies as a function of the order n are obtained. A set of power spectra for $5 \leq l \leq 20$ is given in Fig. 5.10 where, for each l, spectra belonging to different azimuthal or-

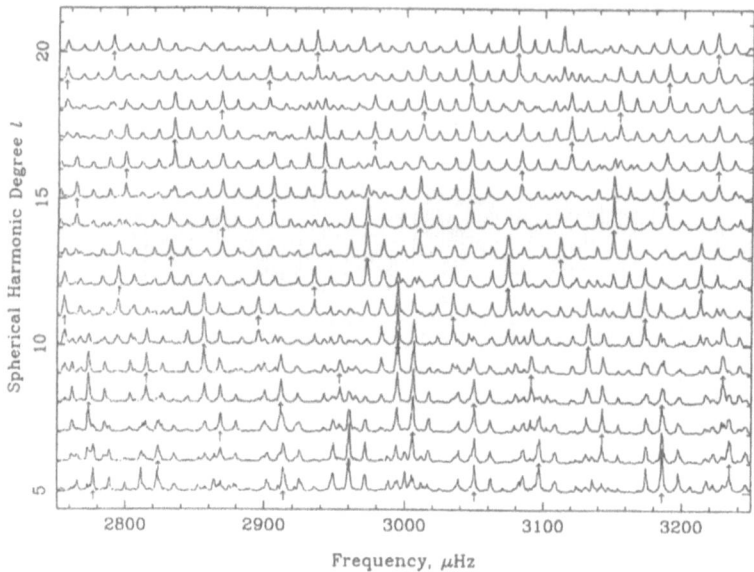

Fig. 5.10. Power spectra of fit coefficients A_{lm} in arbitrary units for $5 \leq l \leq 20$ averaged over the $2l+1$ values of m. All spectral features can be identified either as modes (indicated by arrows), mode aliases, or sidelobes caused by night-time data gaps. [from Libbrecht and Zirin (1986)].

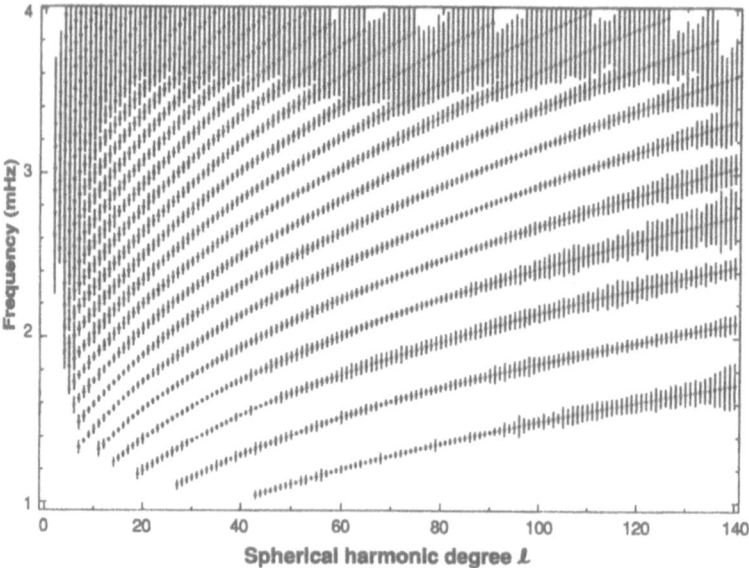

Fig. 5.11. Observed p mode frequencies as a function of degree l. The various ridges correspond to different orders n with $n = 1$ for the lowest ridge. The vertical bars indicate usual $1\,\sigma$ errors magnified by a factor of 10^3 [from Libbrecht and Woodard (1991)].

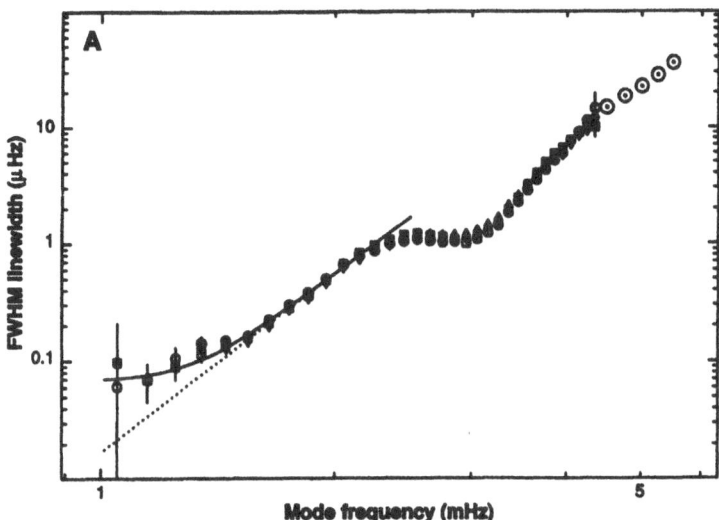

Fig. 5.12. Full-width at half-maximum linewidths of low-degree p modes as a function of frequency. The various symbols indicate different data sets. The dotted line corresponds to a power law ν^5, and the solid line is a modification of this power law due to the finite observation period of about 5 months [from Libbrecht and Woodard (1991)].

ders m have been combined. Note that in a spherically symmetric star the eigenfrequencies are degenerate with respect to m; however, the sun shows (differential) rotation which leads to small frequency splittings that have to be taken into account before averaging over m. With this method, mode frequencies $\nu_{nl} = \omega_{nl}/2\pi$ can be identified with high accuracy. Figure 5.11 shows ν_{nl} for $l \leq 140$, but frequencies of high-degree modes have also been observed; for example, Libbrecht and Kaufman (1988) present tables of ν_{nl} up to $l = 1320$. The observed spectral shape of each mode can be fitted to a given line-profile function in order to determine the linewidths. The results, under the assumption of a Lorentzian profile, are represented in Fig. 5.12 for low degree p modes, and the distribution of mode velocities is shown in Fig. 5.13. The lowest frequency modes are likely to oscillate in a coherent way for several months with velocities of only a few millimeters per second.

As in the case of the Earth (see Sect. 5.3) the seismic data provide the means to explore the internal structure of the oscillating object. Such information can be obtained either by inversion methods, where certain properties of the solar structure are extracted directly from the data, or by direct methods where the observed normal mode frequencies are compared with those calculated from a theoretical model of the sun. The solar model used in this latter approach is based on the assumptions and equations outlined in Sect. 2.1. In order to obtain the structure of the sun at its current age one has to follow

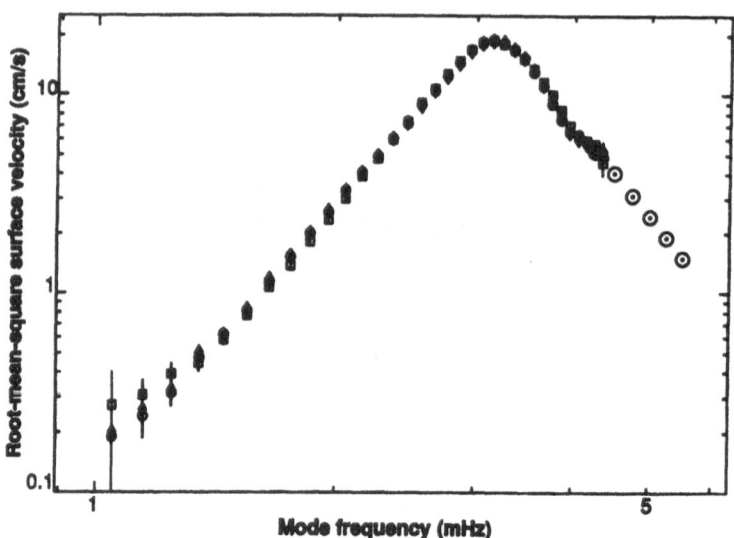

Fig. 5.13. Surface velocities of p modes as a function of frequency averaged over l values between 5 and 60. The various symbols indicate different data sets. [from Libbrecht and Woodard (1991)].

the evolution of a one-solar-mass star from its zero-age-main-sequence state to the present. The initial configuration is assumed to be of uniform chemical composition which then changes due to the nuclear burning process in the center. After a presumed age of 4.6×10^9 y the model must show the correct radius ($R = 6.9626 \times 10^8$ m) and luminosity ($\ell = 3.845 \times 10^{33}$ erg s^{-1}) which is achieved by varying the initial chemical composition and the mixing length for the convective energy transport (Sect. 2.4). The resulting overall stellar structure is usually called the standard solar model (Guenther et al. 1989, 1992; Guzik and Cox 1993, Dziembowski et al. 1994) although there are some differences between models developed by different authors. It contains (by mass) 70% hydrogen, 28% helium, and 2% heavier elements (metals); the central temperature and density are $T_c = 1.58 \times 10^7$ K and $\rho_c = 156$ g cm^{-3}, respectively. Energy is transported by radiative diffusion in the inner parts of the sun and by convection in the outer regimes, with the base of the convection zone being located at $r = 0.713\,R$.

As mentioned above, the observed oscillation patterns at the solar surface can be understood as a manifestation of acoustic waves trapped in the interior. This picture corresponds to the short-wavelenght limit of oscillations discussed at the end of Sect. 5.1. According to (5.57) the cutoff frequencies (5.59) determine the regions of possible wave propagation inside the sun, and the corresponding conditions are given by (5.60). The radial dependence of the cutoff frequencies $\nu_\pm = \omega_\pm/2\pi$ is shown in Fig. 5.14 for a few values of l calculated for a standard solar model. It can be seen that p modes are trapped

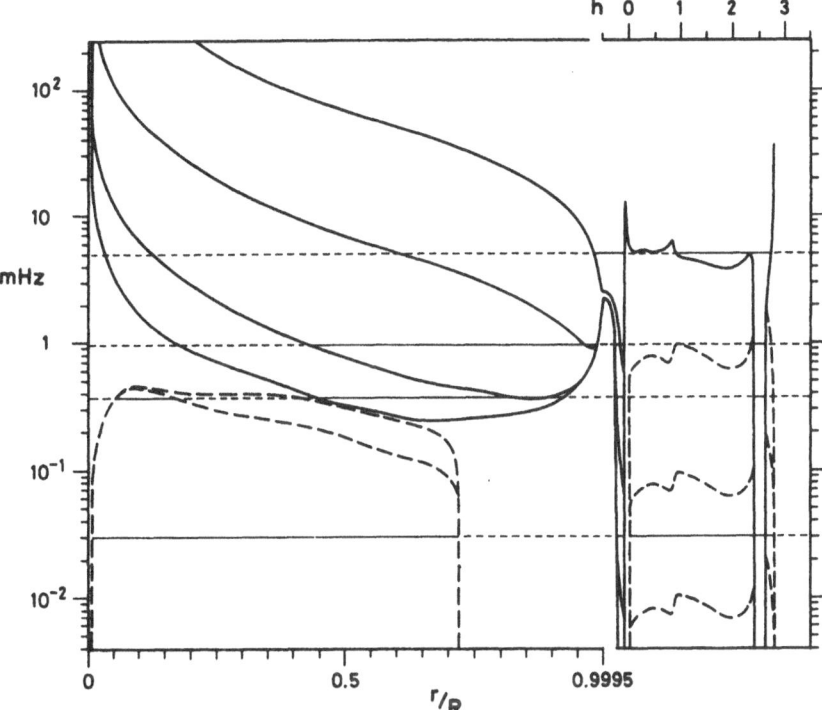

Fig. 5.14. Radial dependence of the cutoff frequencies $\nu_\pm = \omega_\pm/2\pi$ for a standard solar model. The solid and the dashed curves indicate ν_+ and ν_-, respectively. On the lower abscissa the radius extends up to $0.9995\,R$; then the scale is magnified and h denotes the height above the photosphere in units of $10^{-3}\,R$. The various curves are for $l = 1, 5, 50, 500$ and $\nu_\pm$ increases with l at fixed r/R, but in the interior the ν_- curves are indistinguishable for $l = 5, 50, 500$. The horizontal lines indicate various normal modes, they are continuous in regions where the modes can propagate and dashed otherwise. The lowest line represents a g mode with $l > 25$, and the next one corresponds to a $n = l = 1$ mode, which behaves like a g mode in the core and like a p mode in the envelope. The upper two lines denote p modes with $(n = 4, l = 5)$ and $(n = 6, l = 500)$, respectively [from Deubner and Gough (1984)].

in a region between the surface (at $r = R$) and some inner boundary ($r = r_1$) depending on the frequency and on l, and that g modes can propagate in the deep interior with an upper reflection boundary at the base of the convection zone. Note that there is a second possible small regime for g modes located above the photosphere; however, these modes have not yet been observed. Close to the surface we have $\omega_c^2 \gg S_l^2$ and thus $\omega_+ \approx \omega_c$; therefore, p modes are trapped only if $\nu < \nu_c = \omega_c/2\pi \approx 5.5$ mHz. Higher-frequency modes can propagate freely in the solar photosphere and chromosphere. Deep in the solar interior the conditions $\omega_c^2 \ll S_l^2$ and $\omega_V^2 \ll S_l^2$ hold; thus $\omega_+ \approx S_l$, and the reflection radius r_1 is approximately given by

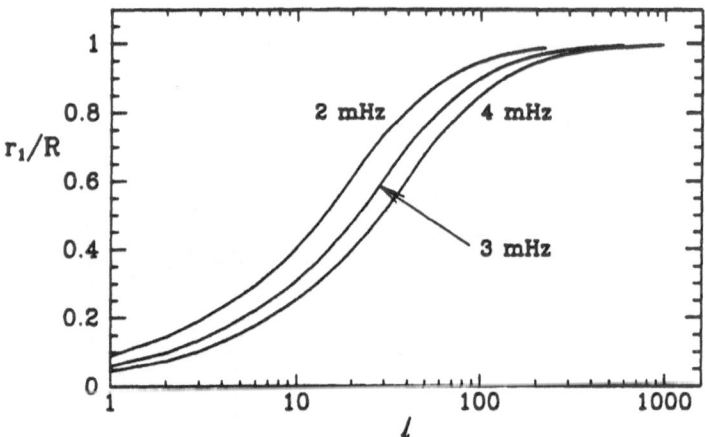

Fig. 5.15. Inner turning radius r_1/R as a function of l for a standard solar model. The curves are determined from the condition $k_r = 0$ for three mode frequencies $\nu = \omega/2\pi$. [from Gough and Toomre (1991)].

$$\frac{\sqrt{l(l+1)}}{r_1} c_s(r_1) = S_l \approx \omega_+ = \omega \;. \tag{5.61}$$

Since the sound speed c_s increases with decreasing radius, smaller l values lead to smaller reflection radii r_1 and low-l modes penetrate deeper into the solar interior than do high-l modes. This dependency is shown in Fig. 5.15 for three different frequencies. The trapping of modes is illustrated in Fig. 5.16 where ray paths of p and g modes are shown for a standard solar model. It can also be seen that the wave motion is mainly vertical at the surface.

In the short-wavelength limit we can derive an equation to determine the eigenfrequencies by requireing that an integral number n of half wavelengths must fit between the turning points r_1 and r_2 (which define the resonant cavity); thus from the expression (5.57) for the radial wave number k_r we get

$$(n + \alpha)\pi = \int_{r_1}^{r_2} k_r \, dr = \int_{r_1}^{r_2} \frac{\omega}{c_s} \left[1 - \frac{\omega_c^2}{\omega^2} - \frac{S_l^2}{\omega^2}\left(1 - \frac{\omega_V^2}{\omega^2} \right) \right]^{1/2} dr \;. \tag{5.62}$$

Here r_1 and r_2 follow from $k_r^2 = 0$, and the phase α is a constant that accounts for the detailed conditions at the turning points where the wave-like solutions of (5.56) change into decaying modes.

For p modes we have $\omega^2 \gg \omega_V^2$ and (5.62) can be approximated by

$$\frac{\pi(n + \alpha)}{\omega} \approx \int_{r_1}^{R} \left[\frac{r^2}{c_s^2} - \frac{l(l+1)}{\omega^2} \right]^{1/2} \frac{dr}{r} \;, \tag{5.63}$$

where also the term ω_c^2/ω^2 has been neglected throughout the cavity. This is reasonable everywhere except near the upper boundary where $\omega \approx \omega_c$. However, this holds only in a very thin layer and its influence may be included in the phase α [see Gough and Toomre (1991)]. From (5.63) we can obtain

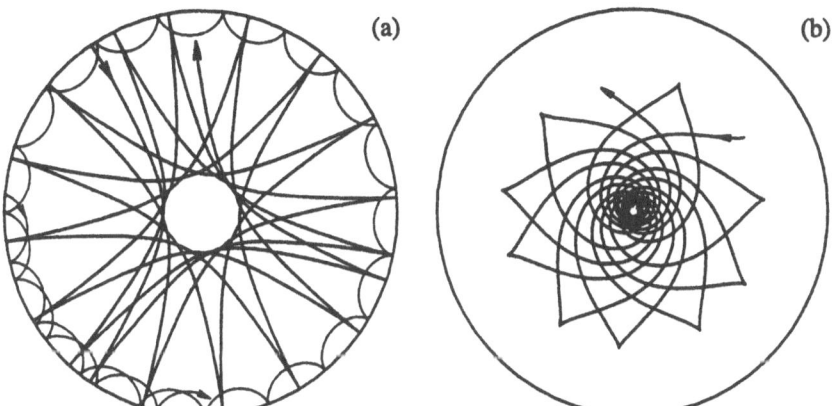

Fig. 5.16. Ray paths for various modes in a standard solar model. The circle represents the surface of the sun; **a** shows the paths of two p_8 waves with $l = 2$ and $l = 100$, respectively. The propagation of a gravity wave with $n = 10$ and $l = 5$ is illustrated in **b** [from Gough and Toomre (1991)].

the remarkable result that the observable quantity $(n+\alpha)\pi/\omega$ on the left-hand side is a function of the surface phase velocity $u = \omega/k_\perp(R) = \omega R/\sqrt{l(l+1)}$ since the integrand contains only this combination of ω and l, and so does the reflection condition (5.61) from which $r_1(u)$ is obtained. The function

$$\frac{\pi(n+\alpha)}{\omega} = R F(u) = R \int_{x_1}^{1} \left[\frac{x^2}{c_s^2} - \frac{1}{u^2} \right]^{1/2} \frac{dx}{x} , \qquad (5.64)$$

where $x = r/R$ and $x_1 = r_1/R$, has been first determined from observations by Duvall (1982), and a more recent presentation is shown in Fig. 5.17. Since $F(u)$ is known from observation, the intergal relation (5.64) can be used to extract the radial dependence of the sound speed $c_s(x)$ directly from the data. Substituting $c_s/x = s$ we obtain from (5.64)

$$u^2 \frac{dF}{du} = - \int_{s_1}^{u} \frac{\Phi(s)}{\sqrt{u^2 - s^2}} \, ds , \qquad (5.65)$$

where $s_1 = s(x=1) = c_s(r=R)$ and $\Phi(s) = d\ln x/d\ln s$. This relation can be converted into an integral equation of Abel's type, which yields an analytic solution [see for example Christensen-Dalsgaard et al. (1985)]

$$x = \frac{r}{R} = \exp\left[-\frac{2s}{\pi} \int_{s_1}^{s} \frac{u}{\sqrt{s^2 - u^2}} \frac{dF}{du} \, du \right] \qquad (5.66)$$

From this $x(s)$ is determined and therefore $c_s = xs(x)$. This procedure is a classical inversion method that was invented by Wichert and Herglotz at the beginning of this century to analyze the internal structure of the Earth from seismic data [see Bullen (1956)]. Other inversion techniques have also been applied to obtain the sound speed in the solar interior with reduced systematic

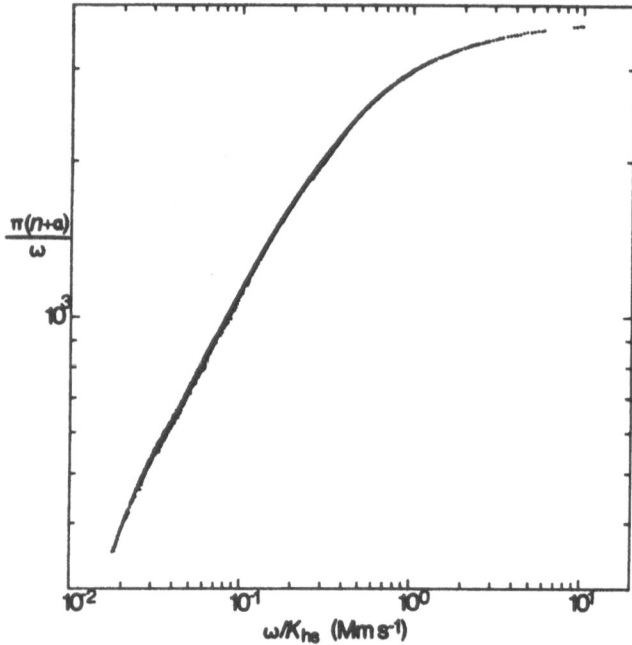

Fig. 5.17. The quantity $\pi(n + \alpha)/\omega$ as a function of the phase velocity $\omega/k_\perp(R)$ (in units of 10^6 m s^{-1}) for $\alpha = 1.58$ determined from 2820 observed p modes with degrees l ranging from 1 to 892 [from Christensen-Dalsgaard et al. (1985)].

errors (Christensen-Dalsgaard et al. 1991). One major achievement of these methods is the measurement of the depth of the convection zone. For that purpose the quantity

$$W(r) = \frac{r^2}{Gm(r)} \frac{dc_s^2}{dr} \tag{5.67}$$

is determined seismologically. As discussed in Sect. 2.4 the convection zone is quite accurately in an adiabatic state, thus $p \propto \rho^{\gamma_{ad}}$, and $W = 1 - \gamma_{ad}$, i.e., $W = -2/3$ for a monoatomic gas. The function W is shown in Fig. 5.18 and it can be clearly seen that it is almost constant in the convective regime but rises abruptly beneath this region where energy is transported by radiative diffusion and the temperature gradient becomes subadiabatic. Thus the radius r_b of the base of the convection zone can be determined quite accurately from an analysis of the normal mode frequencies; Christensen-Dalsgaard et al. (1991) obtain $r_b = (0.713 \pm 0.003)\,R$, and r_b has to be considered as an observed input parameter for modelling the solar structure that is as relevant as the mass, the radius, or the luminosity of the sun. The solar structure in the deep interior, however, cannot be determined very reliably from an anlysis of p modes because information about this regime is contained mainly in the data

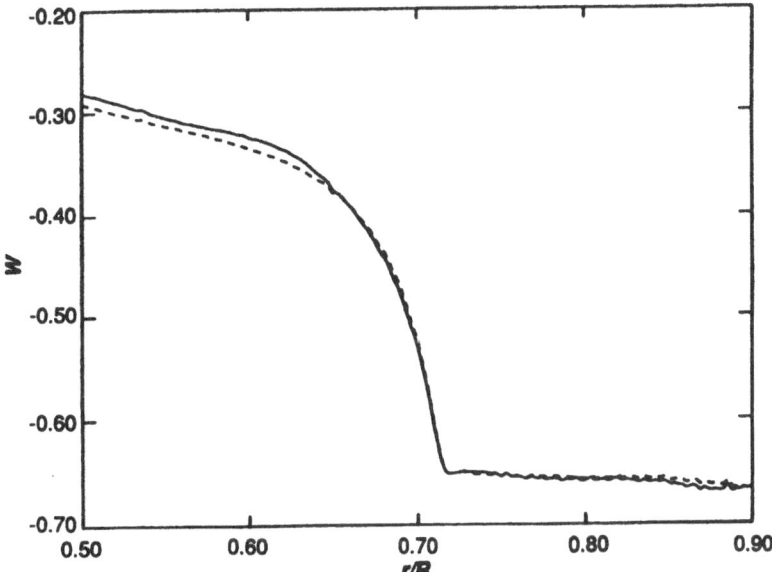

Fig. 5.18. The quantity W as a function of the fractional solar radius r/R determined by an inversion method from helioseismic data. W traces essentially the gradient of the sound speed which changes abruptly at the base of the convection zone (at $r/R = 0.713$) [from Christensen-Dalsgaard et al. (1991)].

from low-l modes, which are currently not availiable with sufficient accuracy for $l < 5$. The situation would of course greatly benefit from a discovery of g-mode frequencies since these waves naturally probe the inner core of the sun. Outside the inner region ($r/R > 0.2$) the agreement between theoretical standard solar models and the results from seismic inversion methods are better than 1% (Dziembowski et al. 1994).

Another essential piece of information that can be obtained from the p mode data is the rotational structure of the sun. Rotation splits the degeneracy of the mode frequencies ω_{nl} into a multiplet ω_{nlm} depending on the azimuthal order m. For a uniformily rotating sun

$$\omega_{nlm} = \omega_{nl} + m\Omega , \qquad (5.68)$$

where $\Omega \approx 2.5 \times 10^{-6}$ Hz is the (average) solar rotation frequency. Since the sun shows differential rotation, Ω depends on the radius and the lattitude (i.e., $\Omega = \Omega(r,\theta)$) and additional terms appear on the right-hand side of (5.68), the next term being of the order of m^3 (note that there are no contributions from even powers of m; Gough and Toomre 1991). Because Ω is much smaller than the p mode frequencies, theoretical estimates of the mode splitting can be performed by considering the rotation as a small perturbation of the non-rotating solar structure. As a consequence, the difference $(\omega_{nlm} - \omega_{nl\,-m})$ can be expressed as

$$\omega_{nlm} - \omega_{nl-m} = \mathcal{F}(\Omega) \; ,$$

where $\mathcal{F}(\Omega)$ is a linear functional of the rotation frequency involving integrals over the entire sun that contain the eigenfunctions of the non-rotating solar model [see Gough and Toomre (1991)]. From this, inversion methods are used to express $\Omega(r,\theta)$ in terms of the observed frequency splittings. Figure 5.19 shows the resulting rotation profiles. It is a surprising result that Ω is a function of latitude throughout the entire convection zone (for $r > 0.7 R$), and that the sun is in a state of solid-body rotation with $\Omega/2\pi \approx 4.3 \times 10^{-7}$ Hz below the convective regime in the range $0.4 R < r < 0.6 R$ (Brown et al. 1989); the rotation profile in the core and underneath the surface are not well resolved yet. The observed solar rotation profile is a result of the transport

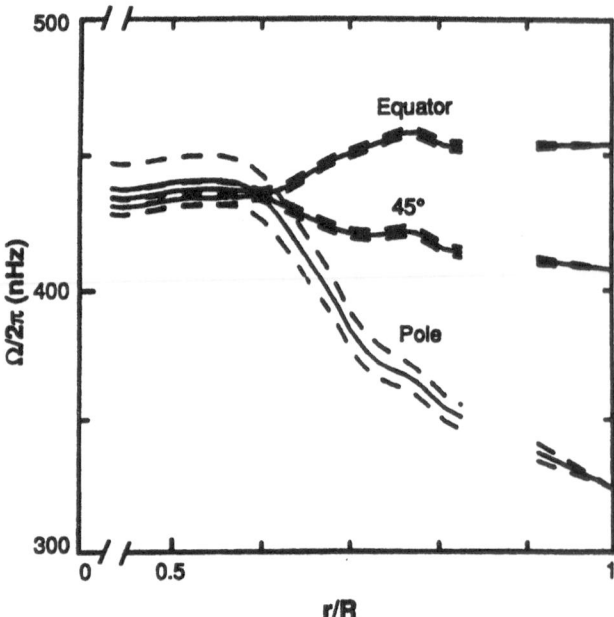

Fig. 5.19. Solar rotation profile as determined from the frequency splitting of p modes. [from Libbrecht and Woodard (1991)].

and redistribution of angular momentum due to convection, a process that is currently not well understood. Together with the density profile the above rotation pattern can be used to estimate the rotational distortion of the sun and to calculate the quadrupole moment J_2 of the external gravitational potential: $J_2 \approx 1.5 \times 10^{-7}$ (Gough and Toomre 1991). This is currently the most accurate method availiable to measure J_2. A solar quadrupol moment leads to an addtional precession of perihelia for the planets, thus an accurate determination of J_2 is essential for testing general relativity in the solar system, and the above value does not even produce a measurable effect for Mercury.

As pointed out in Sect. 1.3 the hydrogen burning processes in the solar core lead to the production of electron neutrinos which carry away about 3% of the energy. These neutrinos have been measured in four different underground experiments, and all show deficits in the neutrino flux of a factor of two or three compared with the predictions based on standard solar models (Bahcall et al. 1995). There are two possible causes for this solar neutrino problem: either there is some error in the solar models, or neutrinos are able to oscillate between three different types (i.e., electron, muon, and τ neutrinos) provided they have a small but finite mass. For the first alternative, the neutrino production rate can be most efficiently reduced by lowering the temperature gradient in the solar core and thus the central temperature. This may be achieved either by reducing the opacity due to a lower helium and heavy-element abundance or by introducing some additional energy transport mechanism, for example, by employing hypothetical so-called WIMPs (weakly interacting massive particles) that have a large mean free path and thus smear out the temperature gradients. In fact, both, models with low helium and metal abundances, and appropriately choosen WIMP models are able to reduce the predicted neutrino fluxes to the observed level. However, detailed comparisons with helioseismic data indicate that such models are inconsistent with the observed p-mode frequencies (Christensen-Dalsgaard 1992), whereas standard solar models are able to reproduce these frequencies with high accuracy. Thus it seems likely that the solution to the solar neutrino problem lies in particle physics rather than in modelling the solar structure.

In this section we have demonstrated the power of helioseismology to probe many details of the internal structure of the sun, and thus to explore the properties of matter under high pressures and temperatures. In the near future there will be a number of promising efforts to improve the accuracy of the observations. The SOHO satellite (Solar and Heliospheric Observatory) has already been launched with three instruments designed especially for helioseismological observations, including a Michelson interferometer Doppler imager with a one-million pixel CCD camera. In order to operate in continuous sunlight this spacecraft will orbit the L_1 Lagrangian point between the Earth and the sun. Ground-based observations will improve through the use of telescope networks around the world. The GONG project (Global Oscillation Network Group) consisting of six sites went into operation recently producing Doppler images from Michelson interferometers continuously for about three years. As the accuracy of the observations increases, the sun could become a laboratory for high-density plasmas and for testing the equation of state under such conditions as described in Sect. 2.2.

5.4 Asteroseismology of White Dwarfs

This area of research is the latest branch of investigations into the properties
of celestial bodies using the equation of state for matter at high densities
together with the observation of eigenmodes. The first observation of a short-
period variability of a white dwarf with a period of about 12 minutes was
reported by Landolt (1968) and related to g modes by Chanmugam (1972)
and Warner and Robinson (1972). The first observation of multi-periodic
variability of a white dwarf was performed by Lasker and Hesser (1969) and
later by McGraw (1979). There are two main difficulties. First, white dwarfs
are objects with low luminousities and, additionally, the light variations have
very small amplitudes. Second, the typical oscillation periods are of the order
of minutes, therefore, in order to obtain a good resolution in frequency space,
long and continuous observations are nessecary. The first problem can be
overcome by using large telescopes and modern high-sensitivity detectors, and
the second one by observing with many telescopes, which should be uniformly
distributed in longitude around the world. The realization of such a concept
– called the Whole-Earth Telescope (WET) – is mainly an organizing and
administrative task, its basic operation is quite simple (Nather et al. 1990,
Nather and Winget 1992).

   The result is a new multimirror ground-based telescope for time-series
photometry of rapid variable stars, designed to minimize or eliminate gaps
in the brightnes record caused by the rotation of our planet. The individual
telescopes involved are distributed in longitude and coordinated from a single
control center (Fig. 5.20), and target stars are tracked as long as darkness
lasts.

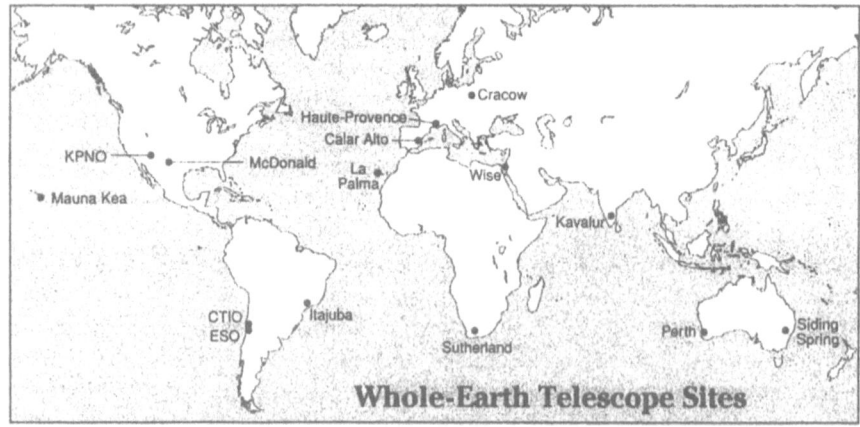

Fig. 5.20. Locations of the observatories that have joined the Whole-Earth Telescope for obtaining long time-series photometry (from Nather and Winget 1992).

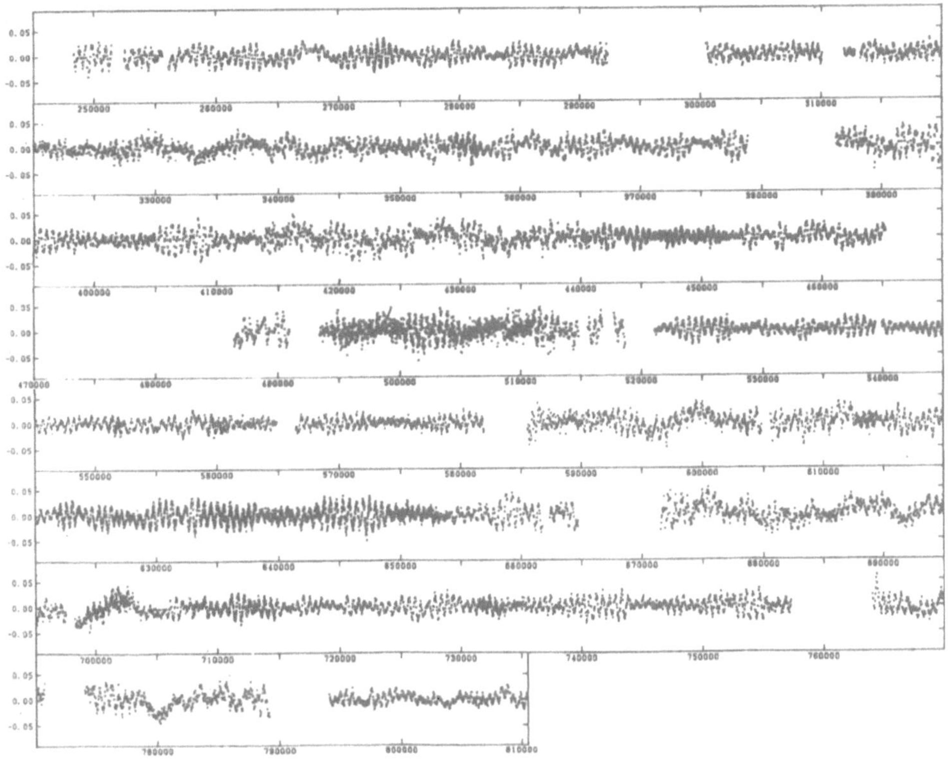

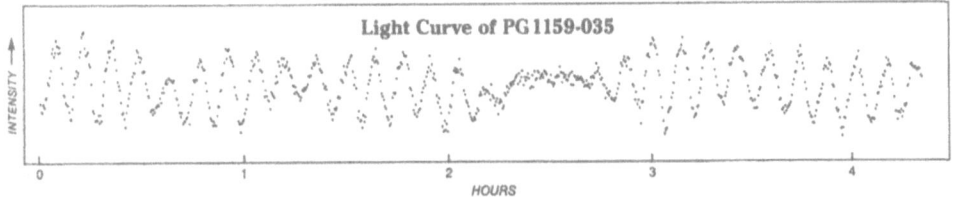

Fig. 5.21. Light curve of the hot white dwarf PG 1159–035 during the central six days of the WET run. The horizontal axis shows the elapsed time in seconds, and the vertical axis shows the fractional intensity. (From Winget et al. 1991)

Fig. 5.22. A 4.5-hour section of the light curve of PG 1159–035. The complex structure is caused by the superposition of oscillations with different periods. (From Nather and Winget 1992)

Under ideal conditions a continuous registration of the light curve over hundreds of hours may be achieved. The data are returned by electronic mail to the control center and analysed in real time. WET is the first instrument to provide continuous data of a quality that permits true high-resolution power spectroscopy of pulsating white dwarfs. Figure 5.21 shows the result of a

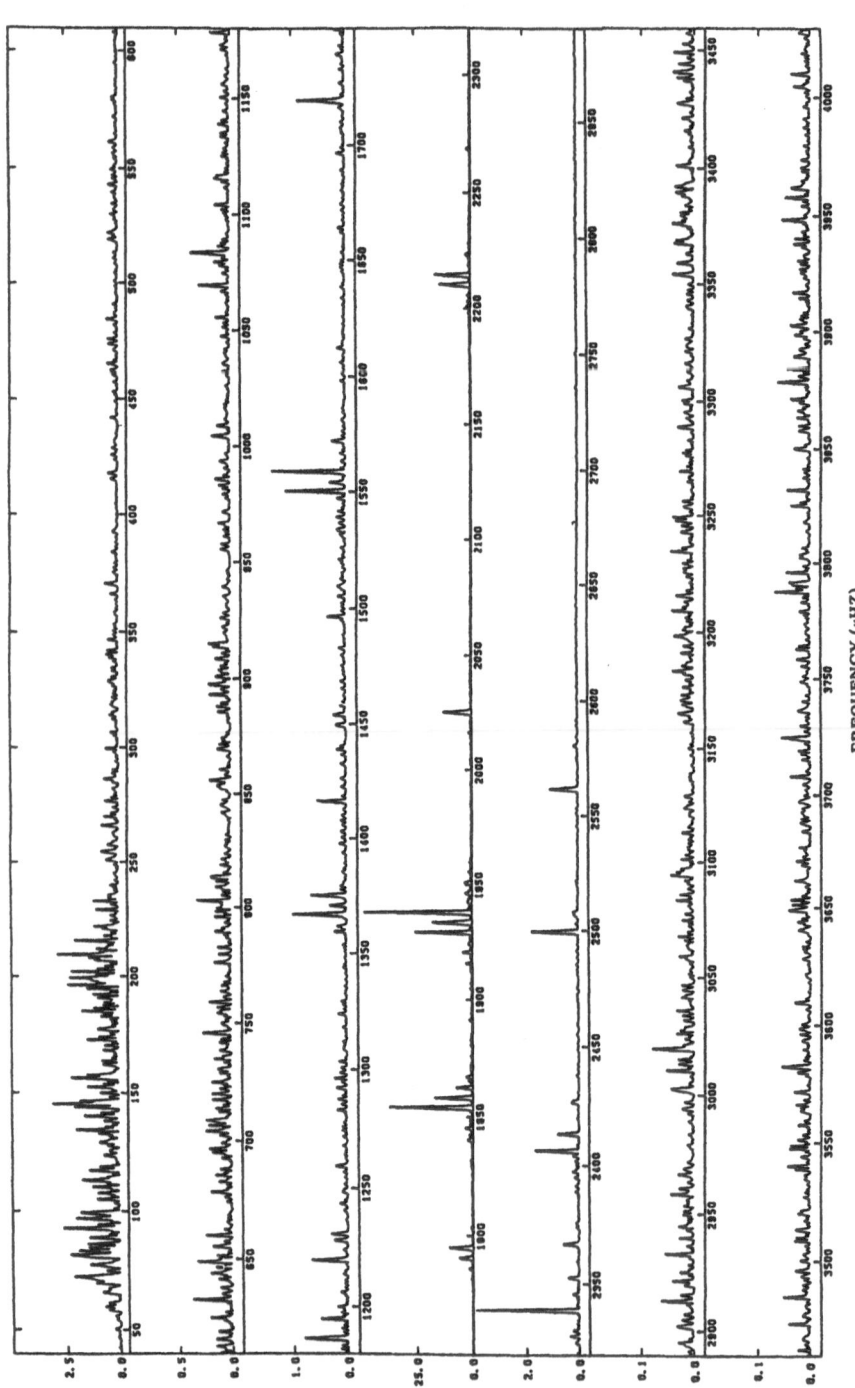

Fig. 5.23. Power spectrum of the intensity pulsations of PG 1159–035. Note the vertical scale differs from panel to panel to accommodate the large dynamic range. (From Winget et al. 1991)

150 hour observing run using a WET consisting of nine telescopes with aper-
tures ranging from 0.6 m to 2.5 m. The target object is the 14th-magnitude
hot white dwarf PG 1159–035 with a temperature of about 140 000 K.

In the 4.5-hour section, shown in Fig. 5.22, the dominant oscillation with
a period of about 8 minutes can clearly be seen, the deviations from a pure
sinoidial wave are caused by the superposition of oscillations with different
periods.

A Fourier transform leads to the power spectrum, a portion of which is
depicted in Fig. 5.23 ranging from 50 to 4000 μHz in frequency or from
4 min to 5.5 h in period. The oscillation periods are determined with a relative accuracy of 10^{-5}. A prominent feature is the occurence of multiplets,
especially triplets. These are rotationally induced m splittings, which open
the possibility of measuring the rotation period of the white dwarf.

To extract detailed information from the precisely measured periods of
the eigenmodes requires very sophisticated models of the radial dependence
of the chemical composition and a detailed knowledge of the corresponding
equation of state. In Fig. 5.24 the mass distribution of helium, carbon, and
oxygen in a model of the interior of a hot white dwarf are shown as obtained
from an evolution calculation starting with a main sequence star of 0.6 solar
mass (Kawaler and Bradley 1994).

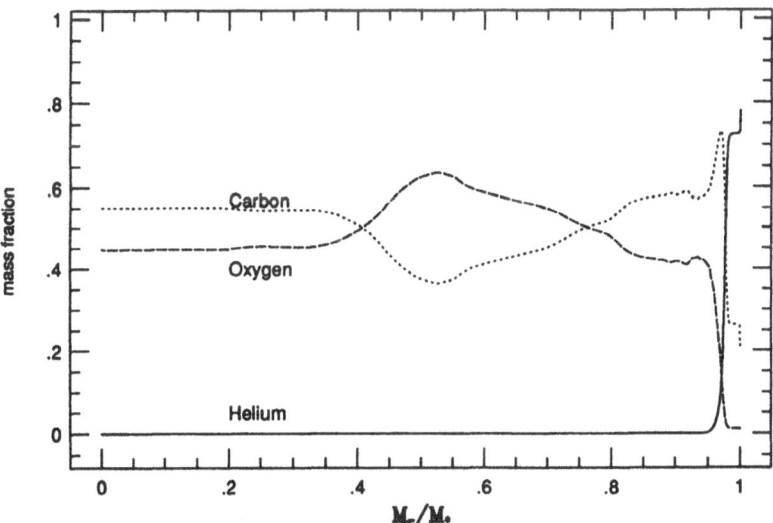

Fig. 5.24. Typical composition profile used in white dwarf evolution codes (from
Kawaler and Bradley 1994)

Once the pulsation spectrum is resolved, and the individual frequencies
have been adapted to modern white dwarf pulsation theory, the astrophysical
information obtainable from WET observations of pulsating white dwarfs is

quite impressive (Nather et al. 1990, Winget et al. 1991, 1994, Bradley 1995, Bradley and Winget 1994, Kawaler and Bradley 1994, Kawaler et al. 1995).

(1) Exact values for the indices l and m and a value for the radial index n within ± 2 are available for DOV stars [variable (V) dwarfs (D) with an effective temperature T_{eff} of about 10^5 K corresponding to the spectral class (O)], and even better for DBV ($T_{\text{eff}} \approx 2.2 \times 10^4$ K) and DAV stars ($T_{\text{eff}} \approx 1.2 \times 10^4$ K).

(2) Rates of change for well-resolved periods can be derived, thus the different values of the radial index n are used as a probe of the internal structure of the star, and the overall change is used as a measure of the stellar cooling rate (Winget et al. 1985, 1991, Kepler et al. 1990, 1991).

(3) The mass of the star can be determined to within ± 0.02 solar mass, e.g., the mass of PG 1159–035 is estimated as 0.59 ± 0.01 solar mass (Kawaler and Bradley 1994).

(4) From the mass determined and the equation of state the radius of the white dwarf can be calculated, and if a value of T_{eff} is measured, e.g., using the Hubble Space Telescope, the star's absolute magnitude is known and from that, together with its apparent magnitude, the distance is obtained. The corresponding values for PG 1159–035 are 17 680 km for the radius, 136 000 K for T_{eff}, and a distance of 440 pc (Bradley 1995, Kawaler and Bradley 1994).

(5) From the m splitting of the oscillation frequencies for a given l the rotation period of the white dwarf can be determined (Cox 1984). Furthermore the amplitude ratios of the multiplets contain information about the inclination angle with respect to the line of sight. Applying this to the white dwarf PG 1159–035 leads to a rotation period of 1.38 ± 0.01 days, and an inclination angle of about 60° provided that the m components of the $l = 1$ multiplets are all excited equally (Winget et al. 1991).

(6) MHD calculations in the linear perturbation approximation (Jones et al. 1989) yield a magnetic m splitting of l multiplets in $l + 1$ components with a frequency spacing proportional to m^2, which, in principle, could be distinguished from rotational m splitting, which is proportional to m. From this, the presence of magnetic fields down to 6 000 G may be detectable – a range that is not accessable through any other method.

In order to obtain a better understanding for the astrophysical importance of white dwarf astereoseismology for the non-specialist as well, we will now briefly describe the life of a white dwarf [for a review see Koester and Chanmugam (1990)]. As Nather and Winget (1992), two of the pioneers in this area of research, clearly express, white dwarfs are the shrunken remains of main-sequence stars with initial mass smaller than about eight solar mass that have used up all their nuclear fuel and are quietly cooling toward oblivion. The birth of a white dwarf as a star in its own right is the moment when it casts off the outer atmosphere of a red-giant in the form of a planetary nebula. A freshly formed white dwarf is extremly hot with a temperature of 140 000 K or more. In this stage it cools very quickly and, therefore, only a few of these hottest white dwarfs can be observed. Later on, the cooling becomes more and more gradual and the number of observable white dwarfs in our galaxy increases as their temperatures reduce. Then, quite abruptly at surface temperatures of about 3 500 K, we find no more. The reason for this sudden cutoff is that the most ancient white dwarf could not have cooled below this temperature in the lifetime of our galaxy. Thus, by searching the oldest white dwarfs and using a sophisiticated cooling theory, we have a powerful method for determining the age of our galaxy totally independent of other assumptions. The best estimate to date is 9.5 billion years for the oldest white dwarfs in our neigborhood, with an uncertainty of perhaps 20 percent, mostly due to the lack of an accurate cooling theory. By further improvements such results may gain fundamental importance for cosmology.

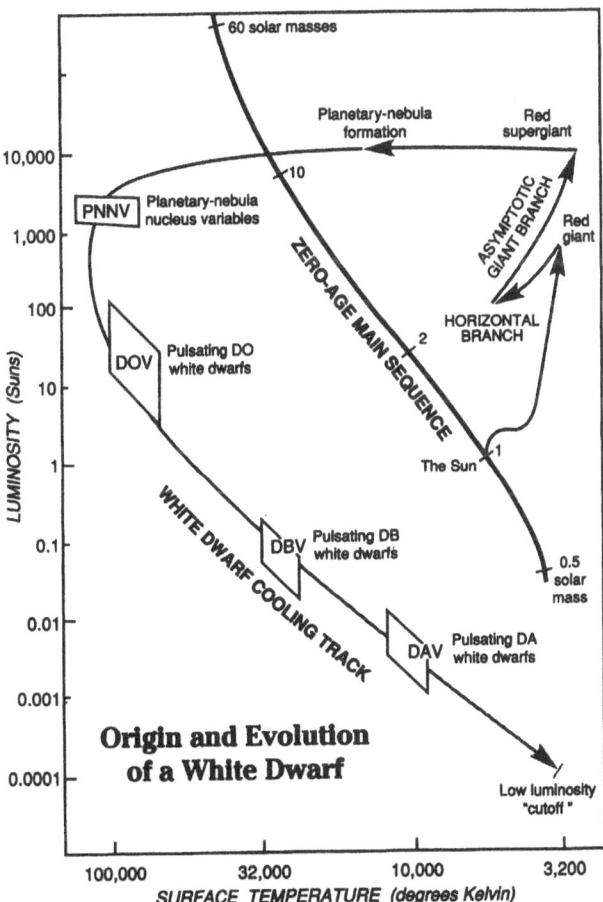

Fig. 5.25. The origin and evolution track of a white dwarf in the Hertzsprung–Russel diagram. The four zones of instability are marked by boxes. (From Nather and Winget 1992)

The standard graphical presentation of the evolution described above is the path of the star in a Hertzsprung–Russell diagram as shown in Fig. 5.25, taken from Nather and Winget (1992). Also shown in this figure are the four locations along the white-dwarf cooling sequence where the stars are variable.

The mechanism for the excitation of these light oscillations is similar to those of the δ-Cepheid variables. In these instability regions the opacity of the outer atmosphere of the star strongly varies with temperature. Below the star's surface, a zone of partially ionized matter modulates the steady flow of heat from the deep interior into a series of puffs.

An intuitive picture of this process is given by Nather and Winget (1992). Imagine a lid on a pot of boiling water. The steam may simply lift the lid slightly to escape in a steady flow, but if the lid's weight and other conditions are just right it will flap up and down, clanking and banging and releasing the steam in bursts. In a white dwarf this activity sets the whole star into vibration. A detailed picture of the oscillation behavior is, of course, highly complex and depends strongly on the radial structure. It turns out that the p modes are confined to the deep interior of the white dwarf star, whereas the g mode amplitudes are large only in a thin surface layer with a stable density stratification. Detailed numerical model calculations show that the frequencies of g modes with small l values and radial wave numbers of a few tens ($n \approx 20, \ldots, 40$ for PG 1159–035) coincide with the observed oscillations, while the periods of p modes are of the order of a few seconds and have not been observed up to now, probably they are not excited.

During the cooling of white dwarfs the interial structure of star will change and therefore corresponding changes in the oscillation periods are expected. From the theory of stellar g-mode pulsations the rate of change of the period is connected with the rates of change of the temperature $\dot{T}$ and the radius $\dot{R}$ (Winget et al. 1983, Kawaler and Bradley 1994):

$$\frac{\dot{P}}{P} \approx -a\frac{\dot{T}}{T} + b\frac{\dot{R}}{R} , \tag{5.69}$$

where a and b are positive constants of the order of unity. For DAV stars the radial changes can be ignored, and the time scale of the change in the period is an immediate measurement of the cooling time scale. Since $\dot{T} < 0$ the periods must increase with time. This has indeed been observed for a few DAV stars, and the most convincing case is G117–B15A (Kepler et al. 1990, 1991, 1993):

$$P = 215.197\,387 \pm 0.000\,001 \text{ s} ,$$
$$\dot{P} = 3.3 \times 10^{-15} \text{ s}\,\text{s}^{-1} , \tag{5.70}$$

which is in reasonable agreement with the predictions from cooling theory.

For hot white dwarfs the situation is much less satisfactory. The measured values of one particularly strongly excited period for PG 1159–035 are

$$P = 516.025\,31 \pm 0.000\,06 \text{ s} ,$$
$$\dot{P} = (-2.49 \pm 0.06) \times 10^{-11} \text{ s}\,\text{s}^{-1} , \tag{5.71}$$

which correspond to a period change of 1 s in 1 270 years (Winget et al. 1985, 1991). The order of magnitude for $\dot{P}$ is in the correct range for models of a fast-cooling hot star; however, it shows the wrong sign. Stellar-evolution calculations indicate that this cannot be attributed to the second term in the relation (5.69) (Winget et al. 1991). A possible explanation is based on the notion of trapped modes (Kawaler and Bradley 1994), i.e., the observed oscillation is a g mode trapped in the very outer layers of the star, and does not therefore mirror the overall stellar structure but is instead a localized feature.

The cooling theory of white dwarfs is currently a most innovative area of research, and is also strongly connected to the physics of matter at high densities and pressures. To achieve the high accuracy necessary for the use of cooling white dwarfs as an independent age indicator for our galaxy, extremely detailed knowledge is required of the stellar chemical composition, the crystallization processes, the opacities in the outer layers, etc. (Wood 1990, 1992, 1994, 1995, Chabrier et al. 1992, Chabrier 1993, Hernanz et al. 1994, Segretain et al. 1994). As an example, in Fig. 5.26 the time behavior of the luminousity is shown, depending on different assumptions concerning the chemical composition.

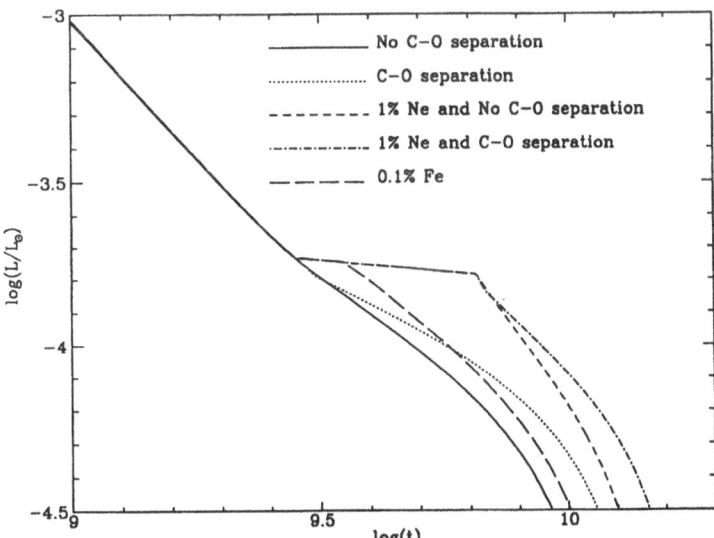

Fig. 5.26. Cooling times of different white dwarf models; Carbon/Oxygen star (dotted line), Carbon/Oxygen/^{22}Neon(1%) star (short-dashed line), Carbon/Oxygen/^{22}Neon(1%) star where the remaining C/O mixture crystallizes after ^{22}Ne has crystallized (dot-dashed line), Carbon/Oxygen/^{56}Fe(0.1%) star (long-dashed line). The full line corresponds to the reference model, when the C/O star is forced to crystallize with no chemical separation. (From Segretain et al. 1994)

The differences in the interesting region around 10^{10} years are about 30 percent. This uncertainty is already relatively small compared with other methods and may be further reduced by improvements in astereoseismology and the equation of state.

5.5 Oscillations of Neutron Stars

In Sect. 4.2 we learned how to tackle the problem of oscillations of spherical distributions of matter. This theory is applicable to a wide range of astrophysical objects such as the Earth, the Sun, or even white dwarfs that can all be adequately described within the Newtonian theory of gravity. Neutron stars, however, have to be treated in the framework of general relativity, which means that oscillations of the matter will couple to corresponding space-time perturbations and these distortions of the metric can act as a source of gravitational waves. Fortunately, the theory of oscillating spheres is quite easily incorporated into general relativity since the basic ideas remain the same as in the Newtonian case and the concept of decomposing the perturbation into spherical harmonics can be extended for tensors in a straightforward manner.

Following the same procedure as in the previous sections we assume that the structure of the static (unperturbed) neutron star is given by solving the TOV equations (4.25). Then the undisturbed metric $g_{\mu\nu}$ and the energy-momentum tensor $T_{\mu\nu}$ are known. According to (5.1) a small perturbation is added to the static structure including the metric:

$$\bar{g}_{\mu\nu} = g_{\mu\nu} + h_{\mu\nu} , \qquad |h_{\mu\nu}| \ll |g_{\mu\nu}| , \qquad (5.72)$$

where the bar denotes a perturbed quantity. This ansatz is inserted into the Einstein equations which are subsequently linearized with respect to the small perturbations. Thus, we get

$$\delta G_{\mu\nu} = 8\pi \delta T_{\mu\nu} , \qquad (5.73)$$

where $\delta T_{\mu\nu}$ is the disturbed energy-momentum tensor and the tensor $\delta G_{\mu\nu}$ contains the metric perturbation $h_{\mu\nu}$. Equation (5.73) represents a system of ten coupled linear partial differential equations. The undisturbed metric of a non-rotating spherically symmetric neutron star is written as (see Chap. 4)

$$ds^2 = -e^{2\nu(r)}dt^2 + e^{2\lambda(r)}dr^2 + r^2(d\theta^2 + \sin^2\theta d\varphi^2) \qquad (5.74)$$

where the two functions $\nu(r)$ and $\lambda(r)$ have to be determined by solving the unperturbed Einstein equations for the neutron star model. We will again assume the star to consist of a perfect fluid specified by the energy density $\bar{\rho}$, the pressure $\bar{p}$ and the four-velocity $\bar{u}^\mu$, hence the energy-momentum tensor reads

$$\bar{T}^{\mu\nu} = (\bar{\rho} + \bar{p})\bar{u}^\mu\bar{u}^\nu + \bar{p}\bar{g}^{\mu\nu} , \qquad (5.75)$$

where $\bar{p}$ and $\bar{\rho}$ are related by the EOS

$$\bar{p} = \bar{p}(\bar{\rho}) \ . \tag{5.76}$$

The disturbed energy-momentum tensor $\delta T^{\mu\nu}$ can be expressed in terms of the perturbations of the metric $h_{\mu\nu}$, the pressure δp, the energy density $\delta\rho$, and the four-velocity δu^μ. Note that the perturbations are functions of all four space-time coordinates x^μ and that δp and $\delta\rho$ are to first order related by

$$\delta p = \frac{\mathrm{d}p}{\mathrm{d}\rho} \, \delta\rho = c_s^2 \delta\rho \ , \tag{5.77}$$

where c_s is the sound velocity of the fluid. Furthermore, δu^0 may be fixed by the normalization condition $\bar{u}^\mu \bar{u}_\mu = -1$. Hence, there remain four perturbation functions for the matter inside the star. As in the non-relativistic case (5.7), we will now introduce a fluid displacement three-vector ξ^k that describes the displacement of a volume element of the fluid with respect to the unperturbed position. It is related to the four-velocity by

$$\delta u^k = \mathrm{e}^{-\nu} \frac{\partial \xi^k}{\partial t}. \tag{5.78}$$

The density perturbation $\delta\rho$ can be expressed in terms of ξ^k by employing the baryon-number conservation,

$$(\bar{n} \, \bar{u}^\mu)_{;\mu} = 0 \ , \tag{5.79}$$

and the first law of thermodynamics. This last relation follows from the condition that the divergence of the energy-momentum tensor must vanish and can be written as a relation between the changes in the energy density $\bar{\rho}$ the baryon-number density $\bar{n}$ (see also (4.34))

$$\frac{\mathrm{d}\bar{\rho}}{\mathrm{d}\bar{n}} = \frac{\bar{\rho} + \bar{p}}{\bar{n}} \ . \tag{5.80}$$

Rewriting those two equations for the perturbations and then combining them by eliminating the baryon number perturbations will yield an equation for the energy-density perturbation as a function of the fluid displacement and the metric perturbations

$$\delta\rho = (\rho + p) \left(-\xi^k_{\,|k} - \frac{1}{2} \frac{\delta g^{(3)}}{g^{(3)}} \right) - \xi^k \rho_{,k} \ . \tag{5.81}$$

Here the vertical bar represents the covariant derivative with respect to the unperturbed spatial three-metric $g_{ik}^{(3)}$, $g^{(3)}$ is its determinant, and $\delta g^{(3)}$ its perturbation. Thus the oscillations are completely described by the ten metric components $h_{\mu\nu}$ and the three compontents ξ^k of the fluid displacement. The problem will still be underdetermined since so far we have not specified any particular gauge. By choosing a gauge we may deliberately fix four components of the perturbation of the metric tensor thus leaving us with six independent components. Since the unperturbed metric is already specified we will not be able to perform a general coordinate transformation but only an infinitesimal one

$$x^\mu_{\text{old}} \to x^\mu_{\text{new}} = x^\mu_{\text{old}} + \eta^\mu(\mathbf{x}) \,, \tag{5.82}$$

where the $\eta^\mu(\mathbf{x})$ is assumed to be a small quantity. This will result in a transformed perturbed metric

$$
\begin{aligned}
\bar{g}^{\text{new}}_{\mu\nu} &= \bar{g}^{\text{old}}_{\mu\nu} + \Delta h_{\mu\nu} \\
\Leftrightarrow \qquad g_{\mu\nu} + h^{\text{new}}_{\mu\nu} &= g_{\mu\nu} + h^{\text{old}}_{\mu\nu} + \Delta h_{\mu\nu} \\
\Leftrightarrow \qquad h^{\text{new}}_{\mu\nu} &= h^{\text{old}}_{\mu\nu} + \Delta h_{\mu\nu} \,,
\end{aligned}
$$

where to first order $\Delta h_{\mu\nu} = \eta_{\mu;\nu} + \eta_{\nu;\mu}$. The most common choice of gauge is due to Regge and Wheeler (1957) who used it first to examine the stability of a black hole. This particular gauge may be obtained by imposing the following conditions on the metric perturbations

$$
\begin{aligned}
h_{\theta\varphi} &= 0 \,, \\
h_{\varphi\varphi} &= \sin^2\theta \, h_{\theta\theta} \,, \\
\frac{\partial}{\partial\theta}(\sin\theta \, h_{t\theta}) &= -\frac{\partial}{\partial\varphi}\left(\frac{1}{\sin\theta} h_{t\varphi}\right) \,, \\
\frac{\partial}{\partial\theta}(\sin\theta \, h_{r\theta}) &= -\frac{\partial}{\partial\varphi}\left(\frac{1}{\sin\theta} h_{r\varphi}\right) \,.
\end{aligned}
$$

The motivation for this requirement will become clear below. We now proceed as in the non-relativistic case by expanding the perturbations into spherical harmonics. Since here we are dealing with four-dimensional vectors and tensors we have to extend the previously described method to four-tensors in a fully covariant way, i.e. we have to construct a set of ten tensorial spherical harmonics which represents a basis for the decomposition of the perturbations $h_{\mu\nu}$. This may be accomplished in a straightforward way by using tensor products of the operators $\mathbf{L}$, ∇ and $\mathbf{n}$. Since $\mathbf{n}$ represents a unit vector in the r direction, we will replace it by its four-dimensional counterpart $\mathbf{e}_r = \frac{1}{\sqrt{g_{rr}}}\frac{\partial}{\partial r}$. The gradient operator ∇ just becomes the covariant derivative $\mathbf{D}$ and the angular momentum operator $\mathbf{L}$ will be written as

$$
\begin{aligned}
L_t &= 0 \,, \\
L_i &= -\mathrm{i}\frac{1}{\sqrt{g^{(3)}}} g^{(3)}_{ik} \epsilon^{klm} x_k D_m \,,
\end{aligned}
$$

where $(x_k) = (r, 0, 0)$. To account for the fourth dimension we have to augment the above set by a time-like unit vector $\mathbf{e}_t = \frac{1}{\sqrt{g_{tt}}}\frac{\partial}{\partial t}$. A set of ten linear independent tensorial harmonics may then be obtained by constructing the various tensor products

$$
\begin{array}{llll}
\mathbf{e}_t \otimes \mathbf{e}_t, & & & \\
(\mathbf{e}_t \otimes \mathbf{e}_r)_s \,, & \mathbf{e}_r \otimes \mathbf{e}_r, & & \\
(\mathbf{e}_t \otimes \mathbf{D})_s \,, & (\mathbf{e}_r \otimes \mathbf{D})_s \,, & \mathbf{D} \otimes \mathbf{D}, & \\
(\mathbf{e}_t \otimes \mathbf{L})_s \,, & (\mathbf{e}_r \otimes \mathbf{L})_s \,, & (\mathbf{D} \otimes \mathbf{L})_s \,, & \mathbf{L} \otimes \mathbf{L}
\end{array}
$$

and applying them to the spherical harmonics Y_{lm}. The subscript s indicates symmetrization, e.g., $(e_t \otimes L)_s = \frac{1}{2}(e_t \otimes L + L \otimes e_t)$, etc. The explicit form of this set may be found in Zerilli (1970a). By forming appropriate linear combinations within this set we may obtain two subsets that are characterized by their behavior with respect to parity transformations. The one consists of seven tensorial harmonics Y_{lm}^{e,λ_e}, $\lambda_e = 1, \ldots, 7$ that transform like $(-1)^l$; the remaining ones Y_{lm}^{o,λ_o}, $\lambda_o = 1, \ldots, 3$ transform like $(-1)^{l+1}$. The former ones are often called even, polar, or electric type while the latter are of odd, axial, or magnetic type. The decomposition of the metric perturbation then reads

$$
h_{\mu\nu}(t, r, \theta, \varphi) \;=\; \sum_{l=0}^{\infty} \sum_{m=-l}^{l} \left(\sum_{\lambda_e} R_{lm}^{e,\lambda_e}(t, r) \left(Y_{lm}^{e,\lambda_e} \right)_{\mu\nu} (\theta, \varphi) \right.
$$

$$
\left. + \sum_{\lambda_o} R_{lm}^{o,\lambda_o}(t, r) \left(Y_{lm}^{o,\lambda_o} \right)_{\mu\nu} (\theta, \varphi) \right) \quad (5.83)
$$

Now the advantage of the Regge–Wheeler gauge becomes evident because it makes four of the ten coefficients $R_{lm}^{e/o,\lambda_e/o}$ vanish. The even parity part of the perturbation may then be written as [see Thorne and Campolattaro (1967)]

$$
h_{\mu\nu} = \begin{pmatrix} e^{2\nu(r)} H_0(t, r) & H_1(t, r) & 0 & 0 \\ H_1(t, r) & e^{2\lambda(r)} H_2(t, r) & 0 & 0 \\ 0 & 0 & r^2 K(t, r) & 0 \\ 0 & 0 & 0 & r^2 \sin^2\theta\, K(t, r) \end{pmatrix} Y_{lm} .
$$

Here and in the following we will consider l and m as being fixed. The fluid displacement vector is much easier to handle since here we can take the usual vector harmonics because when applied to functions the covariant derivative reduces just to the ordinary derivative. Again we only show the even part:

$$
\xi^r \;=\; \frac{e^{-\lambda(r)}}{r^2} W(t, r)\, Y_{lm}(\theta, \varphi) ,
$$

$$
\xi^\theta \;=\; -\frac{V(t, r)}{r^2} \frac{\partial}{\partial\theta} Y_{lm}(\theta, \varphi) ,
$$

$$
\xi^\varphi \;=\; -\frac{V(t, r)}{r^2 \sin^2\theta} \frac{\partial}{\partial\varphi} Y_{lm}(\theta, \varphi) .
$$

These expressions are similar to the non-relativistic displacement vector components (5.7) when only the even parity parts ($\propto E_{lm}$ and Q_{lm}) are considered. Plugging both the even and odd parts in the field equations, we may eliminate the angular dependence, having thus reduced the problem to two dimensions. Moreover, we observe that the different parities do not mix, i.e., we obtain two independent sets of equations, which contain only quantities of the same parity. Besides, all equations are independent of the azimuthal order m as it should be for spherically symmetric systems. As it turns out there are no odd-parity oscillations for a perfect fluid, which should not be surprising since ρ and p are scalar fields and therefore possess even parity (the scalar harmonics transform like $(-1)^l$. The only possible odd parity perturbation is a time-independent differential rotation of the fluid that has to be excluded. Hence we will focus on even parity oscillations only.

As is well known, considering the time dependence of Einstein's equations separates them into the equations of motion (describing the dynamics) and the constraint equations (posing certain conditions the data have to fulfill). In our linearized case, the structure of the equations of motion is that of a general wave equation

$$\ddot{Q} = a\,Q'' + b\,Q' + c\,Q \,, \tag{5.84}$$

where dots denote time derivatives and primes radial derivatives. Q stands for either V, W, K or H_0. The coefficients a, b, and c depend on only the radial coordinate r. The constraints reveal that $H_2 = H_0$ and that H_0 as well as H_1 are related to V, W, and K algebraically or by a time-independent first-order differential equation so that in fact there remain only three true degrees of freedom, two (V, W) associated with the motion of the matter inside the star and one (K) with oscillations of the metric. In order to solve the corresponding initial-value problem one has to specify at an initial time t_0 the radial distributions of W, V, K and $\dot{W}, \dot{V}, \dot{K}$ with appropriate boundary conditions and use the equations of motion to propagate them forward in time. The functions H_2 and H_1 will be fixed by algebraic relations, whilst H_0 can either be propagated via its equation of motion or by solving the constraint equation. Outside the star there are no equations for V and W and one has to solve for only the metric perturbations.

In a realistic problem one could consider a disturbance of the matter with no previous gravitational radiation outside the star (i.e., all metric perturbation functions are set to zero) and observe the appearance of a gravitational wave front outside the star, which should propagate outwards with the speed of light whilst inside the star the oscillations should be damped since the gravitational wave will carry energy out of the system.

As an alternative to the solution of an initial-value problem people tried a different approach to examine oscillations of neutron stars (and black holes) by introducing the concept of quasinormal modes. The starting point is the decomposition of the time-dependent perturbations into normal modes

$$\begin{aligned}
H_0(t,r) &= r^l\,\hat{H}_0(r)\,\mathrm{e}^{\mathrm{i}\omega t}\,, \\
H_1(t,r) &= \mathrm{i}\omega\,r^{l+1}\,\hat{H}_1(r)\,\mathrm{e}^{\mathrm{i}\omega t}\,, \\
H_2(t,r) &= r^l\,\hat{H}_2(r)\,\mathrm{e}^{\mathrm{i}\omega t}\,, \\
K(t,r) &= r^l\,\hat{K}(r)\,\mathrm{e}^{\mathrm{i}\omega t}\,, \\
W(t,r) &= r^{l+1}\,\hat{W}(r)\,\mathrm{e}^{\mathrm{i}\omega t}\,, \\
V(t,r) &= r^l\,\hat{V}(r)\,\mathrm{e}^{\mathrm{i}\omega t}\,.
\end{aligned} \tag{5.85}$$

Due to gravitational wave damping the frequency ω will be complex. Following Detweiler and Lindblom (1985), substitution of the perturbation functions according to (5.85) into the field equations will yield a set of four coupled linear first-order ordinary differential equations in the radial coordinate r,

$$\hat{H}_1' = \frac{e^{2\lambda}}{r}\left[-\left((l+1)e^{-2\lambda} + \frac{2m(r)}{r} + 4\pi r^2 (p-\rho)\right)\hat{H}_1\right.$$
$$\left. + \hat{H}_0 + \hat{K} - 16\pi (p+\rho)\hat{V}\right],\tag{5.86}$$

$$\hat{K}' = \frac{1}{r}\left[\hat{H}_0 + \frac{1}{2}l(l+1)\hat{H}_1 - (l+1-r\nu')\hat{K} - 8\pi (p+\rho)e^{\lambda}\hat{W}\right],$$

$$\hat{W}' = \frac{1}{r}\left[-(l+1)\hat{W} + r^2 e^{\lambda}\left(\frac{e^{-\nu}}{\gamma_{ad} p}\hat{X} - \frac{l(l+1)}{r^2}\hat{V} + \frac{1}{2}\hat{H}_0 + \hat{K}\right)\right],$$

$$\hat{X}' = \frac{p+\rho}{2r}e^{\nu}\left\{(1-r\nu')\hat{H}_0 + \left(r^2\omega^2 e^{-2\nu} + \frac{l(l+1)}{2}\right)\hat{H}_1\right.$$
$$+ (3r\nu'-1)\hat{K} - \frac{2l(l+1)}{r}\nu'\hat{V}$$
$$\left. - \left[8\pi (p+\rho)e^{\lambda} + 2\omega^2 e^{\lambda-2\nu} - 2r^2\left(\frac{e^{-\lambda}}{r^2}\nu'\right)'\right]\hat{W}\right\} - \frac{l}{r}\hat{X}.$$

Here $m(r)$ is the mass function (4.19) which is related to $\lambda(r)$ by (4.20)

$$e^{-2\lambda(r)} = 1 - \frac{2m(r)}{r},\tag{5.87}$$

γ_{ad} denotes the adiabatic index

$$\gamma_{ad} = \frac{p+\rho}{p}\frac{dp}{d\rho},\tag{5.88}$$

and $\hat{X}(r)$ has been introduced in order to make the numerical treatment easier ($\hat{X}$ is proportional to the Lagrangian pressure perturbation)

$$\hat{X} = \omega^2 (p+\rho)e^{-\nu}\hat{V} - \frac{e^{\nu-\lambda}}{r}p'\hat{W} + \frac{1}{2}(p+\rho)e^{\nu}\hat{H}_0.\tag{5.89}$$

The function $\hat{H}_0$ can be expressed in terms of the remaining metric perturbations

$$\hat{H}_0 = \frac{1}{3m(r) + n_l r + 4\pi r^3 p}\left[8\pi r^3 e^{-\nu}\hat{X}\right.$$
$$+ \left(\omega^2 r^3 e^{-2(\lambda+\nu)} - (n_l+1)(m(r) + 4\pi r^3 p)\right)\hat{H}_1$$
$$- \left(\frac{e^{2\lambda}}{r}(m(r) + 4\pi r^3 p)(3m(r) - r + 4\pi r^3 p)\right.$$
$$\left.\left. - n_l r + \omega^2 r^3 e^{-2\nu}\right)\hat{K}\right],$$

where $n_l = \frac{1}{2}(l+2)(l+1)$. Note that $\hat{H}_2 = \hat{H}_0$ still holds. For fixed values of l and ω the system (5.86) admits four independent solutions but only two remain finite at the origin $r = 0$. The initial values at $r = 0$ may be obtained by Taylor expansion of (5.86) and then have to be integrated numerically up to the surface ($r = R$) of the star where a further boundary condition has to be satisfied namely the vanishing of the Lagrangian pressure perturbation, i.e., $\hat{X}(R) = 0$. (The surface of the star is actually defined by zero pressure.) This may be accomplished by an appropriate linear combination of the two regular solutions. Thus for given values of l and ω a unique solution (up to a constant factor) will be determined. To decide whether the chosen frequency ω corresponds to a quasinormal mode, one has to examine the behavior of the perturbations outside the star. There exist various ways to determine the quasinormal modes (QNMs) of a neutron star. In the following we shall outline a method developed by Nollert and Schmidt (1992).

In the region outside the star after the angular parts are split off the even perturbations may be described by just one function Z, the Zerilli function (Zerilli 1970b). This function has to satisfy a wave equation that reads

$$\left[\frac{\partial^2}{\partial x^2} - \frac{\partial^2}{\partial t^2} - V_Z(x)\right] Z(t, x) = 0. \tag{5.90}$$

Again indices l and m are omitted, x is the tortoise coordinate related to the usual Schwarzschild coordinate r by

$$x = r + 2M \log\left(\frac{r}{2M} - 1\right), \tag{5.91}$$

and $V_Z(x)$ is the Zerilli potential

$$V_Z = \frac{2(1 - 2M/r)}{(n_l r + 3M)^2} \left[n_l^2(n_l + 1) + n_l^2 \frac{3M}{r} + n_l \frac{9M^2}{r^2} + \frac{9M^3}{r^3}\right]. \tag{5.92}$$

Here M is the total mass of the star. The usual Fourier ansatz $Z(t, x) = \hat{Z}(x)\, e^{i\omega t}$ for the time behavior then leads to a Schrödinger-like equation for the remaining radial function $\hat{Z}$:

$$\hat{Z}''(x) + \left(\omega^2 - V_Z(x)\right) \hat{Z}(x) = 0. \tag{5.93}$$

(Odd perturbations lead to the same sort of equations, but with a different potential, the Regge–Wheeler potential.) Note that $\hat{Z}$ and $\hat{Z}'$ can be expressed as linear combinations of the two independent metric perturbations $\hat{H}_0$ and $\hat{K}$ [see Lindblom and Detweiler (1983)]. Since $\hat{H}_0$ is algebraically related to $\hat{H}_1$, (5.93) thus is equivalent to the first two equations of (5.86) with the matter perturbations set to zero.

Unfortunately neither the Zerilli nor the Regge–Wheeler potential represent potential wells but potential barriers thus prohibiting the formulation of a well-posed eigenvalue problem with appropriate boundary conditions, which should yield a discrete set of normal modes with eigenfrequencies ω^2. However, there is still a way to postulate some boundary conditions in order to obtain a discrete spectrum of modes. Since $V(x) \to 0$ for $x \to \infty$ a general solution of (5.93) will have the asymptotic form

$$\hat{Z}(x) \approx A(\omega)\, e^{i\omega x} + B(\omega)\, e^{-i\omega x} \tag{5.94}$$

which corresponds to a linear combination of an incoming and an outgoing gravitational wave. This suggests the definintion of the quasinormal modes as being those solutions of (5.93) representing only purely outgoing waves, i.e.

$$\frac{A(\omega)}{B(\omega)} = 0\,, \tag{5.95}$$

$$\text{and} \qquad \hat{Z}_{\text{QNM}} \sim e^{-i\omega x}, \qquad x \to \infty\,. \tag{5.96}$$

This concept, however, is not well defined. Since it can be shown that $\text{Im}(\omega) > 0$ for QNMs, i.e., the modes are exponentially damped (as one expects), the solution (5.96) is exponentially growing for increasing values of x. This makes it numerically difficult to separate the decreasing term that represents the incoming wave. Furthermore the asymptotic behavior is not uniquely defined, which can be seen by stating (5.96) more precisely:

$$\hat{Z}_{\text{QNM}}(x) = e^{-i\omega x}\left(1 + O\left(x^{-1}\right)\right), \qquad x \to \infty\,. \tag{5.97}$$

But then the solution

$$\hat{Z}^{\text{in+out}}(x) = \hat{Z}^{\text{in}}(x) + \hat{Z}^{\text{out}}(x) \tag{5.98}$$

where

$$
\begin{aligned}
\hat{Z}^{\text{in}}(x) &= e^{i\omega x}\left(1 + O\left(x^{-1}\right)\right)\,, \\
\hat{Z}^{\text{out}}(x) &= e^{-i\omega x}\left(1 + O\left(x^{-1}\right)\right)
\end{aligned}
$$

also satisfies (5.96) since

$$
\begin{aligned}
\hat{Z}^{\text{in+out}}(x) &= e^{-i\omega x}\left[1 + O\left(x^{-1}\right) + e^{2i\omega x}\left(1 + O\left(x^{-1}\right)\right)\right] \\
&= e^{-i\omega x}\left(1 + O\left(x^{-1}\right)\right)\,, \tag{5.99}
\end{aligned}
$$

because, for $\text{Im}(\omega) > 0$, $e^{2i\omega x}$ falls off faster than any power of $1/x$. Those difficulties can be avoided by defining QNMs via a Laplace transformation (instead of a Fourier transformation) which is the more appropriate way to tackle a non-stationary problem. The Laplace transform of (5.90) reads

$$
\begin{aligned}
\frac{\partial^2 \psi(s,x)}{\partial x^2} &+ \left[-s^2 - V_{\text{Z}}(x)\right]\psi(s,x) \\
&= -sZ(t{=}0,x) - \frac{\partial Z(t{=}0,x)}{\partial t} \quad =:\ I(s,x)\,. \tag{5.100}
\end{aligned}
$$

We now obtain an inhomogeneous ordinary differential equation for

$$\psi(s,x) := \int_0^\infty e^{-st}\, Z(t,x)\, dt\,, \tag{5.101}$$

with the inhomogeneity constructed from the initial values. To find solutions of (5.100) we will look for the Green's functions $G(s,x,x')$ such that

$$\psi(s,x) = \int G(s,x,x')\, I(s,x')\, dx' \; . \tag{5.102}$$

The Green's function may be calculated explicitly by using two linearly independent solutions $f_1(s,x)$ and $f_2(s,x)$ of the homogeneous differential equation corresponding to (5.100),

$$G(s,x,x') = \frac{1}{W(s)} \left\{ \begin{array}{ll} f_1(s,x')\, f_2(s,x) & (x' < x) \\ f_1(s,x)\, f_2(s,x') & (x < x') \, , \end{array} \right. \tag{5.103}$$

where $W(s)$ is the Wronskian of f_1 and f_2. Since $\psi(s,x)$ is the Laplace transform of a regular function $Z(t,x)$ it is analytic and bounded in the right complex half plane $\mathrm{Re}(s) > 0$. Thus $G(s,x,x')$ also has to be analytic and bounded what means that the Wronskian $W(s)$ must not vanish. If we now try to extend the Green's function analytically to the left half plane there will necessarily be points where $W(s)$ vanishes, i.e., the two functions f_1 and f_2 are then linearly dependent and $G(s,x,x')$ will possess isolated singularities.

For $r \to \infty$ the two independent solutions of the homogeneous equation are of the asymptotic form

$$g_+(s,x) \approx e^{sx} \qquad \text{and} \qquad g_-(s,x) \approx e^{-sx} \; . \tag{5.104}$$

If we require the solution to be a Laplace transform it has to be bounded in the right half plane thus excluding the solution g_+ since

$$\lim_{\substack{\mathrm{Re}(s) \to \infty \\ x > 0}} e^{sx} = \infty \; . \tag{5.105}$$

We may now identify f_2 with g_-, and f_1 has to be chosen such that it will obey the boundary conditions at the surface of the star. The quasinormal modes then just correspond to those values of s for which f_1 and f_2 coincide, i.e., their Wronskian vanishes. If we set $s = i\omega$ this is equivalent to saying that f_1 describes a purely outgoing wave, i.e. $f_1 \sim e^{-i\omega x}$ for $x \to \infty$. The advantage of this definition of the QNMs is that there is no ambiguity in posing the boundary conditions, they just naturally follow from the properties of the Laplace transformation if we require a regular behavior of $Z(t,x)$ for $t \to \infty$. In order to perform numerical calculations we have to know the explicit form of f_2 representing the outgoing wave. Now Nollert and Schmidt (1992) have given an explicit representation of the Wronskian as an absolutely convergent series as a function of f_1 only.

Let us briefly review the whole procedure of calculating QNMs. For a fixed value of l choose a frequency ω and two linearly independent sets of initial values. Then start integrating the system of diffential equations (5.86) from the origin up to the surface of the star. Adjust the obtained solutions by taking such linear combinations that the vanishing of the Lagrangian pressure perturbation at the surface of the star is ensured. Now turn to the exterior region. Convert the metric perturbations at the stellar surface into the function f_1 that enters the series representing the above mentioned Wronskian. The vanishing of the Wronskian tells you that your chosen frequency indeed describes a quasinormal mode of the neutron star. If the Wronskian gives you a non-zero result, try another frequency and go through the whole procedure again.

This method leads to a set of quasinormal mode frequencies ω_{nl} (defined as poles of the Green's function) which may be expressed in terms of the period P_{nl} and the corresponding damping time τ_{nl}:

$$\omega = \omega_{nl} = \frac{2\pi}{P_{nl}} + i\frac{1}{\tau_{nl}} \qquad (n = 0, 1, 2, \ldots) . \qquad (5.106)$$

In contrast to the eigenfrequencies discussed in Sect. 4.2 these modes are complex and therefore describe the wave damping due to energy loss by gravitational waves. For a description of various methods by which the quasinormal modes can be calculated we refer the reader to Leins (1994).

Figure 5.27 shows the positions of the normal-mode frequencies in the complex ω plane for $l = 2$ calculated for a polytropic EOS (2.10) with $\gamma = 2$. There are three different branches corresponding to different types of oscillations. One is equivalent to the well-known p modes, the other two, called w and w_{ii} modes are essentially perturbations of the metric and are strongly damped space-time oscillations [see Leins et al. (1993)]. Note that because of the simple structure of the EOS (5.76) the Brunt–Väisälä frequency (5.55) is zero and therefore g modes do not occur in this case. In Table 5.1, the

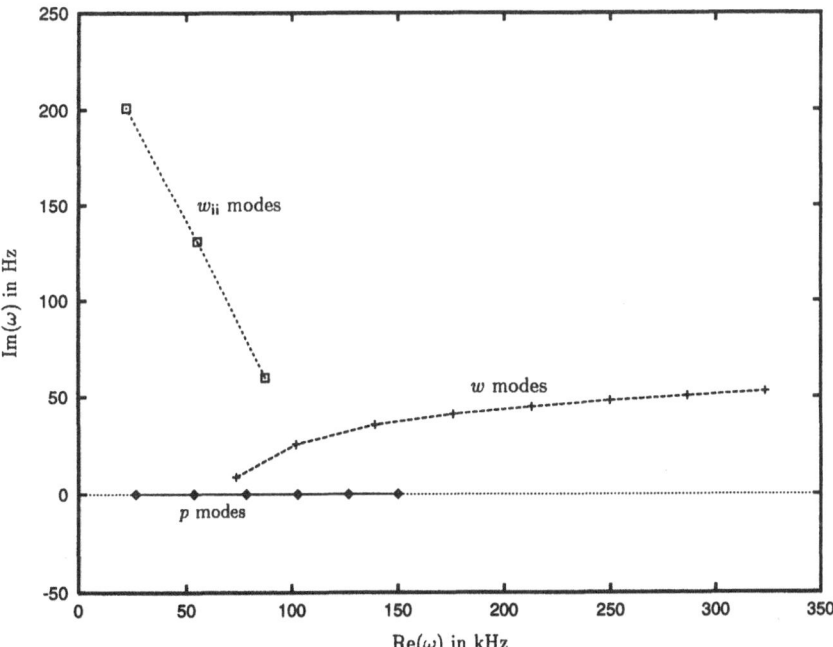

Fig. 5.27. QNM frequencies for $l = 2$ in the complex ω plane for a polytropic neutron star model with polytropic exponent $\gamma = 2$.

periods and damping times for the f mode and the first p modes are given for a neutron star with $M = 1.511\,M_\odot$ and calculated with the MPA equation of state discussed in Chapt. 3. These modes are weakly damped and they can again be labelled according to the number of nodes of the real part of the eigenfunctions, i.e., the f mode which shows no zero in its eigenfunction has $n = 0$, the p_1 mode (with $n = 1$) has one node, etc.

Table 5.1. Oscillation periods and damping times of the f mode and the first p modes for a neutron stars with mass $M = 1.511 M_\odot$ (MPA equation of state)

mode	f	p_1	p_2	p_3
P_{n2} in ms	0.330	0.126	0.0827	0.0638
τ_{n2} in s	0.100	1.7	11	100

The finite damping times τ_{nl} in (5.106) are caused by the emission of gravitational waves during the oscillation of the neutron star. Therefore, this gravitational wave damping occurs only for $l \geq 2$. Most results in the literature have been obtained for quadrupole oscillations, i.e., $l = 2$ [for other angular momentum values see Cutler et al. (1990)], and in this case only for the fundamental mode, the f mode. In Fig. 5.28 the $l = 2$ f mode oscillation periods of neutron stars calculated with different equations of states [EOSs A, B, C, L from Arnett and Bowers (1977) and MPA] as functions of the surface redshift z_0 are compared [see Lindblom and Detweiler (1983)]. For high-mass neutron stars the periods are fractions of a millisecond. The corresponding damping times are presented in Fig. 5.29. It can be seen that these modes will typically oscillate a few hundred times before being damped. Note that Cutler et al. (1990) have also calculated eigenmodes for $l = 0$ which are unaffected by gravitaional wave damping. They decay only through viscous processes in the stellar interior and the corresponding decay times are many orders of magnitude larger than for the $l \leq 2$ modes. Unlike for the sun and white dwarf stars oscilations of neutron stars have not yet been observed. It is very unlikely that the κ mechanism will operate for neutron stars, however, a possible excitation process might be due to quakes in the crust of the neutron star that are observed as glitches in the rotation period of radio pulsars. But even if such oscillations are excited it is unclear whether they will lead to observable effects. One speculative possibility would be that the oscillations couple to the magnetic field structure and thus lead to a modulation of the coherent radio emission.

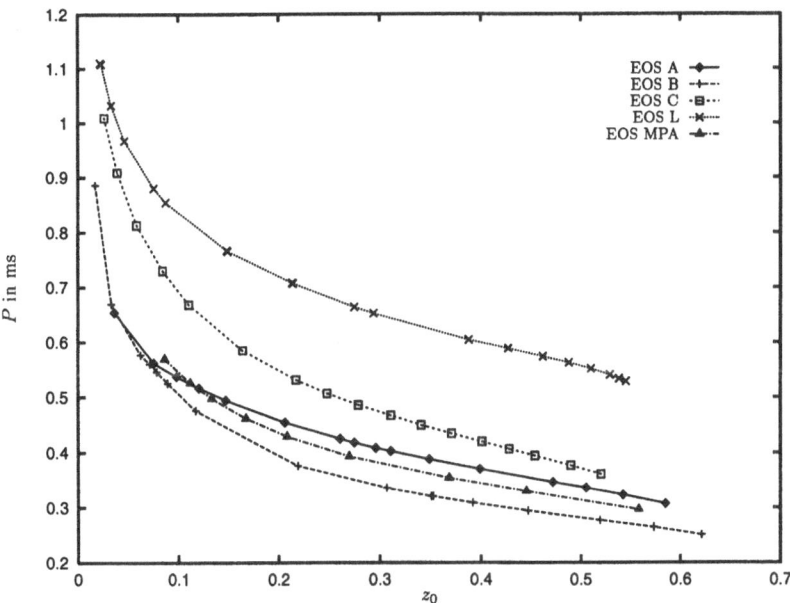

Fig. 5.28. Oscillation period P of the f mode for $l = 2$ as function of the surface redshift z_0 for different equations of state

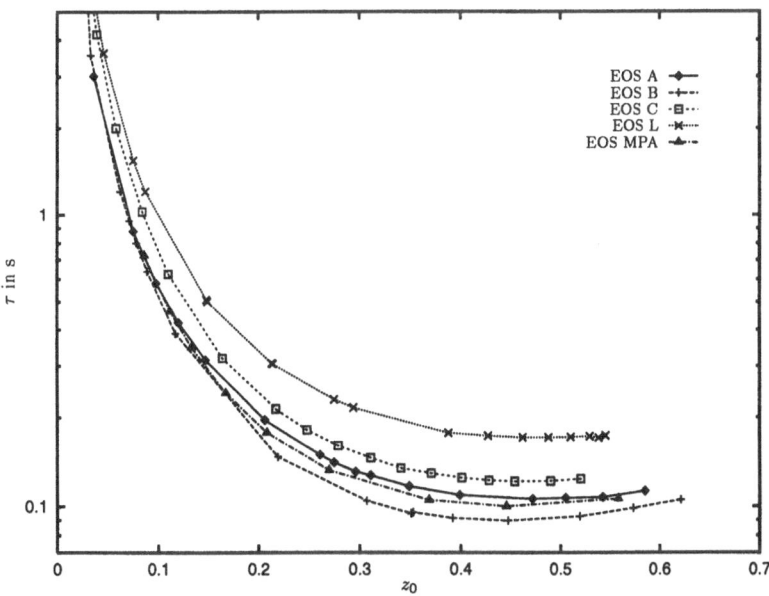

Fig. 5.29. Damping time τ of the f mode for $l = 2$ as function of the surface redshift z_0 for different equations of state

Reprint* of Friedrich Hund's Review Article

Materie unter sehr hohen Drucken und Temperaturen.

von **F. Hund**, Leipzig.

Mit 10 Abbildungen.

Inhaltsverzeichnis.

1. Vorbemerkungen 176
I. Die Zustandsbeziehung 177
  2. Allgemeines 177
  3. Irdische Drucke und Temperaturen............................178
  4. Das Temperaturgebiet der Ionisierung 180
  5. Das Druckgebiet der Atomzerquetschung......................181
  6. Das Elektronengas..182
  7. Das Neutronengas 186
  8. Das Gebiet der Kernumwandlungen..........................187
  9. Abweichungen vom thermodynamischen Gleichgewicht 190
  10. Die Strahlung 191
II. Andere physikalische Eigenschaften 195
  11. Der Energieinhalt...195
  12. Elektrizitäts- und Wärmeleitung.............................198
  13. Absorption von Licht 201
  14. Energietransport 204
III. Das Vorkommen sehr hoher Drucke und Temperaturen 205
  15. Die Planeten 205
  16. Die gewöhnlichen Fixsterne 207
  17. Die dichten Sterne...212
Literaturverzeichnis..214

---

* newly typeset for this edition to include typographical improvements and corrections

Bezeichnungen, die in mehreren Abschnitten gebraucht werden.

$\hbar = \frac{h}{2\pi}$	Wirkungsquantum.	M	Protonenmasse (genähert
c	Lichtgeschwindigkeit.		auch Atomgewichtseinheit
e	Elementarladung		und Neutronenmasse).
	($-e$ Elektronenladung;	μM	mittlere Teilchenmasse.
	vor einem Exponenten,	p	Druck.
	der nicht einfach eine	T	absolute Temperatur.
	Zahl ist, bezeichnet	ϱ	Dichte.
	jedoch e die Basis der	V	Volumen.
	natürlichen Logarithmen).	E	Energie.
k	BOLTZMANNsche Konstante.	ε	Energie pro Teilchen.
G	Gravitationskonstante.	η	Energie pro Masseneinheit.
m	Elektronenmasse.	N	Anzahl.
		n	Anzahl pro Raumeinheit.

1. Vorbemerkungen. Im Laboratorium kann das Verhalten der Stoffe bei Temperaturen bis zu einigen tausend Grad und bei Drucken bis zu einigen zehntausend Atmosphären untersucht werden. Im weiten Bereiche der Natur, insbesondere in den Fixsternen, kommen jedoch Temperaturen und Drucke vor, die eine Anzahl Zehnerpotenzen größer sind. Über das Verhalten der Materie unter solchen Verhältnissen kann die Physik heute weitgehend Auskunft geben. In den Zustandsgebieten, in denen die Materie als aus Atomkernen und Elektronen bestehend angenommen werden kann, kennt sie die Gesetze ihres Aufbaues vollständig. Sie weiß, daß diese Gesetze noch ein Stück weit gelten in den Gebieten extremerer Drucke oder Temperaturen, wo die Atomkerne nicht mehr unverändert bleiben; sie kennt die Kräfte zwischen den Bausteinen der Kerne noch nicht vollständig, wohl aber die Werte der Energie vieler Zustände, die auf diesen Kräften beruhen.

Wesentliche Züge des Verhaltens der Bausteine der Materie sind die Gültigkeit des COULOMBschen Gesetzes zwischen elektrisch geladenen Teilchen bis herab zu Abständen von der Größe der Kerndurchmesser (10^{-13} cm), die durch das Nebeneinanderbestehen der Wellennatur und der Teilchennatur der Materie bedingte, durch das Wirkungsquantum bestimmte Begrenzung der anschaulichen Beschreibung der atomaren Vorgänge, die durch das PAULIsche Ausschließungsprinzip bedingten Abweichungen in der statistischen Beschreibung von Teilchengesamtheiten, der Aufbau der Atomkerne aus Neutronen und Protonen und die Umwandelbarkeit dieser beiden Teilchenarten ineinander.

Sehr hohe Drucke und Temperaturen kommen im Innern der Sterne vor. Die Astronomie vermag aber nur sehr indirekte Angaben über dieses Innere zu machen; die Lage ist heute vielmehr die, daß die Physik verhältnismäßig sichere Aussagen über das Verhalten der Materie dort machen kann, die die Astronomie und Astrophysik zur Deutung ihrer Beobachtungen und zu Schlüssen aus diesen Beobachtungen auf die wirklichen Vorgänge und Zustände benutzt.

Der folgende Bericht beschränkt sich im wesentlichen auf die physikalische Seite der Sache. Der Aufbau der Sterne ist nur soweit herangezogen, als sich aus ihm Hinweise auf das Vorkommen der von uns betrachteten Zustände gewinnen lassen und soweit Andeutungen einer Abbildung der Gesetzmäßigkeiten der Materie auf Regelmäßigkeiten in den beobachtbaren Eigenschaften der Sterne bestehen.

I. Die Zustandsbeziehung.

2. Allgemeines. Wir haben die Aufgabe, die Eigenschaften eines homogenen Stückes Materie im thermodynamischen Gleichgewicht in ihrer Abhängigkeit von den Zustandsvariablen zu untersuchen; als unabhängige Zustandsgrößen wählen wir Temperatur T und Druck p. Zustände, die kein thermodynamisches Gleichgewicht sind, also thermodynamisch unwahrscheinliche Zustände, kommen in der Natur vor, auch lange Zeit unverändert, wenn nämlich die Prozesse, die das Gleichgewicht herstellen helfen, sehr langsam verlaufen. So ist das Vorkommen von Protonen außerhalb der schweren Kerne, also das Vorkommen von Wasserstoff bei tiefen Temperaturen ein solcher „unwahrscheinlicher" Zustand. In weiten Bereichen von Druck und Temperatur können wir aber von der Umwandelbarkeit der Elemente absehen; in diesen Bereichen ist es erlaubt, den Begriff „thermodynamisches Gleichgewicht" in dem etwas weiteren Sinne zu gebrauchen, der durch die Abwesenheit der Kernumwandlungen gegeben ist; wir können dann von einem bestimmten chemischen Element im thermodynamischen Gleichgewicht sprechen.

Wir untersuchen zunächst die *„Zustandsbeziehung"*, d.h. die Abhängigkeit der Dichte ϱ von p und T. Es wird dabei sich zeigen, daß bei einigermaßen hohen Werten von Druck und Temperatur die besondere Natur des Stoffes von recht geringem Einfluß ist, so daß wir von einer Zustandsgleichung der Materie schlechthin sprechen können.

Die ausgezeichneten Marken in unseren ausgedehnten Skalen von Druck, Temperatur, Dichte usw. sind durch die der Natur aufgeprägten absoluten Maßstäbe bestimmt, also durch die Größen: Lichtgeschwindigkeit c, Wirkungsquantum $\hbar$, Elementarladung e, Masse des Elektrons m und des Protons M, BOLTZMANNsche Konstante k. Man benutzt gelegentlich die „atomaren" Einheiten: $\hbar, e, m, k$; die Einheit der Energie ist dann $\frac{me^4}{\hbar^2} = 4,31 \cdot 10^{-11}$ erg ($27,1$ eVolt, doppelte Ionisierungsenergie des Wasserstoffatoms); andere Einheiten dieses Maßsystems gibt Tabelle 1. Da das COULOMBsche Gesetz und damit die Elementarladung im Aufbau der Materie nur bei niedrigen Drucken und Temperaturen eine wesentliche Rolle spielt, sind diese Einheiten für uns verhältnismäßig nebensächlich. Wir benutzen häufiger die Einheiten $\hbar, c, m, k$, also z. B. die Energieeinheit $mc^2 = 8,12 \cdot 10^{-7}$ erg ($0,51$ eM-Volt, Ruheenergie eines Elektrons). Für ganz extreme Verhältnisse kommen noch die Einheiten $\hbar, c, M, k$ in Betracht, also z.B. die Energieeinheit $Mc^2 = 1,49 \cdot 10^{-3}$ erg (Ruheenergie eines Protons, genähert auch eines Neutrons). Eine Zusammenstellung der wichtigsten Einheiten gibt Tabelle 1. Als Dichteeinheit ist dabei in den Maßsystemen mit der Masseneinheit m

die Größe M/Volumen eingeführt, da Elektronen immer nur zusammen mit schweren Teilchen vorkommen.

Wir sehen zunächst vom Vorhandensein eines *Gravitationsfeldes* ab, d.h. wir machen eine begriffliche Trennung zwischen der Energie der Materieteilchen im Gravitationsfeld und ihrem sonstigen Energieinhalt. Da aber sehr hohe Dichten und Drucke wohl nur durch Gravitation erzeugt werden können, ist nicht sicher, ob eine solche begriffliche Trennung für alle Werte der Zustandsvariabeln erlaubt ist. Die Gravitationswirkung läßt sich aber dann begrifflich abtrennen, wenn die Energie der Teilchen im Gravitationsfeld klein ist gegen ihre Ruheenergie oder (anders ausgedrückt) das Gravitationspotential klein gegen c^2.

Tabelle 1. Wichtige Einheiten für den Aufbau der Materie.

Grundeinheiten	$\hbar, e, m, k$	$\hbar, c, m, k$	$\hbar, c, M, k$
Länge.........	$\frac{\hbar^2}{me^2}=0{,}528\cdot10^{-8}$ cm	$\frac{\hbar}{mc}=3{,}84\cdot10^{-11}$ cm	$\frac{\hbar}{Mc}=2{,}09\cdot10^{-14}$ cm
Dichte	$\frac{Mm^3e^6}{\hbar^6}=11{,}3\ \frac{\text{gr}}{\text{cm}^3}$	$\frac{Mm^3c^3}{\hbar^3}=2{,}92\cdot10^7\ \frac{\text{gr}}{\text{cm}^3}$	$\frac{M^4c^3}{\hbar^3}=1{,}82\cdot10^{17}\ \frac{\text{gr}}{\text{cm}^3}$
Energie	$\frac{me^4}{\hbar^2}=4{,}31\cdot10^{-11}$ erg	$mc^2=8{,}12\cdot10^{-7}$ erg	$Mc^2=1{,}49\cdot10^{-3}$ erg
Druck.........	$\frac{m^4e^{10}}{\hbar^8}=2{,}90\cdot10^8$ Atm	$\frac{m^4c^5}{\hbar^3}=1{,}41\cdot10^{19}$ Atm	$\frac{M^4c^5}{\hbar^3}=1{,}62\cdot10^{32}$ Atm
Temperatur ...	$\frac{me^4}{\hbar^2k}=3{,}14\cdot10^5$ Grad	$\frac{mc^2}{k}=5{,}92\cdot10^9$ Grad	$\frac{Mc^2}{k}=1{,}09\cdot10^{13}$ Grad

Man darf also einer Masse $\mathfrak{M}$ nicht auf einen Abstand nahekommen, der mit der Länge $G\mathfrak{M}/c^2$ vergleichbar ist. Wir fügen den Einheiten der Tabelle 1 noch eine Masseneinheit astronomischer Größenordnung zu und eine Längeneinheit, indem wir die Einheiten der Dichte, also der Masse/Volumen, des Druckes, also der Energie/Volumen und damit auch des Potentials, also der Energie/Masse oder des Geschwindigkeitsquadrates aus der letzten Spalte beibehalten und die neuen Einheiten so wählen, daß die Masseneinheit im Abstand der Längeneinheit das Gravitationsfeld Eins hat. Wir erhalten die Masse (G ist die Gravitationskonstante):

$$\left(\frac{\hbar c}{G}\right)^{3/2}\frac{1}{M^2}=3{,}68\cdot10^{33}\,\text{g}\,,$$

das ist eine durchschnittliche Sternmasse, und die Länge

$$\frac{\hbar^{3/2}}{G^{1/2}\,c^{1/2}\,M^2}=2{,}72\cdot10^5\,\text{cm}\,.$$

Solange also Massen von der Größe der Sternmassen sich nicht auf Radien zusammendrängen, die mit dieser Länge vergleichbar sind, ist die Abtrennung der Gravitationswirkung erlaubt.

3. Irdische Drucke und Temperaturen. Unter gewöhnlichen Verhältnissen des Druckes und der Temperatur sind einfache Stoffe entweder im *„kondensierten" Zustand* oder sie bilden ein *Gas aus Molekeln.* Den Unterschied zwischen festem und flüssigem Zustand wollen wir als für unsere

Übersicht geringfügig vernachlässigen. Die Molekeln des Gases zerfallen, wenn sie mehratomig sind, bei höheren Temperaturen in Atome. Auch darauf wollen wir jetzt nicht achten. Das Gas sei also ein ideales einatomiges Gas mit dem Druck

$$p = nkT$$

und der Energie pro Raumeinheit

$$\frac{E}{V} = \frac{3}{2}nkT \, .$$

Für gröbere Betrachtungen können wir den kondensierten Zustand als Zustand konstanter Dichte

$$\varrho = \varrho_k$$

ansehen. Wir übergehen also das ganze interessante Gebiet der hohen Drucke, das BRIDGMAN (6) untersucht. Gas und Kondensat gehen bei höheren Temperaturen und Drucken stetig ineinander über; bei tieferen Temperaturen und Drucken sind sie aneinander grenzende „Phasen", d.h. die Grenzlinie zwischen dem Gasgebiet und dem Kondensatgebiet im p-T-Zustandsdiagramm ist nicht eine Grenze der Existenzmöglichkeit, sondern bis zur Grenze ist eine der Phasen die thermodynamisch wahrscheinliche, jenseits der Grenze ist es die andere Phase. Auf der wärmeren Seite der Grenze ist das Gas wahrscheinlicher wegen des höheren statistischen „Gewichtes" seiner Zustände, auf der kälteren Seite das Kondensat wegen der tieferen Energie. Diesen Wettstreit zwischen statistischem Gewicht und Energie untersucht man bei Benutzung der unabhängigen Variabeln p und T mit Hilfe des „thermodynamischen Potentials" $\Phi = E + pV - TS$ (E, V, S sind Energie, Volumen und Entropie einer gegebenen Stoffmenge). Sind N_k und N_g die Zahl der Atome im kondensierten und im Gaszustand, so ist (einatomiges Gas, Gewicht 1 des Atomzustandes, konstante Atomwärme γ im kondensierten Zustand vorausgesetzt, unter Weglassung kleiner Größen):

$$
\begin{aligned}
E &= N_k \gamma T + N_g \left(\frac{3}{2}kT + Q \right) \\
pV &= N_g kT \\
S &= N_k \gamma + N_g k \left[\frac{5}{2} - \log \frac{(2\pi)^{3/2}\hbar^3 p}{(\mu M)^{3/2}(kT)^{5/2}} \right] .
\end{aligned}
$$

Q ist dabei die Umwandlungswärme pro Atom (auf $T = 0$ extrapoliert) und μM die Atommasse. Das Minimum von

$$\Phi = N_g \cdot \left[kT \log \frac{(2\pi)^{3/2}\hbar^3 p}{(\mu M)^{3/2}(kT)^{5/2}} + Q \right]$$

liegt auf seiten des Gases, wenn $\Phi < 0$ ist, und auf seiten des Kondensats, wenn $\Phi > 0$ ist; die Grenze verläuft bei

$$p = \frac{\mu^{3/2}}{(2\pi)^{3/2}} \cdot \frac{M^{3/2}}{\hbar^3}(kT)^{5/2} e^{-\frac{Q}{kT}} .$$

Unsere Abschätzung wird ungültig, wenn das Gas nicht mehr ideal ist, grob
gesagt, wenn die Formel

$$\varrho = \mu M \frac{p}{kT}$$

eine Gasdichte gäbe, die mit der des Kondensats vergleichbar wäre. In Wirklichkeit haben wir dann stetigen Übergang zwischen Gas und Kondensat mit
keiner einfachen Zustandsgleichung. Für gröbere Betrachtungen idealisieren
wir die Verhältnisse durch die Zustandsgleichung

$$\varrho = \begin{cases} \mu M \frac{p}{kT} & \text{für} \quad p \le \frac{\varrho_k}{\mu M} kT \quad \text{und} \quad p \le \frac{(\mu M)^{3/2} (kT)^{5/2}}{(2\pi)^{3/2} \hbar^3} e^{-\frac{Q}{kT}} \\[2ex] \varrho_k & \text{für} \quad p \ge \frac{\varrho_k}{\mu M} kT \quad \text{oder} \quad p \ge \frac{(\mu M)^{3/2} (kT)^{5/2}}{(2\pi)^{3/2} \hbar^3} e^{-\frac{Q}{kT}}. \end{cases}$$

Die Grenzlinien sind für $\varrho_k = 8 \frac{\text{gr}}{\text{cm}^3}$, $Q \approx 100 \frac{\text{kcal}}{\text{gAtom}}$ und $\mu = 56$ in Abb. 2
eingetragen; sie entsprechen dem Eisen und den ihm benachbarten Elemen-
ten.

4. Das Temperaturgebiet der Ionisierung. Gehen wir zu noch
höheren Temperaturen über, so werden die Atome allmählich ionisiert; das
Gemisch aus Kernen und Elektronen hat dann zwar immer noch höhere Energie, aber viel höheres statistisches Gewicht als das Gas aus Atomen. Daß bei
hohen Temperaturen schließlich ein Gas aus Elektronen und Kernen entsteht
statt des Gases aus Atomen, darauf hat wohl zuerst JEANS (23) hingewie-
sen. Die allmähliche Ionisierung bei steigender Temperatur hat SAHA (37)
im Hinblick auf Anwendungen auf die Sternatmosphären untersucht. Genaue
Angaben über den Ionisierungsgrad unter verschiedenen Bedingungen machen FOWLER und GUGGENHEIM (18). Wir idealisieren den Vorgang durch
die „chemische Formel"

$$\text{Atom} \rightleftharpoons \text{Rest} + Z \text{ Elektronen}$$

und berechnen das thermodynamische Potential

$$\Phi = N_E \left[Q + kT \log \frac{(2\pi)^{3/2} \hbar^3 p_E}{m^{3/2} (kT)^{5/2}} \right] + N_R \cdot kT \log \frac{(2\pi)^{3/2} \hbar^3 p_R}{(\mu M)^{3/2} (kT)^{5/2}} +$$

$$+ N_A \cdot kT \log \frac{(2\pi)^{3/2} \hbar^3 p_A}{(\mu M)^{3/2} (kT)^{5/2}},$$

wo die Zeiger E, R, A sich auf Elektronen, Reste und Atome beziehen und
p_E, p_R, p_A Partialdrucke sind; Q ist die Ionisierungsarbeit pro abgetrenntes
Elektron. Führen wir für die „Konzentration" $\frac{N_E}{N_E + N_R + N_A}$ die Abkürzung $[E]$
und entsprechende Abkürzungen $[R]$ und $[A]$ ein, so liegt das Minimum von
Φ bei

$$[E] \cdot \sqrt[z]{\frac{[R]}{[A]}} = \frac{1}{(2\pi)^{3/2}} \frac{m^{3/2} (kT)^{5/2}}{\hbar^3 p} e^{-\frac{Q}{kT}}.$$

Die Hälfte ist ionisiert, wenn dieser Ausdruck gleich $\frac{Z+2}{Z}$ ist, also von der
Größenordnung 1. Wir erhalten einen verhältnismäßig raschen Übergang von

den Atomen zu dem Gemisch aus Resten und Elektronen in der Nähe der Grenzlinie

$$p = \frac{1}{(2\pi)^{3/2}} \frac{m^{3/2}(kT)^{5/2}}{\hbar^3} e^{-\frac{Q}{kT}}.$$

Sie ist in Abb. 2 gezeichnet für $Q = \frac{1}{2}\frac{me^4}{\hbar^2}$, was für einige leichtere Atome ungefähr der Abtrennung der äußersten Elektronen entspricht und für $Q = \frac{26^2}{2}\frac{me^4}{\hbar^2}$, was der Abtrennung der innersten Elektronen bei Eisen entspricht. Für hohe Temperaturen wird die Grenzlinie unabhängig von Q, in atomaren Einheiten $\hbar, m, k$ (gleichgültig, ob als vierte Einheit e oder c gewählt wird): $p \approx \frac{1}{16}T^{5/2}$.

5. Das Druckgebiet der Atomzerquetschung. Wenn man einen festen Körper oder eine Flüssigkeit ähnlich zusammendrückt oder ausgedehnt denkt, so ändert sich seine Energie in der in Abb. 1 angedeuteten Weise; der Zustand ohne äußere Kraft entspricht dem Minimum der Energie. Allgemein ist, wenn man von Wärmeerscheinungen absieht,

$$-\mathrm{d}E = p\,\mathrm{d}V.$$

Entsprechend dem starken Anstieg der Energie bei Verkleinerung des Volumens sind sehr hohe Drucke nötig, um den Körper merklich zusammenzudrücken. Diesen starken Anstieg der Energie kann man wie in allen Fällen, wo ein atomares System kleiner wird als seine gewöhnliche Ausdehnung, qualitativ so erklären, daß wegen des geringeren Volumens, das den einzelnen Elektronen zur Verfügung steht, die Impulse stark zunehmen. Nähert man

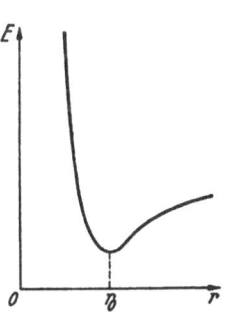

den Zustand des Gebildes durch Zustände der einzelnen Elektronen an und sind die Elektronen hinreichend zahlreich, so kann jede Zelle des Phasenraumes (mit drei räumlichen und drei Impulsdimensionen) nach dem PAULI-Prinzip höchstens zwei Elektronen aufnehmen (FERMI-Statistik). Die auf diese Weise sich vermehrende „Nullpunktsenergie" (weil ohne Temperatur vorhanden) der Elektronen wird schließlich groß gegen die Energie der COULOMBschen Kräfte (die nur mit $\frac{1}{r}$ zunimmt); der Körper kann dann beschrieben werden als ein *ideales Gas aus Elektronen*, deren COULOMBsche Kräfte dadurch unwirksam gemacht sind, daß sie sich *in einer positiven Raumladung* (der Kerne) bewegen. Dieses Gas ist stark „entartet" , d. h. seine Nullpunktsenergie ist groß gegen den Energiebetrag, der von der Temperatur herrührt. Der Beitrag der Kerne zu Druck und Energie ist dann geringfügig. Während das Verhalten der Materie bei gewöhnlichem Druck wesentlich von der betrachteten Stoffart abhängt, gleichen sich die Stoffe bei hohem Druck mehr und mehr einander an. Auf die Möglichkeit solcher Materie hat FOWLER (17) aufmerksam gemacht.

Abb. 1. Energie eines kondensierten Körpers bei ähnlicher Deformation.

Eine genäherte Berechnung der Zunahme der Energie eines Kristallgitters beim Zusammendrücken und damit der Abhängigkeit des Volumens vom

Druck ist mit der vereinfachenden Voraussetzung vieler Elektronen im Kraftfeld der Kerne (THOMAS-FERMIsche statistische Methode) von SLATER und
KRUTTER (39) durchgeführt worden. Für geringe Drucke ist die Annäherung
wegen der groben Form der Berücksichtigung der COULOMBschen Kräfte
nicht gut; sie wird aber um so besser, je geringer der Einfluß der COULOMB-
schen Kräfte ist, also bei hohen Drucken. Den Grenzübergang zum Elektronengas kann man anschließend an SLATER-KRUTTER durch ein geeigne-
tes Entwicklungsverfahren durchführen und hat dann eine Zustandsgleichung
$\varrho = \varrho(p)$ für das Übergangsgebiet. Da sie reichlich verwickelt ist, wollen wir
hier die Verhältnisse so schematisieren, daß wir für niedrige Drucke $\varrho = \varrho_k$,
also konstant, setzen, für hohe Drucke das Verhalten des Elektronengases
(Abschnitt 6) annehmen und die Grenzlinie so ziehen, daß $\varrho(p)$ stetig bleibt.
Von der Temperatur hängt der Übergang erst dann merklich ab, wenn der
Zustand der Elektronen nicht mehr entartet ist, wenn also kT in die Gegend
der atomaren Energieeinheit kommt.

6. Das Elektronengas. In einem sehr großen Gebiet der Temperatur
und des Druckes sind die Eigenschaften der Materie die eines *idealen Gases
aus Elektronen*, dem soviel Kerne beigemischt sind, daß das Ganze elektrisch
neutral ist. Wenn die Kernladung einigermaßen groß ist gegen die Ladung des
Protons, so machen die Kerne wegen ihrer geringen Anzahl in der Zustandsgleichung nicht viel aus. Zu den Voraussetzungen des idealen Gases gehört,
daß die COULOMBsche Energie gegen die kinetische Energie der Teilchen vernachlässigt werden kann; das ist nicht der Fall, wenn gleichzeitig die Temperatur und der Druck zu klein sind. Diese Fälle haben wir schon betrachtet;
es bilden sich Atome (bei niedrigem Druck und höherer Temperatur) oder
das Kondensat (bei niedriger Temperatur und höherem Druck). Bei höheren
Drucken und Temperaturen wird der Einfluß der COULOMBschen Kräfte bald
vernachlässigbar klein, die Elektronenladung tritt nicht mehr in den Gleichungen auf. Auf der anderen Seite wird das Gebiet des idealen Elektronengases da begrenzt, wo die Kerne zerfallen, da dann die Zahl der beigemengten
schweren Teilchen größer wird. Das macht aber noch verhältnismäßig wenig
in der Zustandsgleichung aus (wohl aber im Energieinhalt). Zu Ende ist aber
das Gebiet des Elektronengases, wenn Vorgänge auftreten, die die Zahl der
Elektronen einer betrachteten Materiemenge verändern (Umwandlung von
Elektronen und Protonen in Neutronen, Umwandlung von Strahlung in Elektronen und Positronen). Zu Ende wäre das Gebiet des idealen Elektronengases auch bei so hohen Dichten, daß die Teilchen einander auf die Entfernungen
nahekommen, wo die nichtcoulombschen Kräfte, die im Kernaufbau wirksam
sind, wesentlich werden. Die anderen genannten Begrenzungen treten jedoch
früher auf, so daß das Analogon zur VAN DER WAALSschen Abänderung der
Zustandsgleichung der gewöhnlichen Gase beim Elektronengas nicht auftritt.

Für die Aufstellung der Zustandsgleichung ist das PAULI-Prinzip wichtig;
es hat zur Folge, daß auch bei der Temperatur Null die Elektronen eine
kinetische Energie haben. Bei sehr großer Geschwindigkeit der Elektronen hat
man ferner die relativistische Form der Mechanik zu beachten. Eine für das
ganze Gebiet gültige Zustandsgleichung ist mathematisch sehr verwickelt. Sie

ist einfach in den vier Grenzfällen, wo einerseits die Nullpunktsenergie groß ist gegen den thermischen Energieanteil (entartetes Gas) oder das Umgekehrte der Fall ist (nichtentartetes Gas), wo andererseits die Geschwindigkeit der Elektronen klein gegen die Lichtgeschwindigkeit ist (nichtrelativistisches Gas) oder fast gleich der Lichtgeschwindigkeit ist (relativistisches Gas). Wir geben Zustandsgleichung, Energieinhalt und thermodynamisches Potential $\Phi = E + pV - TS$ für diese vier Fälle an (7, 14, 17, 33, 41, 44).

Nichtrelativistisches nichtentartetes Gas:

$$p = nkT$$
$$E = \frac{3}{2}pV$$
$$\Phi = nV \cdot kT \log \frac{2^{1/2}\pi^{3/2}\hbar^3 p}{m^{3/2}(kT)^{5/2}}.$$

Wenn wir unter n die Zahl der Elektronen in der Raumeinheit verstehen, so geben die Gleichungen den Anteil der Elektronen an Energie, Druck und thermodynamischem Potential an (auch hinter dem log in Φ ist p der Partialdruck). Da die Kerne in unserem Fall auch sicher nichtrelativistisch und nichtentartet sind, können wir die beiden ersten Gleichungen auch so auffassen, daß n die Gesamtteilchenzahl in der Raumeinheit ist, sie geben dann die gesamte Translationsenergie und den gesamten Druck an. Man kann dann n durch die Dichte ausdrücken

$$n = \frac{\varrho}{\mu M},$$

wo μM die im Mittel auf ein Teilchen entfallende Masse ist. Bei einem Gas aus Elektronen und Protonen ist also $\mu = \frac{1}{2}$; bei einem Gas aus schweren Kernen und Elektronen ist μ etwas größer als 2.

Nichtrelativistisches entartetes Gas:

$$p = \frac{3^{2/3}\pi^{4/3}}{5} \cdot \frac{\hbar^2}{m} \cdot n^{5/3}$$
$$E = \frac{3}{2}pV$$
$$\Phi = \frac{5}{2}pV.$$

Dabei ist n die Zahl der Elektronen in der Raumeinheit. Der Beitrag der Kerne ist außer bei Wasserstoff sehr gering; er hängt davon ab, ob die Kerne auch entartet sind oder nicht.

Relativistisches nichtentartetes Gas:

$$p = nkT$$
$$E = 3pV$$
$$\Phi = nV \cdot kT \log \frac{\pi^2 \hbar^3 c^3 p}{2(kT)^4}.$$

In den Gleichungen ist wieder nur der Anteil der Elektronen ausgedrückt. Setzt man

$$n = \frac{\varrho}{\mu M},$$

so beschränkt man sich auf den Fall, wo die Zunahme der Nullpunktsenergie der Elektronen (die größer als mc^2 ist) gegenüber Mc^2 noch nicht ins Gewicht fällt.

Relativistisches entartetes Gas:

$$p = \frac{3^{1/3}\,\pi^{2/3}}{2^2}\cdot \hbar c \cdot n^{4/3}$$
$$E = 3pV$$
$$\Phi = 4pV.$$

Die Zustandsgleichung des Übergangsgebietes vom nichtrelativistischen nichtentarteten zum nichtrelativistischen entarteten Gas hat FERMI (14) angegeben. Die Zustandsgleichung für den Übergang vom nichtrelativistisch entarteten zum relativistisch entarteten Gas gaben STONER (41) und CHANDRASEKHAR (7). Man kann aber ohne allzu große Fehler die Verhältnisse so schematisieren, daß man eine der drei oben angegebenen Zustandsgleichungen als gültig annimmt und die Gültigkeitsgebiete so abgrenzt, daß die Variabeln p, T, ϱ stetig bleiben. Die Grenze zwischen nichtentartetem und nichtrelativistisch entartetem Gebiet verläuft dann da, wo

$$n = \frac{p}{kT} = \left(\frac{5mp}{3^{2/3}\,\pi^{4/3}\,\hbar^2}\right)^{3/5},$$

also

$$p = \frac{5^{3/2}}{3\pi^2}\frac{m^{3/2}}{\hbar^3}(kT)^{5/2}$$

ist. Die Grenze zwischen nichtentartetem und relativistisch entartetem Gebiet erhalten wir entsprechend bei

$$p = \frac{2^6}{3\pi^2}\frac{1}{\hbar^3\,c^3}(kT)^4,$$

die Grenze zwischen den beiden entarteten Gebieten bei

$$p = \frac{5^4}{2^{10}\,3\pi^2}\frac{m^4\,c^5}{\hbar^3}.$$

Der „Tripelpunkt" zwischen den drei Gebieten liegt also bei

$$p = \frac{5^4}{2^{10}\,3\pi^2}\frac{m^4\,c^5}{\hbar^3} = 0,021\cdot\frac{m^4\,c^5}{\hbar^3} = 2,9\cdot 10^{17}\ \text{Atm}$$
$$T = \frac{5}{2^4}\frac{mc^2}{k} = 0,31\cdot\frac{mc^2}{k} = 1,8\cdot 10^9\ \text{Grad}$$
$$\varrho = \frac{5^3\,\mu}{2^6\,3\pi^2}\frac{Mm^3\,c^3}{\hbar^3} = 0,067\cdot\mu\frac{Mm^3\,c^3}{\hbar^3} = \mu\cdot 2,0\cdot 10^6\ \frac{\text{gr}}{\text{cm}^3}$$

in der Nähe der durch $\hbar, c, m, k$ definierten Einheiten (Tabelle 1).

Als Grenze zwischen Kondensat und entartetem Elektronengas (Abschnitt 5) erhalten wir

$$p = \frac{3^{2/3} \, \pi^{4/3}}{5 \mu^{5/3}} \left(\frac{\varrho_k}{M m^3 \, e^6 / \hbar^6}\right)^{5/3} \frac{m^4 \, e^{10}}{\hbar^8} = \frac{1,9}{\mu^{5/3}} \left(\frac{\varrho_k}{M m^3 \, e^6 / \hbar^6}\right)^{5/3} \frac{m^4 \, e^{10}}{\hbar^8} \, .$$

Diese Grenze bestimmt zusammen mit der Grenze gegen das nichtentartete Elektronengas einen weiteren „Tripelpunkt"

$$T \;=\; \frac{3^{2/3} \, \pi^{4/3}}{5 \mu^{2/3}} \left(\frac{\varrho_k}{M m^3 \, e^6 / \hbar^6}\right)^{2/3} \frac{m e^4}{\hbar^2 \, k} = \frac{1,9}{\mu^{2/3}} \left(\frac{\varrho_k}{M m^3 \, e^6 / \hbar^6}\right)^{2/3} \frac{m e^4}{\hbar^2 \, k}$$

$$\varrho \;=\; \varrho_k$$

in der Nähe der durch $\hbar, e, m, k$ definierten Einheiten. Mit $\mu = 2, \varrho_k = 8 \frac{\mathrm{gr}}{\mathrm{cm}^3} = 0,7 \cdot \frac{M m^3 e^6}{\hbar^6}$ liegt er bei $9,6 \cdot 10^7$ Atm und $3,0 \cdot 10^5$ Grad.

Die hier berechneten Grenzen sind in Abb. 2 eingetragen zusammen mit dem früher angegebenen Gebiet der Ionisierung des Gases aus Atomen.

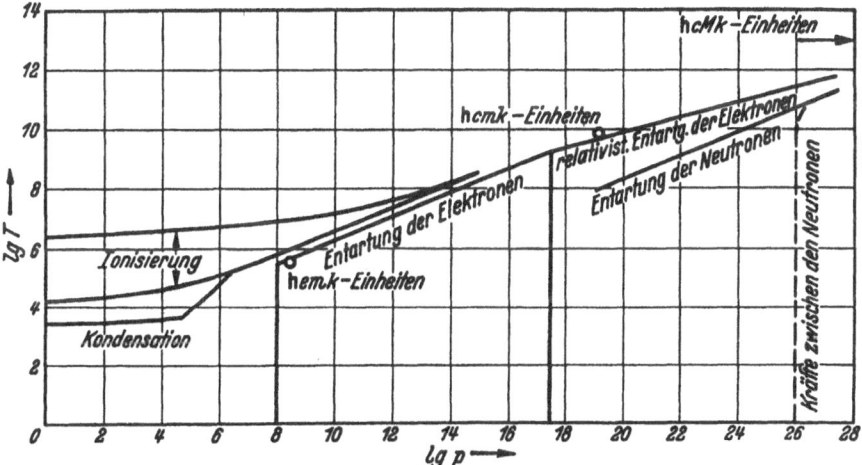

Abb. 2. Wichtige Grenzlinien im p-T-Zustandsdiagramm (logarithmische Skalen).

Man sieht, daß zwischen der berechneten Ionisierungsgrenze und der Entartungsgrenze des Elektronengases (sie unterscheiden sich für hohe Drucke um einen Faktor $\frac{10^{3/2}}{3\pi^{1/2}} = 6,0$ in p) ein schmaler Raum bleibt. In Wirklichkeit haben wir also dort ein Übergangsgebiet zwischen entartetem Elektronengas und teilweise ionisiertem Atomgas. Da wenig ionisiertes Atomgas bei diesen Drucken nicht mehr als ideales Gas angesehen werden kann, haben wir ziemlich verwickelte Verhältnisse. Bei stärkerer, aber nicht vollkommener Ionisation kann man das Ganze als ein Elektronengas mit hochionisierten Atomresten darin ansehen, was gegenüber dem Elektronengas mit Kernen eine kleine Vergrößerung von μ bedeutet. Auch im entarteten Gebiet selbst bezeichnet eine geringe Zunahme von μ bei Annäherung an den gewöhnlichen kondensierten Zustand die Verhältnisse besser als konstantes μ.

7. Das Neutronengas. Das Gebiet, wo die Materie als Elektronengas angesehen werden kann, ist nach der Seite hoher Drucke und Temperaturen dadurch begrenzt, daß schließlich die Kerne umgewandelt werden. Zur Berechnung dieser Verhältnisse brauchen wir die Eigenschaften eines Gases, das aus Elektronen, Kernen, Protonen und Neutronen besteht. Die Eigenschaften lassen sich aus den Angaben des Abschnittes 6 unter Berücksichtigung der anderen Massen entnehmen. Für ein Gas, das nur aus Neutronen besteht, erhalten wir im nichtrelativistischen nichtentarteten Zustand

$$p \;=\; nkT \qquad\qquad n = \frac{\varrho}{M}$$

$$\Phi \;=\; nV \cdot kT \log \frac{2^{1/2}\,\pi^{3/2}\,\hbar^3\,p}{M^{3/2}\,(kT)^{5/2}}\,,$$

im nichtrelativistisch entarteten Zustand

$$p \;=\; \frac{3^{2/3}\,\pi^{4/3}}{5}\frac{\hbar^2}{M}n^{5/3}$$

$$\Phi \;=\; \frac{5}{2}pV\,.$$

Ersetzen wir wieder das Übergangsgebiet durch eine Grenze, wo die eine Gleichung stetig an die andere grenzt, so ist diese Grenze

$$p = \frac{5^{3/2}}{3\pi^2}\frac{M^{3/2}}{\hbar^3}(kT)^{5/2}\,;$$

sie ist in Abb. 2 eingezeichnet. Wir werden nachher sehen, daß es ein Gebiet sehr hoher Drucke gibt, wo die Materie sich wie ein Neutronengas verhält.

Bei so hohen Drucken haben wir darauf zu achten, ob die Voraussetzung des idealen Gases nicht dadurch ungültig wird, daß der Beitrag der Kräfte zwischen den Teilchen nicht mehr unmerklich klein ist. Wir wissen aus dem Bau der Kerne, daß zwischen Protonen und Neutronen Kräfte wirken, wenn diese Teilchen sich auf etwa 10^{-13} cm nahe kommen, und es gibt auch Gründe für die Annahme, daß bei diesen Abständen Kräfte zwischen Neutronen und Neutronen auftreten. Um vorsichtig zu sein, wollen wir unser Gasgebiet da abschließen, wo die Teilchen sich auf solche Entfernungen nahe kommen, beim Neutronengas also mit

$$\left(\frac{M}{\varrho}\right)^{1/3} \approx 10^{-13}\,\mathrm{cm} \quad\text{bis}\quad 10^{-12}\,\mathrm{cm},$$

das entspricht etwa 10^{24} bis 10^{28} Atmosphären Druck. Über das Verhalten bei etwas höheren Drucken wissen wir zunächst nichts, wegen unserer geringen Kenntnis der Kräfte zwischen den Teilchen. Bei extrem hohen Drucken muß aber schließlich die Energiedichte groß werden gegen die Dichte der Ruheenergie nMc^2, so daß wir den bisher als Energie E gerechneten Überschuß über die Ruheenergie gleich der gesamten relativistischen Energie setzen können

$$\frac{E}{V} = \varrho c^2$$

(bei einer Wägung würde ja die gesamte Energie gemessen). Wenn alle Teilchen sich ungefähr mit Lichtgeschwindigkeit bewegen, so ist der Druck

$$p = \frac{1}{3}\frac{E}{V} = \frac{1}{3}\varrho c^2.$$

In diesem Gebiet kennen wir also wieder die Zustandsgleichung, sie ist für jede Art Materie dieselbe [J.v.NEUMANN, in (7)]. Es ist aber zweifelhaft, ob dieses Ergebnis noch einen Sinn hat, da man so hohe Drucke nur mit großen gravitierenden Massen erzeugen kann. Es würde dabei die Energie der Teilchen im Gravitationsfeld von der Größenordnung ihrer Ruheenergie (vgl. Abschnitt 2 und 17).

8. Das Gebiet der Kernumwandlungen. Wenn unter verschiedenen möglichen Zuständen einer ist, dessen Energie tiefer und dessen statistisches Gewicht größer ist als bei den andern, so tritt er im thermodynamischen Gleichgewicht auf. Ebenso wenn er trotz etwas höherer Energie sehr viel größeres statistisches Gewicht oder trotz etwas geringerem statistischem Gewicht sehr viel tiefere Energie hat.

Auf Grund unserer Kenntnis vom Kernbau müssen wir annehmen, daß ein schwerer Kern durch Abgabe von Protonen oder Neutronen in einen leichteren übergehen kann oder ganz in Protonen und Neutronen zerlegt werden kann. Von anderen Prozessen wollen wir zunächst absehen. Bei tiefen Temperaturen kommt es nur auf die Energien an, im thermodynamischen Gleichgewicht kommen also nur die Kerne tiefster Energie vor. Das sind Kerne von Atomgewichten in der Gegend von 100. Daß in Wirklichkeit die Materie bei tiefen Temperaturen auch andere Kerne enthält, ist ein Zeichen der Abweichung vom thermodynamischen Gleichgewicht. In der Tat gibt es bei tiefen Temperaturen keinen Prozeß, der das Gleichgewicht herstellen hilft. Bei hohen Temperaturen wird das statistische Gewicht eines (elektrisch neutralen) Gases aus Protonen, Neutronen und Elektronen zunehmend größer im Vergleich zum statistischen Gewicht eines Gases aus schwereren Kernen und Elektronen. Dieser Vorteil des Gases aus leichteren Teilchen überwiegt schließlich den Nachteil der höheren Energie. Bei hohen Temperaturen werden schließlich im thermodynamischen Gleichgewicht keine schweren Kerne mehr vorkommen.

Aus der Erfahrung des β-Zerfalls von Kernen mußte man schließen, daß sich Protonen in Neutronen umwandeln können unter Aufnahme von Elektronen oder Abgabe von Positronen. Da ein Neutron höhere Energie hat als ein Proton plus Elektron, kommt für eine Betrachtung des thermodynamischen Gleichgewichtes zunächst der Prozeß

$$\text{Proton} + \text{Elektron} \rightleftharpoons \text{Neutron}$$

in Betracht. Bei höheren Temperaturen (wo keine schweren Kerne existieren) ist für die Protonen plus Elektronen die höhere thermische Energie (wegen der doppelten Teilchenzahl) ungünstig, die geringere Eigenenergie und das größere statistische Gewicht günstig. Bei tieferen Temperaturen, wo es nur auf die Energie ankommt, ist die hohe Nullpunktsenergie der Elektronen ungünstig. Es kann bei hohen Drucken günstig sein, wenn die vorhandenen Elektronen mit den vorhandenen Kernen sich zu Neutronen umsetzen [STERNE (40)].

Eine genaue Durchrechnung der Verhältnisse ist sehr verwickelt, da wir eigentlich alle möglichen Kernarten zu berücksichtigen hätten. Wir werden aber das Wesentliche richtig treffen, wenn wir außer Protonen und Neutronen nur eine Art Kerne annehmen, sie sollen aus Z Protonen und Z Neutronen bestehen. Wir schematisieren also durch Annahme der möglichen Reaktionen

$$\text{Kern} \rightleftharpoons Z \text{ Neutronen} + Z \text{ Protonen}$$
$$\text{Proton} + \text{Elektron} \rightleftharpoons \text{Neutron}$$

und damit auch der Reaktionen

$$\text{Kern} \rightleftharpoons 2 Z \text{ Protonen} + Z \text{ Elektronen}$$
$$\text{Kern} + Z \text{ Elektronen} \rightleftharpoons 2 Z \text{ Neutronen.}$$

Da wir die Umwandlungsprozesse in Gebieten verschiedener Zustandsgleichungen der Teilchen erwarten, haben wir unter Weglassung weniger wichtiger Fälle die Rechnung für folgende Fälle durchzuführen:

a) Elektronen und schwere Teilchen nichtrelativistisch nichtentartet;

b) Elektronen und schwere Teilchen nichtentartet, Elektronen relativistisch, schwere Teilchen nichtrelativistisch;

c) Elektronen relativistisch entartet, schwere Teilchen nichtrelativistisch nichtentartet;

d) Elektronen relativistisch entartet, schwere Teilchen nichtrelativistisch entartet.

Im Falle a) ist bei gegebenen Werten von p und T das Minimum des thermodynamischen Potentials

$$\Phi = N_E \left[k T \log \frac{2^{1/2} \pi^{3/2} \hbar^3 p_E}{m^{3/2} (kT)^{5/2}} - Q \right] + N_P \cdot k T \log \frac{2^{1/2} \pi^{3/2} \hbar^3 p_P}{M^{3/2} (kT)^{5/2}} +$$

$$+ N_N \cdot k T \log \frac{2^{1/2} \pi^{3/2} \hbar^3 p_N}{M^{3/2} (kT)^{5/2}} + N_K \left[k T \log \frac{\pi^3 \hbar^6 p}{M_K^3 a^3 (k T)^4} - 2 Z R \right]$$

aufzusuchen. Dabei sind N_E, N_P, N_N, N_K die Anzahlen der Elektronen, Protonen, Neutronen und Kerne, p_E, p_P, p_N, p_K die entsprechenden Partialdrucke, M_K die Kernmasse und a eine Länge von Kerngröße. Im letzten Glied ist berücksichtigt, daß die Kerne bei den in Betracht kommenden Temperaturen rotieren; doch macht dies praktisch nichts aus. Um Q ist die Ruheenergie eines Neutrons höher als die eines Protons plus Elektrons; um $2ZR$ ist die Ruheenergie von Z Neutronen und Z Protonen höher als die eines Kernes. Man erhält für die Anzahlen der einzelnen Teilchen Gleichungen von der Art, wie sie bei gewöhnlichen chemischen Reaktionen gelten. Für Temperaturen $T \geq \frac{R}{k}$ erhält man keine Kerne mehr. Für nicht zu hohe Drucke erhält nur Protonen und Elektronen, für hohe Drucke (und nicht zu hohe Temperaturen) nur Neutronen. Im Falle b) ist das Ergebnis ein ähnliches, wenn man von extrem hohen Drucken absieht. Im Falle c) erstreckt sich das Gebiet der Neutronen bei hohen Drucken auch zu tiefen Temperaturen hinunter. Den interessanten Fall d) wollen wir ausführlicher behandeln. Es ist hier einfacher V und T als unabhängige Variable anzusehen. Dann ist das thermodynamische Gleichgewicht dadurch bestimmt, daß die freie Energie, die hier gleich der Energie ist, ein Minimum hat. Es ist

$$\frac{E}{V} = \frac{3^{4/3} \pi^{2/3}}{2^2} \hbar c n_E^{4/3} - Q n_E + \frac{3^{5/3} \pi^{4/3} \hbar^2}{2 \cdot 5} \left[\frac{n_P^{5/3}}{M} + \frac{n_N^{5/3}}{M} + \frac{n_K^{5/3}}{M_K} \right] - 2 Z R n_K;$$

dabei ist $n_E = n_P + Z n_K$. Beim Aufsuchen des Minimums ist $n_P + n_N + 2 Z n_K$ konstant zu halten. Bei geringeren Werten dieser Zahl ist im Minimum $n_P = n_N = 0$. Neutronen treten zum erstenmal auf, wenn

$$n_E = \frac{1}{3\pi^2} \frac{1}{\hbar^3 c^3} (2R + Q)^3$$

ist, bei noch größerer Dichte verschwinden Kerne und Elektronen rasch. Die Umwandlung der Materie in Neutronen geschieht ziemlich plötzlich bei dem Druck

$$p = \frac{1}{12\pi^2} \left(\frac{2R + Q}{mc^2} \right)^4 \cdot \frac{m^4 c^5}{\hbar^3}.$$

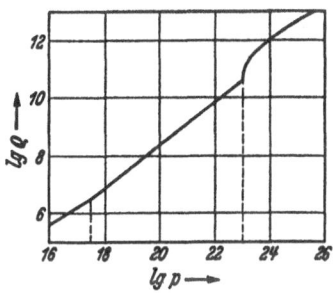

Abb. 3. Verlauf der Dichte in der Nähe der Umwandlung in Neutronen.

Aus den bekannten Atomgewichten folgt R zu etwa $0,008 \cdot Mc^2$ oder $15 \, mc^2$. Der Massenunterschied von Neutron und Proton ist nur ganz unsicher bekannt. Mit dem „Atomgewicht" des Neutrons $1,0085 \pm 0,0005$ (3, 15) und des Protons plus Elektrons $1,0080$ erhalten wir $Q \approx mc^2$ mit einer Unsicherheit, die ebenso groß ist. Wir erhalten so die Umwandlung in Neutronen bei einem Druck von etwa $1 \cdot 10^{23}$ Atmosphären, also bei einem Druck, wo man Elektronen oder Neutronen noch als ideales Gas betrachten kann. Die Abhängigkeit der Dichte vom Druck für die Gleichgewichtsmischung

$$p = \frac{3^{1/3} \pi^{2/3}}{4} \hbar c n_E^{4/3} + \frac{3^{2/3} \pi^{4/3} \hbar^2}{5} \frac{n_N^{5/3}}{M}$$

mit der aus der Gleichgewichtsbedingung folgenden Beziehung zwischen n_E und n_N und

$$\varrho = \mu M n_E + M n_N; \qquad \mu = 2,0$$

gibt Abb. 3 an. Bei Drucken unterhalb der Umwandlung in Neutronen ist die Materie ziemlich kompressibel (aber weniger als das ideale nichtentartete Gas), bei der Umwandlung selbst ist sie ein Stück weit viel stärker kompressibel, um bei noch höheren Drucken etwas weniger kompressibel zu werden als vorher.

Unter Benutzung der vier gerechneten Fälle läßt sich der wirkliche Verlauf der Grenzen zwischen den Gebieten der Elektronen und Kerne, der Elektronen und Protonen, der Neutronen mit geringer Unsicherheit angeben. Die Gebiete sind in Abb. 4 angegeben. Eine Änderung von Q würde die Figur kaum beeinflussen. Für hohe Temperaturen haben wir noch Ergänzungen anzubringen; der Strahlungsdruck darf nicht weggelassen werden, ebenso darf man die Möglichkeiten, Protonen in Neutronen und Positronen umzuwandeln oder aus Strahlung Elektronen, Positronen zu erzeugen, nicht ganz vergessen.

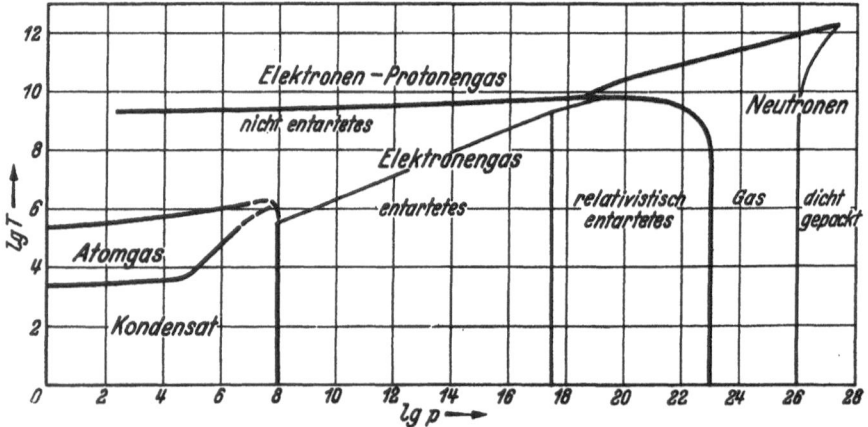

Abb. 4. Zustandsdichte der Materie.

Wir haben bisher das thermodynamische Gleichgewicht betrachtet, ohne auf die Prozesse zu achten, die es herstellen helfen. Bei tiefen Temperaturen wandeln sich erfahrungsgemäß Kerne nicht um; Abweichungen vom Gleichgewicht können sich also (praktisch) unendlich lange halten. Bei den hohen Temperaturen und Drucken, wo das Gleichgewicht auf der Seite der Protonen oder Neutronen liegt, haben aber alle Teilchen so hohe Energien, daß Umwandlungsprozesse häufig stattfinden können. Dort wird also das Gleichgewicht auch hergestellt werden.

9. Abweichungen vom thermodynamischen Gleichgewicht. Bei tiefen Temperaturen und geringen Drucken gibt es keinen Prozeß, der nichtradioaktive Kerne umwandelt. Abweichungen vom thermodynamischen Gleichgewicht in der Zusammensetzung der Kerne, wie das Vorkommen leichter Kerne, können nicht beseitigt werden. Der Grund dafür, daß leichte Kerne sich nicht zu schweren vereinigen, ist die COULOMBsche Abstoßung der Kerne, die verhindert, daß sie sich zu nahe kommen. Erst bei großen Geschwindigkeiten der Teilchen wird das anders. Für ein einfaches Modell des Kernes haben GAMOW (20) sowie ATKINSON und HOUTERMANS (2) die Wahrscheinlichkeit des Eindringens eines geladenen Teilchens in einen schweren Kern der Ladung $Z e$ berechnet. Für den zunächst in Betracht kommenden Geschwindigkeitsbereich wird die Eindringungswahrscheinlichkeit eines Protons pro Stoß

$$\mathrm{e}^{-2Z\frac{e^2}{\hbar v}} = \mathrm{e}^{-\sqrt{2}Z\sqrt{\frac{Me^4/\hbar^2}{\epsilon}}},$$

wo v die Geschwindigkeit des Protons ($\frac{e^2}{\hbar}$ die atomare Geschwindigkeitsein-
heit) oder ϵ seine kinetische Energie ist. Wenn diese Energie das $Z^2 \frac{M}{m}$-fache
der Einheit $\frac{me^4}{\hbar^2}$ oder (falls der Vorgang im Temperaturgleichgewicht stattfindet) wenn die Temperatur das $Z^2 \frac{M}{m}$-fache der Einheit $\frac{me^4}{\hbar^2 k}$ ($3 \cdot 10^{-5}$ Grad)
erreicht, nähert sich die Wahrscheinlichkeit der Eins. Bei einer Temperatur
von $4 \cdot 10^7$ Grad (die als im Innern der Sterne herrschend angenommen wird)
erhalten ATKINSON und HOUTERMANS als mittlere (unter Berücksichtigung
der MAXWELLschen Geschwindigkeitsverteilung) Eindringungswahrscheinlichkeit eines Protons pro Stoß mit einem Li-Kern 10^{-10}, für einen Ne-Kern
nur noch 10^{-24}. Ein eingedrungenes Proton braucht nicht notwendig zur Umwandlung des Kernes in einen schwereren zu führen. Die experimentellen
Ergebnisse über Umwandlung leichter Elemente durch Beschießung mit Protonen scheinen aber zu zeigen, daß die Umwandlungswahrscheinlichkeit nicht
viel kleiner ist als die gerechnete Eindringungswahrscheinlichkeit.

Bei Temperaturen um $4 \cdot 10^7$ Grad würden also Protonen neben Li-Kernen
nur in äußerster Verdünnung längere Zeit bestehen können; neben Ne-Kernen
könnten sie in Dichten der Größenordnung $1\,\mathrm{gr/cm}^3$ über Zeiträume existieren, die mit der Lebensdauer eines Sternes vergleichbar sind. Auf die Bedeutung dieser Umstände für den Energieinhalt und für Energieumsetzungen
in Materie, die nicht im thermodymanischen Gleichgewicht ist, und für die
Energieerzeugung in Sternen kommen wir nachher zurück.

10. Die Strahlung. Ein mit Materie gefülltes oder leeres Volumen enthält auch eine Strahlung. Beim Temperaturgleichgewicht hat sie im Vakuum
die Energiedichte

$$\frac{E}{V} = \frac{\pi^2}{15} \frac{(kT)^4}{\hbar^3 c^3} = aT^4.$$

Die Strahlung übt einen Druck

$$p = \frac{1}{3}\frac{E}{V} = \frac{\pi^2}{45} \frac{(kT)^4}{\hbar^3 c^3} = \frac{a}{3}T^4$$

aus. Die Energie ist nach dem PLANCKschen Gesetz auf die verschiedenen
Frequenzen verteilt. Das Maximum der Energie im Wellenlängenintervall liegt
bei der Wellenlänge

$$\lambda = \frac{2\pi}{4,965} \frac{\hbar c}{kT},$$

für die Temperatur $\frac{mc^2}{k}$ ($5,9 \cdot 10^9$ Grad) also in der Nähe der „COMPTON-
Wellenlänge" $\frac{2\pi\hbar}{mc}$ ($2,4 \cdot 10^{-10}$ cm).

Ist das betrachtete Volumen mit Materie erfüllt, so ist die Trennung in
Energie der Strahlung und Energie der Materie nur bei geringer Wechselwirkung zwischen Strahlung und Materie durchführbar. Wenn die Materie z. B.
nicht absorbiert und ihr optisches Verhalten durch eine Lichtgeschwindigkeit
v oder einen Brechungsindex $n = \frac{c}{v}$ beschrieben werden kann, so ist

$$\frac{E}{V} = \frac{\pi^2}{15} \cdot \frac{c^3}{v^3} \cdot \frac{(kT)^4}{\hbar^3 c^3}.$$

In einem Atomgas ist $n^2 - 1$ von der Größenordnung des Verhältnisses des von den Atomen eingenommenen Volumens zum Gesamtvolumen, also klein. Auch in einem Gas aus Elektronen und Kernen ist die Wechselwirkung von Strahlung und Materie klein, solange es nicht entartet ist, und solange die Temperatur nicht in die Nähe von $\frac{mc^2}{k}$ kommt.

Ziehen wir in unserem p-T-Diagramm (Abb. 2 und 4), das bisher nur den von der Materie herrührenden Druck angibt, die Gerade

$$p = \frac{(kT)^4}{\hbar^3 c^3},$$

so kommt auf dieser Geraden zu dem angegebenen Druck der Materie noch ein Strahlungsdruck, der etwa $1/5$ davon ist. Unterhalb dieser Geraden kann der Strahlungsdruck vernachlässigt werden (da n mit T^4 geht, der Materiedruck höchstens mit T). An der fast auf dieser Geraden liegenden Grenze zwischen nichtentartetem und relativistisch entartetem Elektronengas

$$p = \frac{64}{3\pi^2} \frac{(kT)^4}{\hbar^3 c^3}$$

ist das Verhältnis von Strahlungsdruck zu Materiedruck

$$\frac{p_{\text{rad}}}{p_{\text{mat}}} = \frac{\pi^4}{15 \cdot 64} \approx \frac{1}{10},$$

wenn man dort wie im Vakuum rechnet. Oberhalb der Geraden

$$p = \frac{(kT)^4}{\hbar^3 c^3}$$

überwiegt bald der Strahlungsdruck den Materiedruck. Geben wir im p-T-Diagramm den Gesamtdruck an, so wird das Diagramm durch die Gerade

$$p = \frac{\pi^2}{45} \frac{(kT)^4}{\hbar^3 c^3}$$

begrenzt (Abb. 5), da bei gegebener Temperatur der Gesamtdruck nicht kleiner sein kann. Bei Annäherung an die Grenze nimmt die Dichte der Materie rasch ab. Die Energie der Strahlung hat aber auch eine Masse und damit eine Massendichte:

$$\varrho = \frac{1}{c^2} \frac{E}{V} = \frac{3p}{c^2};$$

das strahlende Vakuum hat also dieselbe Zustandsgleichung wie die relativistische Materie. In dem Gebiet, wo die schweren Materieteilchen Geschwindigkeiten wesentlich unter der Lichtgeschwindigkeit haben, ist der Gesamtdruck

$$p = p_{\text{rad}} + p_{\text{mat}} = \frac{c^2}{3} \varrho_{\text{rad}} + \frac{kT}{\mu M} \varrho_{\text{mat}};$$

in dem Gebiet, wo die schwereren Teilchen im wesentlichen Lichtgeschwindigkeit haben, ist

$$p = p_{\text{rad}} + p_{\text{mat}} = \frac{c^2}{3}(\varrho_{\text{rad}} + \varrho_{\text{mat}}) = \frac{c^2}{3}\varrho.$$

Das Gebiet, wo Strahlungsdruck und Materiedruck vergleichbar sind, ist in unserem p-T-Diagramm sehr schmal. Damit verliert auch die Schwierigkeit an Bedeutung, daß wir bei starker Wechselwirkung von Strahlung und Materie über die Energiedichte der Strahlung nichts Einfaches angeben können.

Nach der Entdeckung des Positrons hat die Folgerung aus der DIRACschen Theorie des Elektrons an Wahrscheinlichkeit zugenommen, daß aus Strahlung genügend hoher Frequenz

$$\hbar\omega \geq 2mc^2; \qquad \lambda \leq \pi\frac{\hbar}{mc}$$

Positronen und Elektronen entstehen können. Unter Berücksichtigung dieser Umwandlungsmöglichkeit gibt es im thermodynamischen Gleichgewicht bei tieferen Temperaturen neben Strahlung und Elektronen keine Positronen; bei höheren Temperaturen bilden sich Paare von Positronen und Elektronen aus Strahlung. Prozesse, die das Gleichgewicht herstellen, gibt es. Das thermodynamische Potential Φ der reinen Strahlung ergibt sich aus der Entropie

$$S = \frac{4}{3}aT^3V$$

zu Null. Wenn N_+ und N_- die Anzahlen der Positronen und der Elektronen sind, und p_+ und p_- die Partialdrucke, so gilt für das Gesamtsystem

$$\begin{aligned}
\Phi \;=\; & N_+\left[mc^2 + kT\log\frac{2^{1/2}\,\pi^{3/2}\,\hbar^3\,p_+}{m^{3/2}\,(kT)^{5/2}}\right] + \\
& + N_-\left[mc^2 + kT\log\frac{2^{1/2}\,\pi^{3/2}\,\hbar^3\,p_-}{m^{3/2}\,(kT)^{5/2}}\right].
\end{aligned}$$

Im thermodynamischen Gleichgewicht ist also

$$p_+p_- = \frac{m^3}{2\pi^3\hbar^6}(kT)^5\,\mathrm{e}^{-\frac{2mc^2}{kT}},$$

d.h. Paarerzeugung tritt wesentlich auf, wenn

$$\frac{p}{(kT)^{5/2}} > \frac{m^{3/2}}{2^{1/2}\,\pi^{3/2}\,\hbar^3}\mathrm{e}^{-\frac{mc^2}{kT}}$$

ist. Wir vergleichen den durch die Paarerzeugung bedingten Druck

$$p_{\text{mat}} = \frac{m^{3/2}}{2^{1/2}\,\pi^{3/2}\,\hbar^3}(kT)^{5/2}\mathrm{e}^{-\frac{mc^2}{kT}}$$

mit dem Strahlungsdruck

$$p_{\text{rad}} = \frac{\pi^2}{45} \frac{(kT)^4}{\hbar^3 c^3}.$$

Das Verhältnis

$$\frac{p_{\text{mat}}}{p_{\text{rad}}} = \frac{45}{2^{1/2} \pi^{7/2}} \left(\frac{m c^2}{kT}\right)^{3/2} e^{-\frac{mc^2}{kT}}$$

beträgt im Maximum etwa 1/10, so daß die Paarerzeugung an unserem Gesamtbild kaum etwas ändert. Andeutungen über einen geringen Einfluß der Paarerzeugung auf die Zustandsgleichung machen CHANDRASEKHAR und ROSENFELD (8).

Für einen hypothetischen Vorgang der Erzeugung von positiven und negativen Teilchen von Protonenmasse können wir entsprechend das Gleichgewicht rechnen. Solche Paare kämen erst dann wesentlich vor, wenn $kT > Mc^2$ geworden ist.

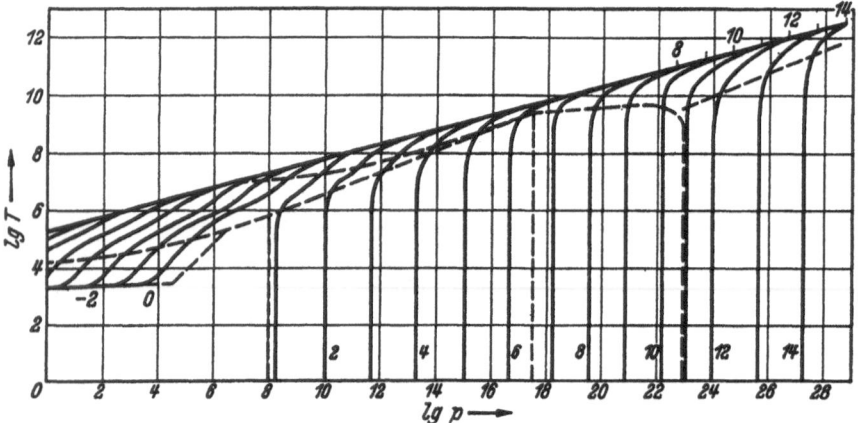

Abb. 5: Dichte der Materie im p-T-Diagramm (mit Berücksichtigung der Strahlung).

In Abb. 5 ist die Dichte ϱ (angegeben ist der Zehnerlogarithmus der in gr/cm³ gemessenen Dichte) als Funktion der Temperatur T und des Gesamtdruckes p (aus Materiedruck und Strahlungsdruck) für Materie und Strahlung im thermodynamischen Gleichgewicht eingezeichnet. Von tieferen Drucken abgesehen, gilt die Abb. 5 für Materie schlechthin; bei tieferen Drucken sind etwa die Verhältnisse des Eisens gezeichnet.

In Abb. 6 ist die Fläche, die zusammengehörige Werte von p, T, ϱ verbindet, als Blockdiagramm perspektivisch dargestellt. Beschreiben wir das Verhalten der Materie in großen Zügen durch *Kompressibilität* und *thermischen Ausdehnungskoeffizienten*, so ist die Materie mit Ausnahme des gewöhnlichen kondensierten Zustandes ziemlich kompressibel, aber im allgemeinen weniger kompressibel als das nichtentartete ideale Gas, die thermische Ausdehnung ist in weiten Bereichen (den entarteten) fast Null, bei höheren Temperaturen ist sie die des nichtentarteten idealen Gases, in den Gebieten der Ionisierung und der Kerndissoziation ist sie noch höher und bei überwiegendem Strahlungsdruck wird sie sehr groß. Es sind im wesentlichen Kompressibilität und

thermische Ausdehnung, die die Bedeutung der einzelnen Zustandsgebiete für
das Gleichgewicht im Innern der Sterne ausmachen.

Wir übersehen ein recht großes Gebiet der Temperatur und des Druckes
dank dem Umstande, daß die Kräfte zwischen den Teilchen eine geringe Rolle
spielen, die Materie also ein ideales Gas ist. Auch die übrigen Eigenschaften
der Materie in diesem Gebiet bilden keine Rätsel mehr (nächste Abschnitte).
Das heißt nicht, daß die für das Innere der Sterne in Betracht kommenden
Vorgänge völlig geklärt wären, denn diese Vorgänge können sehr verwickelt
sein. Nur über die Grundlagen dieser Vorgänge ist in den in Betracht kommenden Gebieten kein Zweifel mehr. Das unbekannte Gebiet liegt bei höheren
Drucken und höheren Temperaturen, als sie in gewöhnlichen Sternen vorkommen, oder bei höheren Gravitationsfeldern.

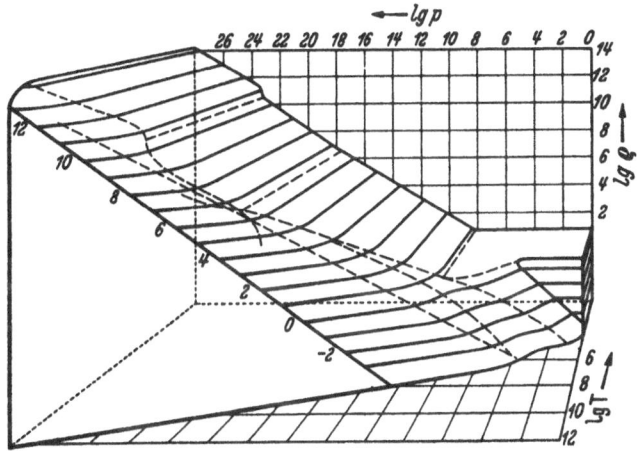

Abb. 6. Zusammenhang von p, T, ϱ.

II. Andere physikalische Eigenschaften.

11. Der Energieinhalt. Für das thermische Verhalten der Materie ist
neben der Zustandsbeziehung (Zusammenhang von Druck, Temperatur und
Dichte) besonders der Energieinhalt wichtig. In einer Darstellung des Ener-
gieinhaltes als Funktion von Druck und Temperatur werden besonders die
Gebiete hervortreten, in denen starke Umwandlungen im Aufbau der Materie
eintreten, während im Zustandsdiagramm besonders die Änderungen sichtbar
wurden, die die Elektronen betrafen.

Wir geben die Energie pro Masseneinheit an

$$\eta = \frac{E}{\varrho V}$$

(sie hat die Dimension c^2); als Nullpunkt nehmen wir die Energie des kondensierten Zustandes bei $p = 0$, $T = 0$. Unter Weglassung der komplizierten
Verhältnisse in den Übergangsgebieten erhalten wir für die einzelnen einfachen Zustände folgendes: Das Kondensat (solange es wirklich inkompressibel,
nach der Regel von DULONG-PETIT schematisiert) hat

$$\eta = \frac{3kT}{\mu M}.$$

Das Atomgas hat

$$\eta = \frac{\epsilon_1}{\mu M} + \frac{3kT}{2\mu M} \qquad \frac{\eta - \eta_1}{c^2} = \frac{3}{2\mu}\frac{m}{M}\frac{kT}{mc^2},$$

wo ϵ_1 bei sehr schwer flüchtigen Stoffen Werte zwischen 1/10 und 1/5 der atomaren Einheit $\frac{me^4}{\hbar^2}$ oder Werte von etwa $\frac{1}{2}10^{-5}$ bis 10^{-5} der Einheit mc^2 annehmen kann. Das nichtrelativistische, nichtentartete Elektronengas hat

$$\eta = \frac{\epsilon_2}{\mu M} + \frac{3kT}{2\mu M},$$

wo μ jetzt nahe bei 2 liegt (außer bei Wasserstoff) und ϵ_2 die Größenordnung $Z^{4/3} \cdot \frac{me^4}{\hbar^2}$ hat [die statistische Behandlung der Elektronenhülle mit der Methode von THOMAS und FERMI gilt nach MILNE (30), davor den Zahlenfaktor 0,77; die Verdampfungsenergie ϵ_1 kann dagegen außer Betracht bleiben]. Für das nichtrelativistische, entartete Elektronengas gilt

$$\eta = \frac{\epsilon_2}{\mu M} + \frac{3^{7/5}\,\pi^{4/5}}{2 \cdot 5^{3/5}\,\mu}\frac{\hbar^{6/5}}{M\,m^{3/5}}p^{2/5} \qquad \frac{\eta - \eta_2}{c^2} = \frac{3^{7/5}\,\pi^{4/5}}{2 \cdot 5^{3/5}\,\mu}\frac{m}{M}\left(\frac{p}{m^4\,c^5/\hbar^3}\right)^{2/5}$$

mit dem gleichen ϵ_2-Wert, für das relativistisch nichtentartete

$$\eta = \frac{\epsilon_2}{\mu M} + \frac{3kT}{\mu M},$$

und für das relativistisch entartete

$$\eta = \frac{\epsilon_2}{\mu M} + \frac{3^{5/4}\,\pi^{1/2}}{2^{3/2}\,\mu}\frac{(\hbar c)^{3/4}}{M}p^{1/4} \qquad \frac{\eta - \eta_2}{c^2} = \frac{3^{5/4}\,\pi^{1/2}}{2^{3/2}\,\mu}\frac{m}{M}\left(\frac{p}{m^4\,c^5/\hbar^3}\right)^{1/4}.$$

Für das nichtrelativistische Neutronengas erhalten wir

$$\eta = \frac{\epsilon_3}{M} + \frac{3kT}{2M} \qquad \frac{\eta - \eta_3}{c^2} = \frac{3}{2}\frac{kT}{Mc^2}$$

und

$$\eta = \frac{\epsilon_3}{M} + \frac{3^{7/5}\,\pi^{4/5}}{2 \cdot 5^{3/5}}\frac{\hbar^{6/5}}{Mm^{3/5}}p^{2/5},$$

wo ϵ_3 unser früheres $\frac{1}{2}(2R + Q)$ ist, die Umwandlungsenergie der Kerne und Elektronen in Neutronen für ein Neutron (die Größe ϵ_2 kann dagegen außer Betracht bleiben); es ist also η_3 etwa $0,008\,c^2$. Für das nichtentartete Protonen-Elektronengas ist entsprechend

$$\eta = \frac{\epsilon_4}{\mu M} + \frac{3kT}{2\mu M}$$

mit $\mu = \frac{1}{2}$ und $\epsilon_4 = \frac{1}{2}(2R - Q)$. Für die vollkommen relativistische Materie wie für die Strahlung im Vakuum gilt schließlich

$$\eta = c^2.$$

Abb. 7 gibt die Energie pro Gramm als Funktion von p und T an; die Zahlen bedeuten Zehnerlogarithmen von η, dieses gemessen in $10^7 \frac{\text{erg}}{\text{gr}}$. Am Übergang vom Kondensat zum entarteten Elektronengas sind die Energien nicht genau angebbar. Die idealisierte Zustandsgleichung des Abschnittes 5 gäbe bis zur Übergangsstelle keine Druckabhängigkeit der Energie, dann ein rasches Ansteigen, das sich schließlich dem oben für das Elektronengas angegebenen Verlauf anschmiegt. Die dichte Folge der η-Kurven in Abb. 7 an dieser Stelle kommt von der hierfür künstlichen Verwendung des logarithmischen Maßstabes für η.

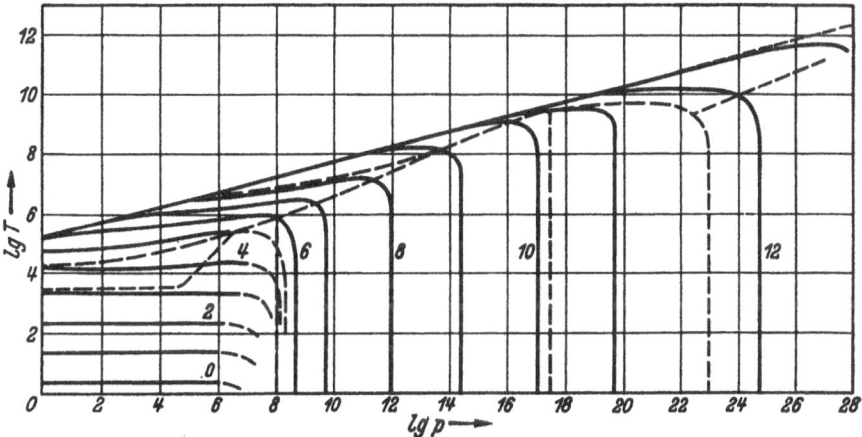

Abb. 7. Energieinhalt der Materie.

Zum Vergleich mit diesen Werten der Energie pro Gramm sei angegeben, daß die Sonne fast $2 \frac{\text{erg}}{\text{gr}}$ in der Sekunde ausstrahlt, die hier benutzte Einheit $10^7 \frac{\text{erg}}{\text{g}}$ also in 2 Monaten. Erniedrigt man also die in Abb. 7 angegebenen Zahlen um den Summanden $\log 6 = 0,78$, so erhält man den Zehnerlogarithmus der Anzahl Jahre, während der die Materie vom angegebenen Anfangszustand aus eine Energiestrahlung von dem gleichen Ausmaß wie die Sonnenstrahlung aufrechterhalten könnte (Gravitationsenergie nicht gerechnet). Man beachte aber, daß in der Abbildung der Energieinhalt der Materie im thermodynamischen Gleichgewicht angegeben ist. Bei Anwesenheit leichter Kerne in Abweichung von diesem Gleichgewicht ist der Energieinhalt viel größer, so geben z. B. Protonen wegen ihrer Umwandelbarkeit in schwere Kerne einen Energieinhalt pro Masseneinheit von $0,008\,c^2$ oder $7 \cdot 10^{11} \cdot 10^7 \frac{\text{erg}}{\text{gr}}$.

In unserer Übersicht über den Energieinhalt tritt die Ruheenergie der schweren Teilchen nicht auf, weil in den betrachteten Gebieten des Druckes und der Temperatur keine Umsetzungen auftreten, die diese Energie angreifen. Die größte Energie pro Masseneinheit, die auftrat, war die der Umwandlung von schweren Kernen in Protonen und Neutronen. Wir haben von Gravitationsenergie abgesehen. In den gewöhnlichen Sternen ist ihre Änderung

auch klein gegen die anderen Energieinhalte. Dies gilt nicht mehr, wenn der
Stern im Laufe seiner Entwicklung einmal sehr klein wird. Dann kann der absolute Betrag der (negativen) Gravitationsenergie eines ganzen Sternes mit
der Ruheenergie der gesamten Sternmaterie vergleichbar werden. Da hört
dann die Möglichkeit der Abtrennung der Gravitation von der Betrachtung
auf, und wir kommen in unbekanntes Gebiet.

12. Elektrizitäts- und Wärmeleitung. Im Zustand des Atomgases
ist die Materie ein Isolator. Im Zustande des Metalls oder Elektronengases
enthält sie bewegliche Elektronen, die die Leitung des elektrischen Stromes
und des Wärmestromes übernehmen können. Wären die Elektronen ganz frei,
so hätten wir eine unendlich gute Leitfähigkeit. Ein Widerstand für Elektri-
zität und Wärme kommt von der Streuung der Elektronen an den Kraftfeldern der Kerne. Von der Wechselwirkung der Elektronen miteinander, soweit
sie sich nicht durch ein statisches Kraftfeld berücksichtigen läßt, das man zu
dem der Kerne schlägt, scheint man nach den Erfahrungen mit der Quantentheorie der metallischen Leitung absehen zu können (außer bei ganz tiefen
Temperaturen, wie die Erscheinung der Supraleitung zeigt).

Wir betrachten zunächst die *elektrische Leitfähigkeit*. Sie läßt sich in der
Form

$$\sigma = \frac{e^2 n l}{m v}$$

schreiben, wo l (für nicht zu tiefe Temperaturen) die Bedeutung einer mitt-
leren freien Weglänge der Leitungselektronen zwischen zwei Streuprozessen
hat, v die mittlere Geschwindigkeit der Elektronen ist und n ihre Anzahl in
der Raumeinheit. Diese Weglänge ist nun nicht etwa durch die Querschnitte
πr^2 der Kernkraftfelder begrenzt, dann wäre

$$l = \frac{1}{\pi r^2 n_K}$$

(n_K Anzahl der Kerne in der Raumeinheit). Vielmehr begrenzt eine vollkommen regelmäßige Gitteranordnung von Kernen die Weglänge überhaupt
nicht und gibt unendliche Leitfähigkeit (5, 38). Die Kernkraftfelder wirken
nur durch ihre Abweichung von der Gittersymmetrie; es liegt nahe, die Querschnitte

$$\pi r^2 \frac{d^2}{a^2}, \qquad \frac{1}{a^3} = n_K,$$

wo d die mittlere Abweichung vom Gitterpunkt und a der Abstand der Git-
terpunkte ist, zur Begrenzung der Weglänge zu benutzen. Wenn die Abweichungen vom Gitter sehr groß sind, also bei ganz ungeordneten Lagen der
Kerne, dürfte πr^2 die richtige Begrenzung geben. Wir erhalten durch diese
rohe Betrachtung

$$l = \frac{1}{\pi r^2 d^2 n_K^{5/3}} \qquad\qquad d \le \frac{1}{n_K^{1/3}}.$$

Haben wir einen idealen Kristall, wo die Abweichungen d der Kerne von den
Gitterpunkten nur durch die Wärmeschwingungen bedingt sind, so ist

$$\frac{1}{2}\overline{\omega^2} M_K d^2 = kT,$$

wo M_K die Kernmasse und $\overline{\omega^2}$ ein Mittelwert des Quadrates der Frequenz ω ist. Schreiben wir, von jetzt ab unter Weglassung von Zahlenfaktoren,

$$\hbar\omega = k\Theta,$$

so wird

$$d^2 = \frac{\hbar^2 T}{M_K k\Theta^2}.$$

Ist ϵ die Energie der Elektronen, so wählen wir als r den Abstand vom Kern, innerhalb dessen die potentielle Energie (dem Betrag nach) größer ist als ϵ, also

$$r = \frac{Ze^2}{\epsilon}$$

(Z ist hier die Anzahl der Leitungselektronen pro Atom, $n = Zn_K$), und erhalten so die Abschätzung

$$l = \frac{M_K \, \epsilon^2 \, k\Theta^2}{Z^{1/3} \, e^4 \, \hbar^2 \, n^{5/3} \, T}.$$

Da die genaue Rechnung von BLOCH bis auf Zahlenfaktoren dasselbe liefert[1], darf unsere rohe Abschätzung als anschauliche Deutung der von BLOCH gerechneten Verhältnisse aufgefaßt werden. Die so gefundene Weglänge benutzen wir für die Temperaturen und Drucke, für die noch eine genäherte Gitterordnung der Kerne vorhanden ist, wo unsere Rechnung

$$\frac{d^2}{a^2} < 1$$

ergibt. Für die höheren Temperaturen und Drucke lassen wir diesen Faktor weg und sind dann im Einklang mit einer Abschätzung von KOTHARI (25). Mit $\epsilon \approx mv^2$ erhalten wir so

$$\sigma = \frac{M_K \, k\Theta^2}{Z^{1/3} \, m^{1/2} \, e^2 \, \hbar^2} \frac{\epsilon^{3/2}}{n^{2/3} \, T} \quad \text{für} \quad n^{2/3} \, kT \leq \frac{Z^{2/3} \, M_K \, (k\Theta)^2}{\hbar^2}$$

$$\sigma = \frac{1}{Z \, m^{1/2} \, e^2} \epsilon^{3/2} \qquad\qquad \text{für} \quad n^{2/3} \, kT \geq \frac{Z^{2/3} \, M_K \, (k\Theta)^2}{\hbar^2}.$$

Wenn wir beachten, daß sich die Energie $k\Theta$ der Gitterschwingungen zur atomaren Energieeinheit $\frac{me^4}{\hbar^2}$ etwa wie $\sqrt{m}$ zu $\sqrt{M_K}$ verhält, so folgt für das Kondensat (soweit es ein Metall ist) und den Übergang zum Elektronengas:

---

[1] Im Falle fast freier Elektronen (38)

$$l = \frac{4}{\pi^3} \frac{M_K \epsilon^2 k\Theta^2}{\hbar^2 C^2 n_K T},$$

wo C die mittlere Tiefe der Potentialmulden, also ungefähr $Ze^2 n_K^{1/3}$ ist.

$$\sigma = \frac{m^{5/2}\, e^6}{Z^{1/3}\, \hbar^6}\, \frac{\epsilon^{3/2}}{n^{2/3}\, kT} \quad \text{für} \quad n^{2/3}\, kT \leq \frac{Z^{2/3}\, m^3\, e^8}{\hbar^6}$$

$$\sigma = \frac{1}{Z m^{1/2}\, e^2}\, \epsilon^{3/2} \quad \text{für} \quad n^{2/3}\, kT \geq \frac{Z^{2/3}\, m^3\, e^8}{\hbar^6}\, .$$

Im nichtentarteten Gas ist natürlich $\Theta = 0$ zu setzen; es gilt also nur die zweite Gleichung. Wir erhalten so schließlich für das nichtentartete Elektronengas mit $\epsilon \approx kT$ die elektrische Leitfähigkeit

$$\sigma = \frac{(kT)^{3/2}}{Z m^{1/2}\, e^2}\, ,$$

wo Z die Zahl der von einem Atom abgetrennten Elektronen ist. Für das entartete Elektronengas mit $\epsilon \approx \frac{\hbar^{6/5}}{m^{3/5}} p^{2/5}$, $n \approx \frac{p}{\epsilon}$ wird die elektrische Leitfähigkeit

$$\varrho = \frac{\hbar^{9/5}\, p^{3/5}}{Z m^{7/5}\, e^2} = \frac{\hbar^3\, \varrho}{Z \mu\, e^2\, m^2\, M}, \quad \text{wenn} \quad p^{2/5}\, kT \geq \frac{Z^{2/3}\, m^{13/5}\, e^8}{\hbar^{26/5}}$$

$$\varrho = \frac{e^6\, m^{6/5}}{Z^{1/3}\, \hbar^{17/5}}\, \frac{p^{1/5}}{kT}, \quad \text{wenn} \quad p^{2/5}\, kT \leq \frac{Z^{2/3}\, m^{13/5}\, e^8}{\hbar^{26/5}}\, ;$$

für das Metall mit $\epsilon \approx \frac{\hbar^2}{m} n^{2/3}$ wird sie

$$\varrho = \frac{e^6\, m}{Z^{1/3}\, \hbar^3}\, \frac{n^{1/3}}{kT}\, .$$

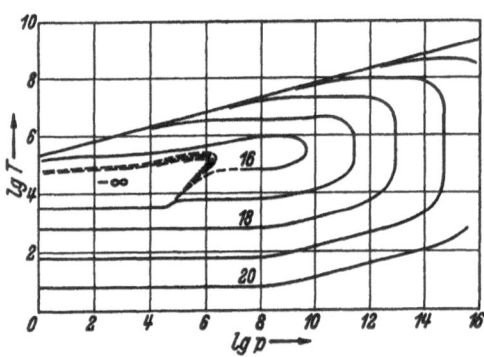

Dabei haben wir die Übergangsgebiete außer acht gelassen. Die aus den gerechneten vier Teilen zusammengesetzte Funktion σ ist also noch etwas abzurunden. Abb. 8 gibt dieses σ als Funktion von p und T an. Das *Atomgas* und, wenn es nicht metallisch ist, das *Kondensat* sind (abgesehen vom Neutronengas der ganz hohen Drukke) *die einzigen Gebiete, wo die Materie nicht leitet.* Die anderen Zustände leiten den elektrischen Strom wie ein Metall oder besser.

Abb. 8. Elektrische Leitfähigkeit der Materie. Der kondensierte Zustand ist als metallisch angenommen.

Bei unseren vereinfachten Ableitungen dürfen wir nicht erwarten, daß unsere Angaben genau gelten. So gibt unser Ausdruck für ein Metall von Zimmertemperatur $\sigma \approx 2 \cdot 10^{18}$, während die besten Leiter nur etwa $\sigma \approx 6 \cdot 10^{17}$ haben. Für das nichtentartete Elektronengas und das entartete Elektronengas mit regelloser Verteilung der Kerne liegen genauere, aber immer noch etwas vereinfachte Rechnungen von KOTHARI vor (25). Die genaueren Werte unterscheiden sich von den hier angegebenen durch Faktoren, die logarithmisch von den Zustandsvariabeln abhängen und in den hier betrachteten Gebieten die Größenordnung 1 haben.

Die Elektronen besorgen auch die Wärmeleitung; die Wärmeleitfähigkeit kann aus der elektrischen Leitfähigkeit mit dem WIEDEMANN-FRANZschen Gesetz berechnet werden

$$\lambda \approx \frac{k^2 T}{e^2} \sigma\,;$$

sie ist (in unserer Näherung) im Metall konstant, geht im nichtentarteten Elektronengas mit $T^{5/2}$ und im entarteten Elektronengas mit $p^{3/5}$ bis $p^{1/5}$.

13. Absorption von Licht. Das Atomgas läßt im allgemeinen sichtbares Licht durch und absorbiert im Ultravioletten die Frequenzen, die der Anregung oder Ionisation der Atome entsprechen. Ein Atomgas unter etwa 10^4 Grad Temperatur stellt der Ausbreitung seiner eigenen Temperaturstrahlung wenig Hindernisse entgegen. Erst bei höheren Temperaturen hält es diese Strahlung einigermaßen bei sich. Das optische Verhalten des kondensierten Zustandes hängt im wesentlichen davon ab, ob er ein Isolator oder ein Metall ist.

Das Elektronengas enthält genähert freie Elektronen, deren Beweglichkeit durch Wechselwirkung mit den Kraftfeldern der Kerne begrenzt ist; das bedingt auch eine Absorption von Strahlung. Ein vollkommener Leiter mit unendlicher Leitfähigkeit wäre durchsichtig. Die Absorption eines Elektronengases ist im Anschluß an eine Untersuchung von KRAMERS über die Emission der Elektronen infolge ihrer Bremsung in der Materie (27) von EDDINGTON, MILNE, OPPENHEIMER, GAUNT, KOTHARI und MAJUMDAR (9, 21, 26, 28, 29, 35) berechnet worden. Man kann aber diese Absorption durch eine einfache Überlegung aus der oben berechneten elektrischen Leitfähigkeit herleiten. Als Bewegungsgleichung der Elektronen nehmen wir an (wie in der Dispersionstheorie der guten Leiter)

$$e\mathfrak{E} = m\ddot{\mathfrak{r}} + g\dot{\mathfrak{r}}\,,$$

wo $\mathfrak{r}$ der Ortsvektor ist und g eine Reibung zum Ausdruck bringt und mit der Leitfähigkeit zusammenhängt. Statt der Bewegungsgleichung können wir auch eine phänomenologische Beziehung zwischen der Stromdichte $\mathfrak{s} = \sum e\dot{\mathfrak{r}}$ und dem Feld angeben

$$ne^2\,\mathfrak{E} = m\dot{\mathfrak{s}} + g\mathfrak{s}\,.$$

Im stationären Fall folgt daraus

$$\mathfrak{s} = \frac{ne^2}{g}\mathfrak{E}\,, \qquad \sigma = \frac{ne^2}{g}\,.$$

Wenn wir außer der Leitfähigkeit σ noch die mit der Trägheit der Elektronen zusammenhängende Größe $\gamma^2 = \frac{ne^2}{m}$ einführen, so wird

$$\mathfrak{E} = \frac{1}{\gamma^2}\dot{\mathfrak{s}} + \frac{1}{\sigma}\mathfrak{s}$$

die neben den allgemeinen MAXWELLschen Gleichungen gültige „Material-
gleichung" (den Zusammenhang zwischen $\mathfrak{E}$ und $\mathfrak{D}$, $\mathfrak{B}$ und $\mathfrak{H}$ wollen wir wie
im Vakuum annehmen). Für ein an die Materie angelegtes Wechselfeld

$$\mathfrak{E} = \mathfrak{E}_0 \, e^{i\omega t}$$

folgt dann für die Stromdichte

$$\mathfrak{s} = \mathfrak{s}_0 \, e^{i\omega t}$$

die Beziehung

$$\mathfrak{s} = \frac{\sigma}{1 + i \frac{\omega\sigma}{\gamma^2}} \mathfrak{E} .$$

Mit der üblichen Rechnung der Optik kann man daraus „Brechungsindex" $\overline{n}$
und „Absorptionsindex" $\overline{k}$ finden

$$(\overline{n} - i\overline{k})^2 = 1 - \frac{4\pi\gamma^2}{\omega^2 - i\frac{\omega\gamma^2}{\sigma}} .$$

Statt dieses in der Optik gebrauchten $\overline{k}$, das die Absorption auf dem Wege
der Vakuumwellenlänge bestimmt, wollen wir eine die Absorption der Masseneinheit auf dem Wege der Längeneinheit bestimmende Größe

$$\kappa = \frac{\omega}{\varrho c} \overline{k}$$

einführen; die Schwächung der Lichtintensität auf der Strecke x ist dann
durch

$$J \sim e^{-\varrho \kappa x}$$

gegeben, und es ist

$$\left(\overline{n} - i\frac{c\varrho}{\omega}\kappa\right)^2 = 1 - \frac{4\pi\gamma^2}{\omega^2 - i\frac{\omega\gamma^2}{\sigma}} .$$

In dem einfachen Grenzfall $\omega \gg \gamma$, $\omega\sigma \gg \gamma^2$ folgt

$$\frac{c\varrho}{\omega}\kappa = 2\pi \frac{\gamma^4}{\sigma\omega^3} .$$

Uns interessiert hauptsächlich die Absorption der Strahlung, die mit
der Materie im Temperaturgleichgewicht steht, die „Opazität" der Materie.
Sie ist (von Zahlenfaktoren abgesehen) gegeben durch die Absorption der
hauptsächlich gestrahlten Frequenz

$$\omega \approx \frac{kT}{\hbar}$$

(Zahlenfaktoren lassen wir jetzt wieder weg).
    Für das nichtentartete Elektronengas galt genähert

$$\sigma = \frac{(kT)^{3/2}}{Z m^{1/2} e^2} .$$

Da in einigem Abstand von der Entartungsgrenze

$$\frac{\omega^2}{\gamma^2} = \frac{m(kT)^2}{\hbar^2\,e^2\,n} \gg 1\,, \qquad \frac{\sigma\omega}{\gamma^2} = \frac{m^{1/2}\,(kT)^{5/2}}{Z\hbar e^4\,n} \gg 1$$

ist, folgt

$$\kappa = \frac{Z\hbar^2\,e^6}{c\,m^{3/2}\,(\mu M)^2} \frac{\varrho}{(kT)^{7/2}}$$

in Übereinstimmung mit den oben erwähnten Berechnungen (die auch nur unter Vereinfachungen durchgeführt sind). Für das entartete Gas galt bei höheren Drucken und Temperaturen

$$\sigma = \frac{\hbar^3}{Z m^2\,e^2} \frac{\sigma}{\mu M}\;;$$

es kann dort

$$\frac{\sigma\omega}{\gamma^2} = \frac{\hbar^2\,kT}{Z m\,e^4}$$

als groß angesehen werden; dagegen ist

$$\frac{\omega^2}{\gamma^2} = \frac{m(kT)^2}{\hbar^2\,e^2\,n}$$

nicht sehr von 1 verschieden. Bei Anwendung unserer oben angegebenen einfachen Formel für κ machen wir einen Fehler, der aber die Größenordnung nicht ändert. Es wird

$$\kappa = \frac{Z e^6}{\hbar c \mu M} \frac{1}{(kT)^2}$$

in Übereinstimmung mit dem Ergebnis, das MAJUMDAR (28) auch unter einigen Vereinfachungen findet. Für das relativistisch entartete Gas findet MAJUMDAR eine sehr geringe konstante Opazität.

Die Streuung von Licht an Licht, die bei Licht der COMPTON-Wellenlänge $\frac{\hbar}{mc}$ bei einer Dichte von vielen Lichtquanten in einem Würfel mit dieser Länge als Kante wesentlich wird, wird bei Temperaturen unter $\frac{mc^2}{k}$ keine Rolle spielen. Darüber jedoch gibt sie eine sehr starke Opazität.

Der streuende Querschnitt von Lichtquanten der Frequenz ω ist nach EULER (13) von der Größenordnung

$$\left(\frac{e^2}{\hbar c}\right)^4 \cdot \left(\frac{\hbar}{mc}\right)^8 \left(\frac{\omega}{c}\right)^6 = \frac{1}{137^4} \left(\frac{\hbar}{mc}\right)^8 \left(\frac{\omega}{c}\right)^6\;;$$

das gibt im Temperaturgleichgewicht eine Absorption der Größenordnung

$$\varrho\kappa = \frac{1}{137^4} \frac{mc}{\hbar} \left(\frac{kT}{mc^2}\right)^9\,,$$

die für tiefere Temperaturen als $\frac{mc^2}{k}$ ganz schnell abnimmt. In der Nähe dieser Temperatur geben jedoch unsere früheren Betrachtungen Absorptionswerte der Größenordnung

$$Z \cdot \left(\frac{e^2}{\hbar c} \right)^3 \cdot \frac{mc}{\hbar} = \frac{Z}{137^3} \frac{mc}{\hbar} \, ,$$

so daß die Streuung von Licht an Licht bei höheren Temperaturen bald der ausschlaggebende Vorgang wird.

14. Energietransport. Für die Untersuchung der Verhältnisse im Inneren eines Sternes ist die Frage wichtig, ob der Transport der Energie vom Inneren an die Oberfläche durch Bewegung der Materie, durch Wärmeleitung oder durch Strahlung stattfindet. Die Wirksamkeit der beiden letztgenannten Vorgänge können wir vergleichen. Ein Temperaturgefälle ruft einen Wärmestrom

$$\mathfrak{Q}_{\mathrm{cal}} = -\lambda_{\mathrm{cal}} \operatorname{grad} T$$

hervor. Er ruft aber auch einen Strom von Strahlungsenergie hervor. Wenn die Strahlung genähert im Temperaturgleichgewicht ist, das Temperaturgefälle also so gering, daß es in der Reichweite der Strahlung nicht sehr ins Gewicht fällt, so kommt von der Seite höherer Temperatur mehr Energie als von der anderen auf eine senkrecht zum Temperaturgefälle stehende Flächeneinheit. Ist $\mathfrak{Q}_{\mathrm{rad}}$ der Energiestrom, so wird $\varrho\kappa\mathfrak{Q}_{\mathrm{rad}}$ in der Raumeinheit absorbiert und übt einen Impuls aus; es ist also ein Gefälle des Strahlungsdruckes

$$-\operatorname{grad} p_{\mathrm{rad}} = \frac{\varrho\kappa}{c} \mathfrak{Q}_{\mathrm{rad}}$$

vorhanden. Also ist

$$\mathfrak{Q}_{\mathrm{rad}} = -\frac{c}{\varrho\kappa} \operatorname{grad} \frac{a}{3} T^4 = -\frac{4acT^3}{3\varrho\kappa} \operatorname{grad} T \, ;$$

d. h. der Strahlungsstrom kann durch eine „Leitfähigkeit"

$$\lambda_{\mathrm{rad}} = \frac{4\pi^2 \, k^4 \, T^3}{45\hbar^3 \, c^2 \, \varrho\kappa}$$

beschrieben werden. Das Verhältnis der Leitfähigkeit für Wärme zu der für Strahlung [KOTHARI (25)] ist (nach den Ergebnissen der Abschnitte 12 und 13) für nichtentartetes Gas bis auf unbedeutende Faktoren

$$\begin{aligned}
\frac{\lambda_{\mathrm{cal}}}{\lambda_{\mathrm{rad}}} &= \frac{\hbar^5 \, ce^2}{m^2} \frac{\varrho^2}{(\mu M)^2 \, (kT)^4} = \frac{\hbar^5 \, ce^2}{m^2} \frac{p^2}{(kT)^6} \\
&= \frac{1}{137} \frac{\hbar^6 \, c^2}{m^2} \frac{p^2}{(kT)^6} = 137 \frac{\hbar^4 \, e^2}{m^2} \frac{p^2}{(kT)^6} \, x,
\end{aligned}$$

also in einigem Abstand von der Entartungsgrenze klein. Für ein entartetes Gas ist bei höheren Temperaturen ebenfalls

$$\frac{\lambda_{\mathrm{cal}}}{\lambda_{\mathrm{rad}}} = \frac{\hbar^5 \, ce^2}{m^2} \frac{\varrho^2}{(\mu M)^2 \, (kT)^4} \, ,$$

in einigem Abstand von der Entartungsgrenze ist das groß. *Im Gebiet des nichtentarteten Elektronengases überwiegt der Energietransport durch Strah-*

lung, im Gebiet des entarteten Elektro-nengases überwiegt der Transport durch Wärmeleitung. Bei tieferen Temperaturen greift das Gebiet der Wärmeleitung ein wenig über die Entartungsgrenze hinweg, bei höheren Temperaturen das Gebiet der Strahlung. Die für den Energietransport wichtige Summe der beiden Leitfähigkeiten ist in Abb. 9 dargestellt. Die Zahlen geben die Zehnerlogarithmen der in erg/cm sec Grad gemessenen Energieleit-fähigkeit.

Neutronen könnten eine sehr starke Leitfähigkeit für Wärmeenergie geben (16), sie dürften aber, abgesehen von sehr hohen Drucken, nicht in nennenswerter Anzahl vorkommen.

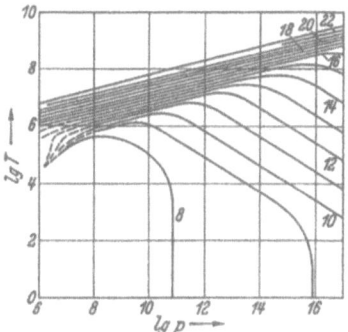

Abb. 9. Leitfähigkeit für Wärmeenergie (Transport durch die Elektronen und durch Strahlung).

III. Das Vorkommen sehr hoher Drucke und Temperaturen

In den folgenden Abschnitten soll kein Bericht über den inneren Bau der Himmelskörper gegeben werden[1]; vielmehr werden nur einige einfache Überlegungen angegeben, die größenordnungsmäßig etwas über die vorkommenden Drucke und Temperaturen aussagen, und die zeigen, wie etwa in großen Zügen die empirischen Eigenschaften der Sterne mit den Eigenschaften der Materie, die wir uns klar gemacht haben, zusammenhängen.

15. Die Planeten. Im *Laboratorium* kann man Temperaturen von einigen tausend Grad herstellen. Bestimmte Wirkungen sehr viel höherer Temperaturen kann man jedoch durch die Versuche mit schnellbewegten Teilchen (radioaktiven Ursprungs oder durch Felder beschleunigt) hervorrufen. Mit der Herstellung hoher Drucke ist BRIDGMAN fast bis 10^5 Atmosphären gelangt (6). Er ist also vom Übergang der festen Körper in den Zustand mit der Kompressibilität des entarteten Elektronengases noch ziemlich entfernt.

Vom *Erdinneren* wissen wir, daß es keine oder geringe Transversalelasti-zität hat; wir können also einen skalaren Druck p im Gleichgewicht mit den Kräften der Schwere annehmen

$$\frac{dp}{dr} = -G\varrho\frac{\mathrm{m}}{r^2} \qquad (\mathrm{I})$$

$$\frac{d\mathrm{m}}{dr} = 4\pi\varrho r^2 \qquad (\mathrm{II})$$

[r ist Abstand vom Erdmittelpunkt, $\mathrm{m}(r)$ Masse vom Erdmittelpunkt bis zum Abstand r, G ist die Gravitationskonstante]. Die beiden Gleichungen zusammengefaßt geben

$$\frac{d}{dr}\left(\frac{r^2}{\varrho}\frac{dp}{dr}\right) = -4\pi G\varrho r^2 . \qquad (1)$$

---

[1] Hierüber hat H. VOGT im Bd. 6 der Ergeb. exakt. Naturwiss. (46) berichtet.

Wenn $\varrho(r)$ oder $\varrho(p)$ bekannt ist, hat diese Gleichung eine einparametrige Schar von Lösungen (die eine der beiden Integrationskonstanten ist nämlich durch $\mathfrak{m}(0) = 0$, d. h. durch Verschwinden von $\frac{dp}{dr}$ im Mittelpunkt bestimmt). Machen wir eine Überschlagsrechnung mit konstantem ϱ, so wird

$$p = \frac{2\pi}{3} G\varrho^2 \left(R^2 - r^2 \right).$$

Mit dem empirischen Wert $\varrho = 5,5\,\mathrm{g/cm}^3$ erhält man im Erdmittelpunkt einen Druck von fast $2 \cdot 10^6$ Atmosphären. Ein Modell aus zwei Schalen, einer inneren mit der Dichte $8\,\mathrm{g/cm}^3$ und einer äußeren mit der Dichte $3\,\mathrm{g/cm}^3$, deren Verhältnis so bestimmt ist, daß die richtige mittlere Dichte herauskommt, gäbe im Erdmittelpunkt $3 \cdot 10^0$ Atmosphären. Das Erdinnere ist also im gewöhnlichen kondensierten Zustand, einem Gebiet, wo die durch die Elektronen bedingten Eigenschaften der Materie nicht wesentlich von der Temperatur abhängen.

Der größte der Planeten, *Jupiter*, hat eine mittlere Dichte von $1,3\,\mathrm{g/cm}^3$. Die Überschlagsrechnung des Druckes im Mittelpunkt mit der Annahme konstanter Dichte ($R = 7 \cdot 10^9$ cm) gäbe $1,2 \cdot 10^7$ Atmosphären; wegen der Dichtezunahme zum Mittelpunkt hin muß der Druck dort in Wirklichkeit größer sein. Die geringe Dichte des Jupiter ist nicht verträglich mit der Annahme, daß die Materie ganz im kondensierten Zustand sei, die Materie des Jupiter liegt im Diagramm der Abb. 5 höher. Dann hängt aber die Dichte ϱ nicht eindeutig vom Druck ab, sondern auch von der Temperatur, und wir können darum die Gleichung des Aufbaues nicht integrieren. Wir nehmen darum unsere Zuflucht dazu, zunächst Jupitermodelle aufzubauen, bei denen Druck und Dichte in einer einfachen vorgegebenen Weise zusammenhängen. Wir wählen EMDENs (12) „polytropen" Zusammenhang

$$\frac{p}{p_0} = \left(\frac{\varrho}{\varrho_0} \right)^{1+\frac{1}{n}}, \tag{2}$$

wo p_0 und ϱ_0 Druck und Dichte im Mittelpunkt sein sollen und n eine feste Zahl ist. Für ein Gas im „adiabatischen" Gleichgewicht (auf- oder absteigende Gasströme behalten darin ihren Temperaturunterschied gegen die Umgebung) gilt z. B. eine solche Beziehung, wo $1 + \frac{1}{n}$ gleich dem Verhältnis c_p/c_v der spezifischen Wärmen ist. Von Dissoziation oder Ionisation abgesehen, wäre dann n zwischen 3/2 und 3. Mit einem solchen Ansatz läßt sich die Aufbaugleichung (1) auf eine einfache Gleichung zurückführen, die außer n nur Zahlenkoeffizienten enthält und von EMDEN für eine Reihe von n-Werten numerisch gelöst worden ist (12, 31).

Tabelle 2. Dichte und Druck im Mittelpunkt von Jupiter-Modellen.

n	ϱ_0 (g/cm^3)	p_0 (Atm)
1,5	8	$0,8 \cdot 10^8$
2	14	$1,5 \cdot 10^8$
2,5	31	$4 \ \cdot 10^8$
3	71	$11 \ \cdot 10^8$

Mit einigen n-Werten erhalten wir folgende Werte der Dichte und des Druckes im Mittelpunkt des Jupiter.

Ein Blick auf Abb. 5 zeigt uns, daß nur die obersten dieser Druck- und Dichtewerte zusammen vorkommen können. Jupiter ragt danach mit seinem Innern in das Übergangsgebiet zwischen Atomgas, Kondensat und entartetem Elektronengas hinein.

Für *die anderen großen Planeten* erhalten wir durch eine solche Abschätzung etwa die gleichen Mittelpunktsdichten und etwas geringere Mittelpunktsdrucke.

16. Die gewöhnlichen Fixsterne. *Die Sonne* hat die mittlere Dichte $1,4\,\mathrm{g/cm}^3$ und den Radius $6,95 \cdot 10^{10}$ cm. Die rohe Abschätzung des Druckes im Sonnenmittelpunkt mit der Annahme konstanter Dichte gäbe hier $1,3 \cdot 10^9$ Atmosphären. Dem entspräche ein Punkt im Zustandsdiagramm (Abb. 5) im Gebiet des nichtentarteten Elektronengases mit einer Temperatur von etwa $2 \cdot 10^7$ Grad.

Die genaueren Abschätzungen machen Gebrauch von Annahmen über die Energieerzeugung und den Energietransport (4, 11, 24, 31, 36, 42, 44, 46). Wenn man von Energietransport durch Bewegung der Materie absieht, darf man im nichtentarteten Elektronengas Transport durch Strahlung, genähert mit einem Absorptionsgesetz

$$\kappa \sim \varrho T^{-7/2} \tag{3}$$

annehmen. Zu den beiden Differentialgleichungen erster Ordnung des Aufbaues (I, II) tritt eine dritte, die das Temperaturgefälle mit dem Energiestrom verknüpft (s. Abschnitt 14):

$$\frac{dT^4}{dr} = -\frac{3}{ac}\varrho\kappa\,\frac{l}{4\pi r^2}\,, \tag{4}$$

wo l die aus einer der Sonne konzentrischen Kugel vom Radius r herauskommende Energiemenge ist. Die drei Gleichungen reichen nicht aus, um die vier unbekannten Funktionen p, T, $\mathfrak{m}$, l zu bestimmen (ϱ sei mit der Zustandsgleichung durch p und T bestimmt). Es muß noch etwas über die Funktion l ausgesagt werden. Da liegt es nahe, anzunehmen, daß die Wirklichkeit zwischen einem Grenzfall liegt, wo nur im Mittelpunkt der Sonne die Materie mehr Energie abgibt als aufnimmt (Punktquellenmodell) und einem anderen Grenzfall, wo jedes Gramm Materie gleich viel mehr Energie abgibt als aufnimmt (gleichförmige Energieerzeugung). Das Modell der gleichförmigen Energieerzeugung führt unter nicht sehr gefährlichen Vereinfachungen auf eine uns schon bekannte Differentialgleichung. Aus [vgl. (I) und (4)]

$$\frac{dT^4}{dp} = \frac{3}{4\pi ac}\kappa\,\frac{l}{\mathfrak{m}} = \frac{3}{4\pi ac}\kappa\,\frac{L}{\mathfrak{M}}\,,$$

wo L und $\mathfrak{M}$ Gesamtstrahlung und Gesamtmasse bedeuten, folgt nämlich durch Integration

$$p \sim T^{17/4} \sim \varrho^{17/13},$$

wenn wir die Zustandsgleichung des nichtentarteten Gases

$$p \sim \varrho T$$

zugrunde legen und die Integrationskonstante so bestimmen, daß T mit p verschwindet (da die Temperatur an der Sternoberfläche vernachlässigbar klein ist gegen die Temperatur im Inneren). Das gibt aber eine „polytrope" Zustandsgleichung (2) mit $n = 3,25$. In die Gleichung (1) des Aufbaues gehen jetzt wieder nur ϱ_0 und p_0 als Parameter ein und wir können sie so bestimmen, daß der gerechnete Stern den Radius und die Masse der Sonne erhält. Man bekommt so im Mittelpunkt der Sonne den Druck $2,3 \cdot 10^{11}$ Atmosphären, die Dichte $120\,\mathrm{g/cm^3}$ und die (mit $\mu = 2$) Temperatur $4,5 \cdot 10^7$ Grad; man ist noch im Gebiet des nichtentarteten Elektronengases. Das Punktquellenmodell liefert Werte, die nicht so sehr davon verschieden sind (4, 11, 42). Genauere Rechnungen berücksichtigen, daß κ etwas anders lautet, da ja weiter außen das Gas nur teilweise ionisiert ist. Auch sie geben nichts wesentlich anderes (44).

Wir wissen aus Altersbestimmungen der Gesteine, daß die Sonne seit etwa 10^9 Jahren ziemlich unverändert den gleichen Energiebetrag strahlt. Sie hat seitdem gegen 10^{18} erg/g verausgabt. Aus ihrer Gravitationsenergie kann sie ihn nicht entnommen haben, selbst die Zusammenziehung aus einem unendlich ausgedehnten Nebel zu ihrer heutigen Größe gäbe viel weniger. Um aus ihrer inneren Energie Beträge von 10^{18} erg/g oder $10^{11} \cdot 10^7$ erg/g im thermodynamischen Gleichgewicht verausgaben zu können, müßte die Sonne (Abb. 7) Materie von einigen 10^9 Grad Temperatur oder einigen 10^{19} Atmosphären Druck enthalten haben. Dann wäre sie aber vor 10^9 Jahren in einem vom heutigen Zustand sehr verschiedenen Zustand gewesen. Wir müssen also annehmen, daß die Energieerzeugung in der Sonne nicht im thermodynamischen Gleichgewicht vor sich geht; sie muß dann im wesentlichen geschehen durch Umwandlung leichter Atomkerne in schwerere bei Temperaturen, die wesentlich niedriger sind als die Temperaturen, wo dieser Vorgang im thermodynamischen Gleichgewicht stattfindet. Die Abschätzungen von ATKINSON und HOUTERMANS (2, 22), die oben erwähnt wurden, zeigen, daß solche Vorgänge bei den in Betracht kommenden Temperaturen hinreichend rasch erfolgen. Auf Schwierigkeiten in der Deutung der Häufigkeit der verschiedenen Elemente, die sich aber vielleicht beheben lassen, sei hier nicht eingegangen (1, 22)[1].

Während ein einzelner Fixstern den Physiker wenig zu interessieren braucht, haben wir in der *Gesamtheit der Fixsterne* ein großes empirisches Material, und es ist denkbar, daß empirische Regeln über die Eigenschaften der Fixsterne irgendwie ein Abbild sind des Verhaltens der Materie bei hohen Drucken und Temperaturen. Wir scheiden die veränderlichen Sterne aus; dann dürfen wir aus der Ähnlichkeit mit der Sonne schließen, daß die Fixsterne insofern im Gleichgewicht sind, als sie sehr lange Zeit hindurch ihre Eigenschaften nicht stark ändern; ihre Energie muß also aus einem Vorrat stammen, dessen Inanspruchnahme nur geringe Rückwirkung auf die Zustandsvariablen hat. Wahrscheinlich ist es die Umwandlung leichter Atomkerne in

---

[1] Über die Energiequellen der Sterne berichtete in der Ergeb. exakt. Naturwiss., Bd.6, E. FREUNDLICH (19)

schwerere, der Übergang von einem thermodynamisch unwahrscheinlichen Zustand in einen wahrscheinlicheren.

Man kennt Radius R und Energiestrahlung L vieler Sterne, bei einer An-

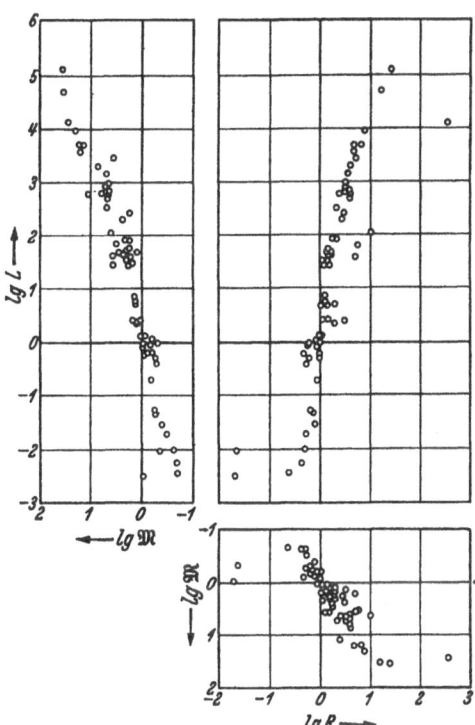

zahl von Sternen auch die Masse $\mathfrak{M}$. Die gemessenen Werte sind mit einer gewissen Regelmäßigkeit im R-$\mathfrak{M}$-L-Koordinatenraum verteilt. Abb. 10 gibt die Lagen für die Sterne, bei denen die drei Koordinaten bekannt sind [nach PILOWSKI; Angaben in (34)]. Für R und L allein hat man sehr viel mehr Material; die entsprechenden Punkte liegen nicht viel anders wie die gezeichneten. Die Sterne zeigen eine Häufung in der Nachbarschaft einer Linie im R-$\mathfrak{M}$-L-Raum, die ungefähr durch die Beziehungen $L \sim \mathfrak{M}^4$ (für große $\mathfrak{M}$ ist die Zunahme von $\mathfrak{M}$ etwas schwächer) und $L \sim R^5$ bis $L \sim R^4$ bezeichnet ist. Die R-L-Beziehung ist ein anderer Ausdruck für das bekannte HERTZSPRUNG-RUSSEL-Diagramm der Spektraltypen und der Leuchtkräfte. Es gibt aber auch zahlreiche Sterne, die der

Abb. 10. Verteilung der R-,$\mathfrak{M}$-,L-Werte der Fixsterne (nach den Zahlen PILOWSKIS).

Beziehung nicht folgen, vor allem die sog. Riesensterne (rechts im R-L-Diagramm). Die $\mathfrak{M}$-L-Beziehung ist ziemlich scharf, die wenigen Riesensterne, deren Masse bekannt ist, fügen sich ein. Einige Sterne sehr geringer Leuchtkraft und ganz kleiner Radien, die sog. weißen Zwerge (links unten im $\mathfrak{M}$-L- und R-L-Diagramm, links oben im R-$\mathfrak{M}$-Diagramm), passen weder in die R-L-Beziehung noch in die $\mathfrak{M}$-L-Beziehung; wir wollen sie nicht zu den gewöhnlichen Fixsternen rechnen.

Zusammenfassend kann man sagen, daß für die gewöhnlichen Fixsterne eine anscheinend *allgemeine Beziehung zwischen $\mathfrak{M}$ und L* ziemlich gut erfüllt ist (es gibt Anzeichen dafür, daß L bei gleichem $\mathfrak{M}$ noch schwach von R abhängt und mit R zunimmt) und daß *für die große Mehrzahl* dieser Sterne (die „Hauptreihe") *eine zweite Beziehung* leidlich gut erfüllt ist (diese können wir nach Belieben als R-L-Beziehung oder R-$\mathfrak{M}$-Beziehung ansehen).

Wir wollen sehen, ob unsere Kenntnis vom Verhalten der Materie Anhaltspunkte zum Verständnis solcher Regeln gibt. Um einen Einblick in das Wesen der Schlüsse zu bekommen, vereinfachen wir etwas, ohne uns aber sehr von den wirklichen Überlegungen von EDDINGTON, VOGT u. a.(10, 11, 24, 31, 36, 44, 45, 46) zu entfernen. Die drei Gleichungen des Aufbaues und Energietransportes

$$\frac{dp}{dr} = -G\varrho\frac{m}{r^2} \tag{I}$$

$$\frac{dm}{dr} = 4\pi\varrho r^2 \tag{II}$$

$$\frac{dT^4}{dp} = \frac{3}{4\pi Gac}\kappa\frac{l}{m} \tag{III}$$

bestimmen, auch wenn die Zustandsgleichung $\varrho = \varrho(p,T)$ und das Absorptionsgesetz $\kappa = \kappa(p,T)$ bekannt ist, den Aufbau nicht eindeutig, da es nur drei Gleichungen für vier unbekannte Funktionen sind. Es fehlt noch eine Beziehung für die Energieerzeugung. Da man über sie von der Physik her wenig weiß, behandelt man statt der wirklichen Sterne Sternmodelle, indem man eine hypothetische Funktion l hinschreibt. Die folgenden Gleichungen beschreiben einige der vorgeschlagenen oder naheliegenden Modelle (f bedeutet eine für alle Modellsterne angenommene Funktion, σ und τ für alle Modellsterne angenommene Zahlen; a und b sind individuelle Parameter). Mit (IVa) gibt es eine einparametrige Schar von Lösungen (T an der Oberfläche Null gesetzt), mit (IVb) und (IVc) eine zweiparametrige Schar und mit (IVd) eine dreiparametrige Schar von Lösungen.

$$\frac{dl}{dm} = f(p,T) \quad \text{z.B.} \quad f = p^\sigma T^\tau \tag{IV a}$$

$$\frac{dl}{dm} = f(a,p,T) \quad \text{z.B.} \quad f = ap^\sigma T^\tau \tag{IV b}$$

$$l = f(a,m) \quad \text{z.B.} \quad l = L \text{ (Punktquellenmodell)} \tag{IV c}$$

$$l = \frac{L}{\mathfrak{M}}m \text{ (gleichförmige Energieerzeugung)}$$

$$l = L\left(\frac{m}{\mathfrak{M}}\right)^\sigma$$

$$l = f(a,b,m) \quad \text{z.B.} \quad l = \left\{\begin{array}{ll} \frac{L}{b\mathfrak{M}}m & \text{für} \quad m \le b\mathfrak{M} \\ L & \text{für} \quad m \ge b\mathfrak{M} \quad (32) \end{array}\right\} \tag{IV d}$$

Wir beschränken uns auf Sternmaterie im Zustand des nichtentarteten Elektronengases, bei dem der Strahlungsdruck noch nicht wesentlich sei

$$p \sim \varrho T$$

und nehmen ein Absorptionsgesetz

$$\kappa \sim \varrho^\alpha T^\beta,$$

z.B. das früher genannte Gesetz

$$\kappa \sim \varrho T^{-7/2}, \tag{3}$$

an. Aus einer Lösung der Gleichungen (I), (II), (III) und dieser beiden Beziehungen kann man dann durch eine einfache Ähnlichkeitstransformation neue Lösungen finden, und zwar eine zweiparametrige Schar, deren Mitglieder z. B. durch $\mathfrak{M}$ und R unterschieden werden können. Man setze

$$\frac{r'}{r} = \frac{R'}{R}; \quad \frac{\mathfrak{m}'}{\mathfrak{m}} = \frac{\mathfrak{M}'}{\mathfrak{M}}; \quad \frac{\varrho'(r')}{\varrho(r)} = \frac{\mathfrak{M}' R^3}{R'^3 \mathfrak{M}}; \quad \frac{p'(r')}{p(r)} = \frac{\mathfrak{M}'^2 R^4}{R'^4 \mathfrak{M}^2}$$

usw. Wir wollen die Transformationen durch die abgekürzte, leicht verständliche Schreibweise

$$r \sim R, \quad \mathfrak{m} \sim \mathfrak{M}, \quad \varrho \sim \frac{\mathfrak{M}}{R^3}, \quad p \sim \frac{\mathfrak{M}^2}{R^4}, \quad T \sim \frac{\mathfrak{M}}{R}$$

und

$$L \sim l \sim \frac{\mathfrak{M}^3}{\kappa} \tag{A}$$

beschreiben, wo die Transformation von κ durch den Zusammenhang mit ϱ und T gegeben ist. $\kappa \sim \varrho T^{-7/2}$ gibt $L \sim \mathfrak{M}^{11/2} R^{-1/2}$. Wir erhalten so aus einer Lösung von (I), (II), (III) natürlich nicht alle Lösungen, und aus einer Lösung von (I), (II), (III) und einer der Gleichungen (IV) erhalten wir durch die Transformation vielleicht überhaupt keine Lösung von (IV). Nehmen wir das Modell (IV a) in der rechts stehenden besonderen Form, so erhalten wir aus einer Lösung durch die angegebene Transformation eine neue Lösung, wenn L wie $p^\sigma T^\tau$ transformiert wird. Die Transformation liefert dann aus einer Lösung nur eine einparametrige Schar von Lösungen, damit aber alle Lösungen, die es überhaupt gibt. Für die Sterne des Modells (IV a) in der angenommenen besonderen Form gelten zwei R-$\mathfrak{M}$-L-Beziehungen; alle Sterne des Modelles liegen auf einer Linie im R-$\mathfrak{M}$-L-Raum; eine der beiden Beziehungen ist (A). Bei den Modellen (IV a) und (IV b) in einer der rechts angegebenen Formen erhalten wir durch die Transformation wieder Lösungen, und zwar gerade die zweiparametrige Schar aller Lösungen. Alle Sterne dieser Modelle liegen auf einer durch (A) bezeichneten Fläche im R-$\mathfrak{M}$-L-Raum; der noch nicht bestimmte Faktor in (A) kann nur durch wirkliche Auflösung der Gleichungen bestimmt werden. Im Modell (IV d) können wir nicht auf so einfache Weise aus einer Lösung neue bilden.

Die für die Modelle (IV a), (IV b) und (IV c) in der rechts stehenden Form gültige R-$\mathfrak{M}$-L-Beziehung

$$L \sim \mathfrak{M}^{11/2} R^{-1/2} \tag{A}$$

liegt in der Nachbarschaft der empirischen Beziehung $L \sim \mathfrak{M}^4$. Genauere Rechnungen mit einem besseren κ-Gesetz, das die nicht vollständige Ionisierung der Atome berücksichtigt (4, 11), bringt die Abhängigkeit von $\mathfrak{M}$ der beobachteten näher; dagegen scheint die Abhängigkeit von $R^{-1/2}$ nicht der Beobachtung zu entsprechen; die Abweichung der empirischen Abhängigkeit kann auf verschiedenen Wasserstoffgehalt der Sterne (also verschiedenes μ der Zustandsgleichung) geschoben werden (4).

Es sieht also so aus, als hätten wir *in der empirischen $\mathfrak{M}$-L-Beziehung* der gewöhnlichen Sterne *einen Ausdruck für die weitgehende Gleichförmigkeit der Sternmaterie in der Zustandsbeziehung und im Gesetz des Energietranspor-tes*. Die zweite empirische Beziehung, die ja nicht so allgemein und nicht so scharf ist, könnte dann mit den Umständen der Energieerzeugung zusam-menhängen. So bedeutet die genähert erfüllte Beziehung $\mathfrak{M} \sim R$ in der oben gegebenen Ähnlichkeitsbetrachtung die Übereinstimmung der Temperaturen an entsprechenden Stellen der Sterne.

17. Die dichten Sterne. Es gibt einige wenige Sterne, deren R-$\mathfrak{M}$-L-Werte aus der Reihe fallen, die „*weißen Zwerge*". Sie fallen durch ihre Klein-heit bei fast normaler Masse (soweit bekannt) auf und haben mittlere Dichten um 10^5 g/cm^3 und in ihrem Mittelpunkt noch höhere Dichten. Solche Dichten sind an sich auch im Zustand des nichtentarteten Elektronengases möglich, die Sterne müßten aber dann größere Radien haben. Man muß daher anneh-men, daß ihre Materie zum Teil im Zustand des entarteten Elektronengases ist. Sie waren es, die die Aufmerksamkeit der Physiker auf diesen Zustand als Zustand der Materie unter hohen Drucken lenkten [FOWLER (17)].

Wenn der größte Teil der Sternmaterie im entarteten Zustand (der ge-wöhnliche kondensierte Zustand sei auch zugelassen) ist, so ist die Tem-peratur ganz unwesentlich für das Verhalten. Für den Aufbau eines Ster-nes im Gleichgewicht ist also auch Energieerzeugung und Energietransport gleichgültig. Die Gleichung des Aufbaues (I) läßt eine einparametrige Lö-sungsschar zu, deren Mitglieder etwa durch die Sternmasse $\mathfrak{M}$ unterschieden werden können. Dabei ist allerdings vorausgesetzt, daß wir die Zustandsglei-chung wirklich kennen, wozu etwa genügt, daß der Stern nicht viel Wasserstoff enthält. Der Radius des Sterns ist eine eindeutige Funktion der Sternmasse, und zwar ist für kleine Massen natürlich $R^3 \sim \mathfrak{M}$. Für mittlere Massen, wo der Hauptteil der Materie nichtrelativistisches entartetes Elektronengas ist, folgt aus der Zustandsgleichung $p \sim \varrho^{5/3}$ und der Aufbaugleichung durch eine Ähnlichkeitsbetrachtung $R^3 \sim \frac{1}{\mathfrak{M}}$. Für noch größere Massen wird der innen gelegene Teil der Materie relativistisch entartet. Die alleinige Annahme der Zustandsgleichung $p \sim \varrho^{4/3}$ gäbe einen singulären Zusammenhang zwischen $\mathfrak{M}$ und R.

Die wirklich beobachteten weißen Zwerge sind noch etwas größer, als solchen Abschätzungen entspricht. Ihre äußeren Teile sind ja auch, wie man an der hohen Oberflächentemperatur sieht, gasförmig. Nach plausiblen Abschätzungen hat der Siriusbegleiter eine Mittelpunktsdichte von etwas über 10^6 g/cm^3 und einen Mittelpunktsdruck von etwa 10^{17} Atmosphären, er kommt also gerade an das Übergangsgebiet zum relativistisch entarteten Elektronengas. VAN MAANENs Stern dürfte wesentlich höhere Mittelpunkts-dichte und höheren Mittelpunktsdruck haben, also zum Teil aus relativistisch entartetem Elektronengas bestehen. Dafür, daß die Umwandlung in Neutro-nen bei den beobachteten Sternen erreicht ist, besteht kein Anzeichen.

CHANDRASEKHAR (7) hat den Aufbau eines Sterns, dessen Temperatur nichts mehr ausmacht, mit einer genaueren Zustandsgleichung, die für das nichtrelativistische und das relativistische Elektronengas gilt, berechnet. Bei zunehmender Masse nimmt der Radius ab, bei Sonnenmasse ist er etwa gleich dem Erdradius, bei noch größerer Masse nimmt er schließlich rasch gegen Null ab; der Radius Null wird bei einer endlichen Masse bald oberhalb der Sonnenmasse erreicht. Dieses letzte Ergebnis ist zunächst nicht ganz wörtlich zu nehmen, denn es ist bei der Berechnung die Zustandsgleichung als für hohe Drucke unbegrenzt gültig angenommen. Dieses Zusammensinken zum Radius Null (oder dem, was ihm in der allgemeinen Relativitätstheorie entspricht) rührt von der großen Kompressibilität der Materie im Zustand des relativistisch entarteten Elektronengases her. Wenn ein Stern oberhalb der Grenzmasse endlichen Radius hätte, so würde der Druck seiner Materie nicht das Gewicht der darüberliegenden Materie aushalten; der Stern würde zusammensinken. Dabei würde zwar der Druck zunehmen, aber nicht rasch genug, um dem jetzt vergrößerten Gewicht der darüberliegenden Schichten gewachsen zu sein.

Die Umwandlung der Materie in Neutronen bei hohen Drucken bedeutet zunächst in einem gewissen Gebiet eine etwas geringere Kompressibilität als die des relativistisch entarteten Elektronengases; die Grenzmasse für endlichen Radius wird also noch ein wenig hinausgeschoben. Wenn wir aber zu sehr kleinen Radien gekommen sind, wird die Gravitationsenergie der Teilchen von der Größenordnung ihrer Ruheenergie und die Grundlagen unserer Rechnungen versagen. Wir dürfen aus den Rechnungen von CHANDRASEKHAR oder ähnlichen mit Berücksichtigung der Neutronenumwandlung angestellten nur da etwas schließen, wo die Sternradien noch nicht so klein sind, daß sie mit der (Abschnitt 2) kritischen Länge von einigen 10^5 cm vergleichbar werden; wir können also schließen, daß Sterne hinreichender Masse so klein werden können.

Es ist nicht ganz ausgeschlossen, daß die Sterne, die jetzt im gewöhnlichen Zustand sind, sich zu einem diffusen, kalten Nebel ausdehnen; sie müßten aber dann damit anfangen, ehe ihr Energievorrat der Abweichung der chemischen Zusammensetzung aufgebraucht ist. Wenn sie das nicht tun, müssen sie sich schließlich zusammenziehen und dichte Sterne werden. Die Sterne großer Masse könnten dabei die oben erwähnten kleinen Radien dadurch vermeiden, daß sie große Teile der beim Zusammenziehen freiwerdenden Gravitationsenergie abstrahlen. Bei Zusammenziehung auf Erdradius wird diese Energiemenge vergleichbar mit $\mathfrak{M}c^2$, der Stern verringert also seine Masse beträchtlich. Sie können vielleicht auch infolge hohen Strahlungsdruckes an der Oberfläche Materie verlieren (7). Als *mögliches Endstadium der Sternentwicklung* dürfen wir also Sterne mäßiger Masse und sehr hoher Dichten erwarten.

Literaturverzeichnis[1]

1. ATKINSON, R. D'E.: (Entstehung der Kerne.) Astrophysic. J. **73**, 250, 308 (1931).
2. — u. F.G. HOUTERMANS: (Eindringen von Protonen in Kerne.) Z. Physik **54**, 656 (1929).
3. BETHE, H.: (Kernenergien.) Physic. Rev. **47**, 633 (1935).
4. BIERMANN, L.: (Sternmodelle.) Z. Astrophysik **3**, 116 (1931).
5. BLOCH, F.: (Metallische Leitung.) Z. Physik **52**, 555 (1928).
6. BRIDGMAN, P.W.: Theoretically interesting aspects of high pressure phenomena. Rev. mod. Physics **7**, 1 (1935).
7. CHANDRASEKHAR, S.: (Sterne aus entarteter Materie.) Monthly not. **91**, 456 (1923); **95**, 207, 226, 676 (1935).
8. — and L. ROSENFELD: (Paarerzeugung.) Nature (Lond.) **135**, 999 (1935)
9. EDDINGTON, A. S.: (Opazität.) Monthly not. **83**, 32, 98 (1922), 431 (1923); **84**, 104, 308 (1924).
10. — ($\mathfrak{M}$-L-Beziehung.) Monthly not. **84**, 308 (1924).
11. — The internal constitution of the stars. Cambridge 1926. Der innere Aufbau der Sterne. Berlin 1928.
12. EMDEN, R.: Gaskugeln. Leipzig u. Berlin 1907. Thermodynamik der Himmelskörper. Enzyklop. math. Wiss. **6/2**, 373, (1926).
13. EULER, H.: (Streuung von Licht an Licht.) Ann. Physik (5) **26**, 398 (1936).
14. FERMI, E.: (Ideales Gas.) Z. Physik **36**, 902 (1926).
15. FLEISCHMANN, R. u. W. BOTHE: Künstliche Kernumwandlung. Erg. exakt. Naturwiss. **14**, 1 (1935).
16. FLÜGGE, S.: (Neutronen im Stern.) Z. Astrophysik **6**, 272 (1933).
17. FOWLER, R.H.: (Dichte Materie aus entartetem Elektronengas.) Monthly not. **87**, 114 (1926).
18. — and E.A. GUGGENHEIM: (Ionisierung der Sternmaterie.) Monthly not. **85**, 939, 961 (1925).
19. FREUNDLICH, E.: Die Energiequellen der Sterne. Erg. exakt. Naturwiss. **6**, 27 (1927).
20. GAMOW, G.: (Eindringen von α-Teilchen in Kerne.) Z. Physik **52** , 510 (1928).
21. GAUNT, J.A.: (Strahlung der Elektronen.) Z. Physik **59**, 508 (1930).
22. HOUTERMANS, F.G.: Neuere Arbeiten über Quantentheorie des Atomkerns. Erg. exakt. Naturwiss. **9**, 123 (1930).
23. JEANS, J.H.: (Ionisation der Sternmaterie.) Observatory **40**, 43 (1917). (War mir nicht zugänglich.)
24. — Astronomy and Cosmogony. Cambridge 1928.
25. KOTHARI, D. S.: (Leitfähigkeit eines Elektronengases.) Philosophic. Mag. (7) **13**, 361 (1932). — Monthly not. **93**, 61 (1932).
26. — u. R. C. MAJUMDAR: (Opazität.) Astron. Nachr. **244**, 65 (1931).
27. KRAMERS, H. A.: (Strahlung der Elektronen.) Philosophic. Mag. **46**, 836 (1923).
28. MAJUMDAR, R. C.: (Opazität.) Astron. Nachr. **243**, 5 (1931); **247**, 217 (1932).
29. MILNE, E. A.: (Opazität.) Monthly not. **85**, 750 (1925).
30. — (Energie der Elektronen eines Atoms.) Proc. Cambridge philos. Soc. **23**, 794 (1927).
31. — Thermodynamics of the stars. Handbuch der Astrophysik, Bd 3/1, S.64 1930, insbes. S.183f.
32. — (Sternmodelle.) Monthly not. **90**, 769; **91**, 1 (1930); **92**, 610 (1922). -Z. Astrophysik **4**, 75; **5**, 337 (1932).

---

[1] Die in Klammern gesetzten Angaben sollen kurz den für uns wichtigen Inhalt der Arbeit kennzeichnen

33. Møller, C. and S. Chandrasekhar: (Relativistisch entartetes Gas.) Monthly not. **95**, 673 (1935).
34. Nernst, W.: (Sternentwicklung.) Z. Physik **97**, 511 (1935).
35. Oppenheimer, J. R.: (Strahlung der Elektronen.) Z. Physik. **52**, 725 (1929).
36. Rosseland, S.: Astrophysik auf atomtheoretischer Grundlage. Berlin 1931.
37. Saha, M. N.: (Ionisierung in Sternen.) Z. Physik **6**, 40 (1921). — Proc. roy. Soc. A **99**, 135 (1921).
38. Sommerfeld, A. u. H. Bethe: Elektronentheorie der Metalle. Handbuch der Physik, 2. Aufl., Bd. 24/2, S.333. 1933.
39. Slater, J. C. and H. M. Krutter: (Feste Körper unter hohem Druck.) Physik. Rev. **47**, 559 (1935).
40. Sterne, T. E.: (Elementumwandlung im thermodynamischen Gleichgewicht.) Monthly not. **93**, 736, 767, 770 (1933).
41. Stoner, E. C.: (Entartetes Elektronengas.) Monthly not. **92**, 651 (1932).
42. Strømgren, B.: (Sternmodell.) Z. Astrophysik **2**, 345 (1931).
43. — (Wasserstoffgehalt der Sterne, Opazität.) Z. Astrophysik **4**, 118 (1932); **7**, 222 (1933).
44. — Thermodynamik der Sterne und Pulsationstheorie. Handbuch der Astrophysik, Bd.7, S.121 (1936).
45. Vogt, H.: ($\mathfrak{M}$-L-Beziehung.) Astron. Nachr. **226**, 301 (1936).
46. — Der innere Aufbau und die Entwicklung der Sterne. Erg. exakt. Naturwiss. **6**, 1 (1927).

An English Translation*
of Friedrich Hund's Review Article

Matter Under Very High Pressures and Temperatures.

by **F. Hund**, Leipzig.

With 10 Figures.

Contents.

1. Preliminaries 218
I. The Equation of State 219
   2. General Remarks 219
   3. Pressures and Temperatures Occurring on Earth 220
   4. The Ionization Temperature Range 222
   5. The Pressure Region of Atomic Deformation 222
   6. The Electron Gas 224
   7. The Neutron Gas 227
   8. The Region of Nuclear Transformation 229
   9. Deviations From Thermodynamic Equilibrium 232
   10. Radiation 233
II. Other Physical Properties 237
   11. The Energy Content 237
   12. Electrical and Thermal Conductivity 239
   13. Absorption of Light 242
   14. Energy Transport 245
III. The Occurrence of Very High Pressures and Temperatures 247
   15. The Planets 247
   16. Ordinary Stars 248
   17. The Compact Stars 253
References 256

---

* including minor corrections

List of Frequently Used Symbols.

$\hbar = \frac{h}{2\pi}$	Quantum of Action.	μM	Mean Particle Mass.
c	Speed of Light.	p	Pressure.
e	Elementary Charge.	T	Absolute Temperature.
k	BOLTZMANN's Constant.	ϱ	Density.
G	Gravitational Constant.	V	Volume.
m	Electron Mass.	E	Energy.
M	Proton Mass (also	ε	Energy per Particle.
	approximately equal to the	η	Energy per Unit Mass.
	Unit of Atomic Weight	N	Number.
	and to the Neutron Mass).	n	Number per Unit Volume.

1. Preliminaries. In the laboratory it is possible to investigate the behavior of matter at temperatures of up to a few thousand degrees and at pressures of up to some ten thousand atmospheres. However, in many areas of nature, particularly in the stars, one encounters temperatures and pressures that are several orders of magnitude higher. Today physics provides a good deal of information about the behavior of matter under such conditions. In the phase regions in which matter can be assumed to exist as atomic nuclei and electrons, we know the laws governing the construction of matter in their entirety. We know that these laws retain some of their validity when we move into the area of extreme pressures or temperatures where the atomic nuclei are no longer immutable. We do not yet have a complete knowledge of the forces between the components of the nuclei, but we do know the energy values of many states which are based on these forces.

Some essential features of the behavior of the components of matter are the following: the validity of the COULOMB law between electrically charged particles at separations as small as the diameter of the nucleus (10^{-13} cm); the wave–particle duality of matter; the limitations on our intuitive description of atomic events that are imposed by the quantum of action; the dicrepancies that arise in the statistical description of particle ensembles due to PAULI's exclusion principle; the construction of atomic nuclei from neutrons and protons; and the fact that these two types of particles can be transformed into one another.

Extremely high pressures and temperatures exist in the interior of stars. Astronomy, however, is only able to give us very indirect information about what goes on inside stars; indeed, the situation today is that physics can make some relatively certain statements about the behavior of matter under such conditions, information which astronomy and astrophysics employs to interpret observations and to draw conclusions from these observations about the actual processes and states.

The following report restricts itself largely to the physical aspects. The contruction of stars is only mentioned to the extent that it provides information about the actual existence of the states that we consider, and where regularities in the observed properties of stars provide support for the relevant laws of matter.

I. The Equation of State.

2. General Remarks. Our task is to investigate the properties of a homogeneous piece of material in thermodynamic equilibrium as a function of the state variables. As independent quantities characterizing the state we choose temperature T and pressure p. States that do not correspond to thermodynamic equilibrium, i.e. thermodynamically improbable states, also exist in nature. They even remain unchanged over long periods of time when the processes that help produce equilibrium only occur very slowly. An example of such an "unlikely" state is the occurence of protons outside the heavy nuclei, and thus the occurence of hydrogen at low temperatures. Over large ranges of pressure and temperature we can ignore the transmutability of elements. In these ranges one can legitimately extend the concept of "thermodynamic equilibrium" to the situation where nuclear transformation is absent. We can then speak of a particular chemical element in thermodynamic equilibrium.

We begin by considering the *"equation of state"*, i.e. the dependence of the density ϱ on p and T. It will emerge that, when relatively high values of temperature and pressure are reached, the particular nature of the material plays only a very minor role, allowing us to speak of a general equation of state of matter.

The special points on our extended scales of pressure, temperature, density, etc., are determined by certain absolute scales that are found in nature, i.e. by the quantities: speed of light c, quantum of action $\hbar$, elementary charge e, electron mass m and proton mass M, and BOLTZMANN's constant k. Occasionally one uses "atomic" units: $\hbar, e, m, k$; the unit of energy is then $\frac{me^4}{\hbar^2} = 4.31 \times 10^{-11}$ erg (27.1 eVolt, twice the ionization energy of the hydrogen atom). Other units within this system are given in Table 1. Since the COULOMB law and thus the elementary charge only plays an important role in the structure of matter at low pressures and temperatures, these units are largely irrelevant for our purposes. We will more frequently be using the units $\hbar, c, m, k$, thus, for example, the energy unit $mc^2 = 8.12 \times 10^{-7}$ erg (0.51 eM-Volt, the rest mass energy of an electron). For very extreme situations the units $\hbar, c, M, k$ will also be needed, thus, for example the energy unit $Mc^2 = 1.49 \times 10^{-3}$ erg (rest mass energy of a proton, also approximately that of the neutron). A summary of the most important units can be found in Table 1. In the system with the mass unit m, the quantity $M/$volume is introduced as a density unit, since electrons only ever occur together with heavy particles.

We assume initially that there is no gravitational field, i.e., we make a conceptual separation between the energy of the material particles in the gravitational field and the other contributions to their energy. However, since very high densities and pressures can only be generated through gravitation, it is not clear whether such a conceptual separation is allowed for all values of the state variables. But it is possible to treat the effect of gravitation separately whenever the energy of the particles in the gravitational field is small in comparison to their restmass energy, or (expressed differently) whenever the gravitational potential is small compared to c^2.

Table 1. Important Units of the Structure of Matter.

Basic Units	$\hbar, e, m, k$	$\hbar, c, m, k$	$\hbar, c, M, k$
Length	$\frac{\hbar^2}{me^2}=0.528\times10^{-8}$ cm	$\frac{\hbar}{mc}=3.84\times10^{-11}$ cm	$\frac{\hbar}{Mc}=2.09\times10^{-14}$ cm
Density.....	$\frac{Mm^3e^6}{\hbar^6}=11.3\,\frac{g}{cm^3}$	$\frac{Mm^3c^3}{\hbar^3}=2.92\times10^7\,\frac{g}{cm^3}$	$\frac{M^4c^3}{\hbar^3}=1.82\times10^{17}\,\frac{g}{cm^3}$
Energy	$\frac{me^4}{\hbar^2}=4.31\times10^{-11}$ erg	$mc^2=8.12\times10^{-7}$ erg	$Mc^2=1.49\times10^{-3}$ erg
Pressure....	$\frac{m^4e^{10}}{\hbar^8}=2.90\times10^8$ atm	$\frac{m^4c^5}{\hbar^3}=1.41\times10^{19}$ atm	$\frac{M^4c^5}{\hbar^3}=1.62\times10^{32}$ atm
Temperature	$\frac{me^4}{\hbar^2 k}=3.14\times10^5$ deg	$\frac{mc^2}{k}=5.92\times10^9$ deg	$\frac{Mc^2}{k}=1.09\times10^{13}$ deg

Thus, one may not approach a mass $\mathfrak{M}$ closer than a distance that is comparable to the length $G\mathfrak{M}/c^2$. To the units listed in Table 1 we add a mass unit of astronomic magnitude and a length unit. We do this by retaining the units of density (mass/volume), of pressure (energy/volume), and thus also of potential (energy/mass or velocity squared) from the last column; and we choose the new units so that at a separation of one length unit the mass unit has a gravitational field of one. We obtain the mass (G is the gravitational constant):

$$\left(\frac{\hbar c}{G}\right)^{3/2}\frac{1}{M^2}=3.68\times10^{33}\,\text{g}\,,$$

which corresponds to the mass of a typical star, and the length

$$\frac{\hbar^{3/2}}{G^{1/2}c^{1/2}M^2}=2.72\times10^5\,\text{cm}\,.$$

Thus provided that masses of the order of star masses do not approach closer to one another than distances comparable with this length, one can legitimately separate out the effect of gravitation.

3. Pressures and Temperatures Occurring on Earth. Under normal conditions of pressure and temperature simple materials exist either in the *"condensed" state* or they form a *gas of molecules*. The difference between the solid and the fluid state is of no great importance for our discussion here. If the molecules of a gas are polyatomic, they may dissociate at high temperatures into the component atoms. This too is of no great importance for the moment. Thus we assume that we have an ideal monatomic gas with the pressure

$$p=nkT$$

and the energy per unit volume

$$\frac{E}{V}=\frac{3}{2}nkT\,.$$

For approximate purposes we may consider the condensed state to be a state of constant density

$$\varrho=\varrho_k\,.$$

Thus we will not address any of the interesting area of high pressures that is examined by BRIDGMAN (6). At high temperatures and pressures gas and condensed state transform continuously into one another; at lower temperatures and pressures they are separate "phases" which border upon one another. This means that the borderline between the gaseous region and the condensed region in the p-T phase diagram does not represent a limit on the existence of these phases: it separates regions in which one or other of the phases is thermodynamically more likely. On the warmer side of the boundary, the gas is the more likely phase because of the higher statistical "weight" of its states. On the colder side the condensed state is favored because of the lower energy. This competition between statistical weight and energy can be investigated by using the independent variables p and T with the help of the "thermodynamic potential" $\Phi = E + pV - TS$ (E, V, S are energy, volume, and entropy of a given amount of material). If N_k and N_g are the numbers of atoms in the condensed and in the gaseous state, then (for a monatomic gas, weight 1 of the atomic state, assuming a constant specific heat γ in the condensed state, and neglecting small quantities):

$$E \quad = \quad N_k \gamma T + N_g \left(\frac{3}{2} kT + Q \right)$$

$$pV \quad = \quad N_g kT$$

$$S \quad = \quad N_k \gamma \ + N_g k \left[\frac{5}{2} - \log \frac{(2\pi)^{3/2} \hbar^3 p}{(\mu M)^{3/2} (kT)^{5/2}} \right].$$

Here Q is the heat of transformation per atom (extrapolated to $T = 0$) and μM is the atomic mass. The minimum of

$$\Phi = N_g \cdot \left[kT \log \frac{(2\pi)^{3/2} \hbar^3 p}{(\mu M)^{3/2} (kT)^{5/2}} + Q \right]$$

lies on the side of the gas if $\Phi < 0$, and in the region of condensed material if $\Phi > 0$. The boundary is at

$$p = \frac{\mu^{3/2}}{(2\pi)^{3/2}} \cdot \frac{M^{3/2}}{\hbar^3} (kT)^{5/2} e^{-\frac{Q}{kT}}.$$

This estimate loses its validity if we no longer have an ideal gas, that is, roughly speaking, when the formula

$$\varrho = \mu M \frac{p}{kT}$$

gives a gas density that is comparable with that of the condensed phase. In reality we then have a continuous transition between the gaseous and condensed phase, which does not have a simple equation of state. For approximate considerations we idealize the situation by means of the equation of state

$$\varrho = \begin{cases} \mu M \frac{p}{kT} & \text{für} \quad p \leq \frac{\varrho_k}{\mu M} kT \quad \text{und} \quad p \leq \frac{(\mu M)^{3/2} (kT)^{5/2}}{(2\pi)^{3/2} \hbar^3} e^{-\frac{Q}{kT}} \\ \\ \varrho_k & \text{für} \quad p \geq \frac{\varrho_k}{\mu M} kT \quad \text{oder} \quad p \geq \frac{(\mu M)^{3/2} (kT)^{5/2}}{(2\pi)^{3/2} \hbar^3} e^{-\frac{Q}{kT}}. \end{cases}$$

The boundaries are indicated in Fig. 2 for $\varrho_k = 8 \frac{g}{cm^3}$, $Q \approx 100 \frac{kcal}{gAtom}$, and $\mu = 56$; they correspond to the case of iron and its neighboring elements.

4. The Ionization Temperature Range. Moving now to even higher temperatures, we approach the point where the atoms become ionized. The mixture of atomic nuclei and electrons still has a higher energy but is made favorable because it has a much higher statistical weight than the gas of atoms. The first person to point out that at high temperatures a gas of electrons and nuclei exists in place of the gas of atoms was JEANS (23). The gradual ionization that occurs as the temperature rises was investigated by SAHA (37) in connection with applications to the stellar atmospheres. Precise information about the degree of ionization under various conditions is given by FOWLER and GUGGENHEIM (18). We will idealize the process by means of the "chemical formula"

$$\text{atom} \rightleftharpoons \text{remainder} + Z \text{ electrons}$$

and calculate the thermodynamic potential

$$\Phi = N_E \left[Q + kT \log \frac{(2\pi)^{3/2} \hbar^3 \, p_E}{m^{3/2} \, (kT)^{5/2}} \right] \; + \; N_R \cdot kT \log \frac{(2\pi)^{3/2} \hbar^3 \, p_R}{(\mu M)^{3/2} \, (kT)^{5/2}} \; +$$

$$+ \; N_A \cdot kT \log \frac{(2\pi)^{3/2} \hbar^3 \, p_A}{(\mu M)^{3/2} \, (kT)^{5/2}} \; ,$$

where the indices E, R, A refer to the electrons, the remainders, and the atoms, and p_E, p_R, and p_A are partial pressures. Q is the ionization energy per electron that is removed. For the "concentration" $\frac{N_E}{N_E + N_R + N_A}$, we introduce the abbreviation $[E]$ together with the corresponding abbreviations $[R]$ and $[A]$. With this notation the minimum of Φ lies at

$$[E] \times \sqrt[z]{\frac{[R]}{[A]}} = \frac{1}{(2\pi)^{3/2}} \frac{m^{3/2}(kT)^{5/2}}{\hbar^3 \, p} e^{-\frac{Q}{kT}} .$$

Half of the atoms are ionized when this expression is equal to $\frac{Z+2}{Z}$, i.e. when it is of the order of 1. We obtain a relatively rapid transition from the atoms to the mixture of ionized remainders and electrons in the vicinity of the dividing line

$$p = \frac{1}{(2\pi)^{3/2}} \frac{m^{3/2}(kT)^{5/2}}{\hbar^3} e^{-\frac{Q}{kT}} .$$

This boundary is shown in Fig. 2 for $Q = \frac{1}{2} \frac{me^4}{\hbar^2}$, which for light atoms corresponds approximately to the removal of the outermost electrons, and for $Q = \frac{26^2}{2} \frac{me^4}{\hbar^2}$, which corresponds to the removal of the innermost electrons of iron. At high temperatures the boundary becomes independent of Q: in atomic units $\hbar, m, k$ (with either e or c as the fourth unit), $p \approx \frac{1}{16} T^{5/2}$.

5. The Pressure Region of Atomic Deformation. If we suppose that a solid body or a fluid is compressed or expanded, then its energy changes in the manner indicated in Fig. 1. The state with no external force corresponds to the minimum energy. Ignoring thermal phenomena, we have the general relation

$$-\mathrm{d}E = p\, \mathrm{d}V.$$

The steep increase in the energy which accompanies a decrease in the volume means that extremely high pressures are necessary to compress the body by any significant amount. As in all cases where an atomic system becomes smaller than its normal size, this steep increase in the energy can be explained qualitatively in terms of the reduced volume that is available to the individual electrons. This leads to a strong increase in the momenta.

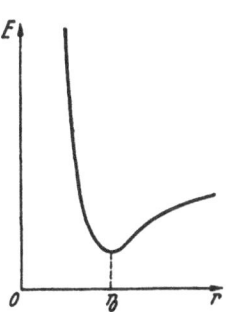

If we approximate the state of the whole system by means of single electron states and assume that the electrons are sufficiently numerous, then every cell of the phase space (with three spatial and three momentum dimensions) can, according to the PAULI principle, contain at most two electrons (FERMI statistics). The compression thus leads to an increase in the "zero point energy" (so called because it exists at zero temperature) of the electrons. Eventually the zero point energy becomes large in comparison to the COULOMB energy (which only increases with $\frac{1}{r}$). It is then possible to describe the body as an *ideal gas of electrons*, whose COULOMB forces have become ineffective due to the fact that the electrons are moving in a positive space charge (the nuclei). This gas is highly "degenerate", i.e., its zeropoint energy is large in comparison to the energy contribution that stems from the temperature.

Fig. 1. Schematic Diagram of the energy of a condensed body as a function of its volume.

The contribution of the nuclei to the pressure and energy is then insignificant. Whereas the behavior of matter at ordinary pressures is significantly influenced by the type of material considered, at high pressures all materials become more and more similar. The possibility that such states of matter exist was first suggested by FOWLER (17).

An approximate calculation of the increase in energy of a crystal lattice under compression, and thus the dependence of the volume on pressure, was carried out by SLATER and KRUTTER (39) under the simplifying assumption of many electrons moving in the force field of the nuclei (THOMAS–FERMI statistical method). This is not a good approximation at low pressures on account of the crude form in which the COULOMB forces are taken into account; the approximation improves as the influence of the COULOMB forces decreases, in other words, at high pressures. Using the SLATER–KRUTTER method and a suitable expansion procedure it is possible to make the transition to the electron gas. One then has an equation of state $\varrho = \varrho(p)$ for the transition region. Since the situation is actually rather complicated, we will schematize it as follows: at low pressures we will assume ϱ to be constant, $\varrho = \varrho_k$; at high pressures we will assume the behavior to be that of the electron gas (Sect. 6) and will take the border such that $\varrho(p)$ remains continuous. The transition only becomes significantly dependent on temperature when the states of the electrons is no longer degenerate, i.e., when kT becomes of the order of the atomic energy unit.

6. The Electron Gas. In a very wide range of temperature and pressure, the properties of matter are those of an *ideal gas of electrons* mixed with just the right number of nuclei to ensure that the whole system is electrically neutral. If the nuclear charge is fairly large in comparison to the charge on the proton, then the nuclei, because they are relatively few in number, do not have much influence in the equation of state. One of the prerequisites for an ideal gas is that the COULOMB energy is negligible in comparison to the kinetic energy of the particles. When both the temperature and the pressure are too low this is not the case. We have already considered the latter situation: at low pressures and high temperatures atoms are formed and at low temperatures and high pressures the matter exists in the condensed state. At higher pressures and temperatures the influence of the COULOMB forces soon becomes negligible and the charge on the electrons no longer appears in the equations. The other limit to the region of the ideal electron gas is the point at which the nuclei decay, because here the number of heavy particles included in the system becomes larger. This still makes relatively little difference in the equation of state (but certainly in the energy content). The region of the electron gas finally comes to an end when processes occur that change the number of electrons in a particular amount of material (transformation of electrons and protons into neutrons; transformation of radiation into electrons and positrons). Another limit to the region of the ideal electron gas would be densities so high that the particles approach close enough to one another that the noncoulombic forces, which determine the structure of the nucleus, become significant. However the previously mentioned limitations occur earlier so that this latter effect, analogous to the VAN DER WAALS modification of the equation of state of ordinary gases, does not occur for the electron gas.

The PAULI principle is important for setting up the equation of state; it has the consequence that even at zero temperature the electrons have a kinetic energy. At very high electron velocities one also has to take account for relativistic effects. To achieve a valid equation of state for the entire region is mathematically very complex. There are four limiting cases in which the equation is much simpler: when the zero point energy is large compared to the thermal energy (degenerate gas); when the reverse situation holds (nondegenerate gas); when the electron velocity is small compared to the speed of light (nonrelativistic gas); or when the electron velocity is almost equal to the speed of light (relativistic gas). We will give the equation of state, energy, and thermodynamic potential $\Phi = E + pV - TS$, for these four cases (7, 14, 17, 33, 41, 44).

Nonrelativistic nondegenerate gas:

$$p = nkT$$
$$E = \frac{3}{2}pV$$

$$\Phi = nV \times kT \log \frac{2^{1/2} \pi^{3/2} \hbar^3 p}{m^{3/2} (kT)^{5/2}} .$$

If we assume that n is the number of electrons per unit volume, then these equations give the electron contributions to the energy, pressure, and thermodynamic potential (in the logarithm in the last equation p is the partial pressure). Since in our case the nuclei are certainly not relativistic and not degenerate, we can also interpret the first two equations by taking n to be the total number of particles per unit volume. The equations then give the total translational energy and the total pressure. One can express n in terms of the density:

$$n = \frac{\varrho}{\mu M} ,$$

where μM is the mean mass of the particles. Thus for a gas containing electrons and protons, $\mu = \frac{1}{2}$. For a gas containing heavy nuclei and electrons, μ is somewhat larger than 2.

Nonrelativistic degenerate gas:

$$p = \frac{3^{2/3} \pi^{4/3}}{5} \times \frac{\hbar^2}{m} \times n^{5/3}$$

$$E = \frac{3}{2} pV$$

$$\Phi = \frac{5}{2} pV .$$

Here n is the number of electrons per unit volume. With the exception of hydrogen, the contribution of the nuclei is extremely small. It depends on whether or not the nuclei are also degenerate.

Relativistic nondegenerate gas:

$$p = nkT$$

$$E = 3pV$$

$$\Phi = nV \times kT \log \frac{\pi^2 \hbar^3 c^3 p}{2(kT)^4} .$$

These equations again express only the contribution of the electrons. If we take

$$n = \frac{\varrho}{\mu M} ,$$

we restrict ourselves to the case where the increase in the zero point energy of the electrons (which is larger than mc^2) remains insignificant in comparison to Mc^2.

Relativistic degenerate gas:

$$p = \frac{3^{1/3} \pi^{2/3}}{2^2} \times \hbar c \times n^{4/3}$$

$$E = 3pV$$

$$\Phi = 4pV .$$

The equation of state of the transition region from the nonrelativistic non-degenerate to the nonrelativistic degenerate gas has been given by FERMI (14). The equation of state for the transition from the nonrelativistic degenerate to the relativistic degenerate gas is to be found in STONER (41) and CHANDRASEKHAR (7). One can schematize the situation with relatively good accuracy by assuming that one of the above three equations of state is valid and choosing the boundaries of its validity so that the variables p, T, ϱ remain continuous. The boundary between nondegenerate and nonrelativistic degenerate regions is then to be found at

$$n = \frac{p}{kT} = \left(\frac{5mp}{3^{2/3}\, \pi^{4/3}\, \hbar^2} \right)^{3/5},$$

and thus at

$$p = \frac{5^{3/2}}{3\pi^2}\, \frac{m^{3/2}}{\hbar^3} (kT)^{5/2}.$$

The boundary between nondegenerate and relativistic degenerate regions is found correspondingly at

$$p = \frac{2^6}{3\pi^2}\, \frac{1}{\hbar^3\, c^3} (kT)^4,$$

and the boundary between the two degenerate regions at

$$p = \frac{5^4}{2^{10}\, 3\pi^2}\, \frac{m^4\, c^5}{\hbar^3}.$$

The "triplepoint" between the three regions thus lies at

$$p = \frac{5^4}{2^{10}\, 3\pi^2}\, \frac{m^4\, c^5}{\hbar^3} = 0.021 \times \frac{m^4\, c^5}{\hbar^3} = 2.9 \times 10^{17}\ \text{atm}$$

$$T = \frac{5}{2^4}\, \frac{mc^2}{k} = 0.31 \times \frac{mc^2}{k} = 1.8 \times 10^9\ \text{degree}$$

$$\varrho = \frac{5^3\, \mu}{2^6\, 3\pi^2}\, \frac{Mm^3\, c^3}{\hbar^3} = 0.067 \times \mu \frac{Mm^3\, c^3}{\hbar^3} = \mu \times 2.0 \times 10^6\ \frac{\text{g}}{\text{cm}^3}$$

in the vicinity of the units defined by $\hbar, c, m$, and k (Table 1).

For the boundary between the condensed state and the degenerate electron gas (Sect. 5) we obtain

$$p = \frac{3^{2/3}\, \pi^{4/3}}{5\mu^{5/3}} \left(\frac{\varrho k}{Mm^3\, e^6/\hbar^6} \right)^{5/3} \frac{m^4\, e^{10}}{\hbar^8} = \frac{1.9}{\mu^{5/3}} \left(\frac{\varrho k}{Mm^3\, e^6/\hbar^6} \right)^{5/3} \frac{m^4\, e^{10}}{\hbar^8}.$$

This boundary, together with that confining the nondegenerate electron gas, determines a further "triplepoint" :

$$T = \frac{3^{2/3}\, \pi^{4/3}}{5\mu^{2/3}} \left(\frac{\varrho k}{Mm^3\, e^6/\hbar^6} \right)^{2/3} \frac{me^4}{\hbar^2\, k} = \frac{1.9}{\mu^{2/3}} \left(\frac{\varrho k}{Mm^3\, e^6/\hbar^6} \right)^{2/3} \frac{me^4}{\hbar^2\, k}$$

$$\varrho = \varrho k$$

in the vicinity of the units defined by $\hbar, e, m$, and k. With $\mu = 2, \varrho_k = 8 \frac{g}{cm^3} = 0.7 \times \frac{Mm^3e^6}{\hbar^6}$, the boundary lies at 9.6×10^7 atm and at 3.0×10^5 degrees.

The boundaries that we have calculated here are shown in Fig. 2 along with the earlier discussed region of ionization of the gas of atoms.

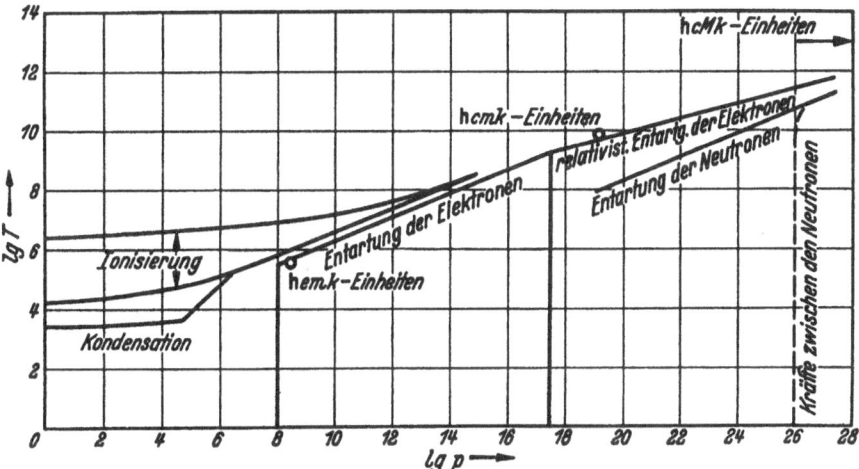

Fig. 2. Important boundaries in the p-T phase diagram (logarithmic scale).

One sees that a narrow region remains between the calculated ionization boundary and the boundary at which the electron gas becomes degenerate (these lines differ at high pressure by a factor $\frac{10^{3/2}}{3\pi^{1/2}} = 6.0$ in p). Thus in reality we have here a transition region between a degenerate electron gas and a partially ionized gas of atoms. Since, at these pressures, a slightly ionized gas of atoms can no longer be considered as an ideal gas, we have here a rather complicated situation. At somewhat stronger, but still incomplete, ionization one can view the whole system as an electron gas containing highly ionized atomic cores (remainders), which, in comparison to the electron gas containing nuclei, represents a small increase in μ. Even in the degenerate region itself, a small increase in μ upon approaching the normal condensed state is a better description of the situation than a constant μ.

7. The Neutron Gas. The area in which the matter can be described as an electron gas is limited on the side of high pressure and temperature by the fact that the nuclei themselves are finally transformed. To describe this situation mathematically we need to know the properties of a gas consisting of electrons, nuclei, protons, and neutrons. One can deduce these properties from the information given in Sect. 6 taking into account the fact that we now have different masses. For a gas that consists only of neutrons, we obtain in the nonrelativistic nondegenerate state

$$p = nkT \qquad\qquad n = \frac{\varrho}{M}$$

$$\Phi = nV \times kT \log \frac{2^{1/2} \pi^{3/2} \hbar^3 p}{M^{3/2} (kT)^{5/2}},$$

and in the nonrelativistic degenerate state

$$p = \frac{3^{2/3} \pi^{4/3} \hbar^2}{5} \frac{\hbar^2}{M} n^{5/3}$$

$$\Phi = \frac{5}{2} pV.$$

If we once more replace the transition region by a boundary where one equation continuously adjoins the other, then this boundary is at

$$p = \frac{5^{3/2}}{3\pi^2} \frac{M^{3/2}}{\hbar^3} (kT)^{5/2};$$

it is drawn on Fig. 2. Later on we will see that there is a region of very high pressures where matter behaves as a neutron gas.

At such high pressures we have to be careful that the prerequisites for the ideal gas are not invalidated due to a nonnegligible contribution of the forces between the particles. We know from the structure of nuclei that forces act between protons and neutrons when these particles approach one another to within about 10^{-13} cm, and there are good reasons to believe that at such separations there are also forces between one neutron and another. To be quite sure, we will take the boundary of our gas region to be at the point where the particles reach such separations; thus for the neutron gas at

$$\left(\frac{M}{\varrho}\right)^{1/3} \approx 10^{-13} \text{ cm} \quad - \quad 10^{-12} \text{ cm}.$$

This corresponds to pressures of $10^{24} - 10^{28}$ atm. At present we know nothing about the behavior at somewhat higher pressures because of our incomplete knowledge of the forces between the particles. At extremely high pressures, however, the energy density will finally become large in comparison to the density of the rest mass energy nMc^2, so that the energy E calculated thus far as the excess above the rest mass energy can now be set equal to the entire relativistic energy

$$\frac{E}{V} = \varrho c^2$$

(by weighing the system one would measure its entire energy). If all particles are moving approximately at the speed of light, then the pressure is

$$p = \frac{1}{3} \frac{E}{V} = \frac{1}{3} \varrho c^2.$$

Thus in this region we know the equation of state once more. It is identical for every type of material [J.V.NEUMANN in (7)]. However it is doubtful whether this result really makes any sense since such high pressures can only be generated with the help of the gravitational forces of large masses. And in such a high gravitational field the energy of the particles would be of the same order of magnitude as their rest mass energy (cf. Sects. 2 and 17).

8. The Region of Nuclear Transformation. If, among several possible states, there is one state whose energy is lower and whose statistical weight is larger than those of the others, then this is the state that occurs in a thermal equilibrium. The same remains true if the state has a somewhat higher energy but nonetheless a much greater statistical weight; likewise if, despite a somewhat lower statistical weight, it has a much lower energy.

From our knowledge of the structure of nuclei, we must assume that a heavy nucleus can be transformed into a lighter nucleus by its losing protons or neutrons. It could even be transformed entirely into protons and neutrons. For the time being we will restrict ourselves to these processes. At low temperatures only the energies are relevant and, in thermodynamic equilibrium, only the nuclei of lowest energy exist. These are nuclei with atomic weights in the vicinity of 100. The fact that, in reality, even at low temperatures, materials also contain other nuclei, is an indication of the deviation from thermodynamic equilibrium. Indeed, at low temperatures, there is no process which helps to produce the equilibrium. At high temperatures the statistical weight of an (electrically neutral) gas of protons, neutrons, and electrons becomes progressively larger in comparison to the statistical weight of a gas composed of heavy nuclei and electrons. This advantage of the gas comprising lighter particles finally overtakes the disadvantage of its higher energy. At high temperatures and in thermodynamic equilibrium there will eventually no longer be any heavy nuclei.

From what is known about the β decay of nuclei, one can conclude that protons can transform into neutrons by absorbing electrons or emitting positrons. Since a neutron has a higher energy than a proton plus an electron, when we consider the thermodynamic equilibrium we first have to look at the process

$$\text{proton} + \text{electron} \rightleftharpoons \text{neutron}.$$

At higher temperatures (where no heavy nuclei exist) the higher thermal energy of the protons plus electrons (because the number of particles is twice as high) is unfavorable, whereas the lower particle energy and the larger statistical weight is favorable. At lower temperatures, where only the energy is important, the high zero point energy of the electrons is unfavorable. At high pressures it can prove to be favorable for the electrons and the nuclei together to transform into neutrons [STERNE (40)].

An exact description of the situation would be very complicated, since we would really need to consider all the possible types of nucleus. However, the essential ingredients should be correctly described if, in addition to protons and neutrons, we simply consider one type of nucleus; it is assumed to contain Z protons and Z neutrons. Our simplification thus assumes the possible reactions

$$\text{nucleus} \rightleftharpoons Z \text{ neutrons} + Z \text{ protons}$$
$$\text{proton} + \text{electron} \rightleftharpoons \text{neutron}$$

and thus also the reactions

$$\text{nucleus} \quad \rightleftharpoons \quad 2\,Z \text{ protons} + Z \text{ electrons}$$
$$\text{nucleus} + Z \text{ electrons} \quad \rightleftharpoons \quad 2\,Z \text{ neutrons.}$$

Since we expect the transformation processes to occur in regions of different equations of state of the particles, we will be omitting only a few important cases if we carry out the calculation for the following systems:

a) nonrelativistic nondegenerate electrons and heavy particles;

b) nondegenerate electrons and heavy particles, relativistic electrons, non-relativistic heavy particles;

c) relativistic degenerate electrons, nonrelativistic nondegenerate heavy particles;

d) relativistic degenerate electrons, nonrelativistic degenerate heavy particles.

In case a), for given values of p and T, the minimum of the thermodynamic potential is to be expected at

$$\Phi = N_E \left[kT \log \frac{2^{1/2}\,\pi^{3/2}\,\hbar^3\,p_E}{m^{3/2}\,(kT)^{5/2}} - Q \right] + N_P \times kT \log \frac{2^{1/2}\,\pi^{3/2}\,\hbar^3\,p_P}{M^{3/2}\,(kT)^{5/2}} +$$

$$+ N_N \times kT \log \frac{2^{1/2}\,\pi^{3/2}\,\hbar^3 p_N}{M^{3/2}\,(kT)^{5/2}} + N_K \left[kT \log \frac{\pi^3\,\hbar^6\,p}{M_K^3\,a^3\,(kT)^4} - 2ZR \right].$$

Here N_E, N_P, N_N, N_K are the numbers of electrons, protons, neutrons, and nuclei; p_E, p_P, p_N, and p_K are the corresponding partial pressures; M_K is the nuclear mass, and a is a length of the order of the nuclear size. The last term in the equation takes into account the fact that the nuclei are rotating at the temperatures that we consider. However this makes hardly any difference practically. The quantity Q is the amount by which the rest-mass energy of a neutron exceeds that of a proton plus electron, and $2ZR$ is the amount by which the rest mass energy of Z neutrons and Z protons exceeds that of a nucleus. For the numbers of individual particles one obtains equations of the type that are valid for ordinary chemical reactions. For temperatures $T \geq \frac{R}{k}$ one no longer obtains any nuclei. At pressures that are not too high one obtains only protons and electrons; for high pressures (and temperatures that are not too high) only neutrons are found. In case b) the result is similar except in the region of extremely high pressures. In case c) the region of neutrons extends, at high pressures, also to low temperatures. The interesting case d) will be discussed in greater detail. Here it is simpler to take V and T as the independent variables. The thermodynamic equilibrium is then determined by the fact that the free energy, which is here the same as the energy, has a minimum. It is given by

$$\frac{E}{V} = \frac{3^{4/3}\,\pi^{2/3}}{2^2} \hbar c n_E^{4/3} - Q n_E + \frac{3^{5/3}\,\pi^{4/3}\,\hbar^2}{2 \times 5} \left[\frac{n_P^{5/3}}{M} + \frac{n_N^{5/3}}{M} + \frac{n_K^{5/3}}{M_K} \right] - 2ZRn_K,$$

where $n_E = n_P + Zn_K$. In searching for the minimum it is necessary to hold constant $n_P + n_N + 2Zn_K$. At low values of this number, $n_P = n_N = 0$ at the minimum. Neutrons occur for the first time when

$$n_E = \frac{1}{3\pi^2}\frac{1}{\hbar^3 c^3}(2R + Q)^3.$$

At even higher densities the nuclei and electrons rapidly disappear. The tranformation of matter into neutrons occurs quite suddenly at the pressure

$$p = \frac{1}{12\pi^2}\left(\frac{2R + Q}{mc^2}\right)^4 \times \frac{m^4 c^5}{\hbar^3}.$$

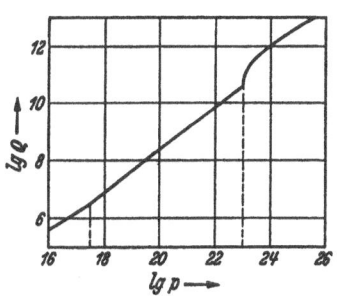

Fig. 3. The behavior of the density in the vicinity of the transformation into neutrons.

From the known atomic weights it follows that R is about $0.008\,Mc^2$ or $15\,mc^2$. The difference in the mass of neutrons and protons is only known to within a considerable uncertainty. With the "atomic weight" of the neutron as 1.0085 ± 0.0005 (3, 15) and that of the proton plus electron as 1.0080, we obtain $Q \approx mc^2$ with an uncertainty which is equally large. Thus we obtain the transformation into neutrons at a pressure of about 1×10^{23} atm, i.e., at a pressure where one can still consider electrons or neutrons to behave as an ideal gas. Figure 3 shows the dependence of the density on pressure for the equilibrium mixture

$$p = \frac{3^{1/3}\,\pi^{2/3}}{4}\hbar c n_E^{4/3} + \frac{3^{2/3}\,\pi^{4/3}}{5}\frac{\hbar^2}{M}n_N^{5/3},$$

the relation between n_E and n_N following from the equilibrium condition, and

$$\varrho = \mu M n_E + M n_N; \qquad \mu = 2.0.$$

At pressures below the neutron transformation the material is fairly compressible (but less so than the ideal nondegenerate gas). At the transformation point itself the material becomes significantly more compressible, and at even higher pressures somewhat less compressible than it was previously.

Using the results for the four calculated cases described above it is possible with only a small uncertainty to plot the boundaries between the areas of electrons and nuclei, electrons and protons, and neutrons. These areas are shown in Fig. 4. A change in the value of Q would hardly influence this diagram. At high temperatures some additions have to be made: one can no longer neglect radiation pressure and should not forget the possibilities that protons transform into neutrons and positrons or that electron–positron pairs are created by radiation.

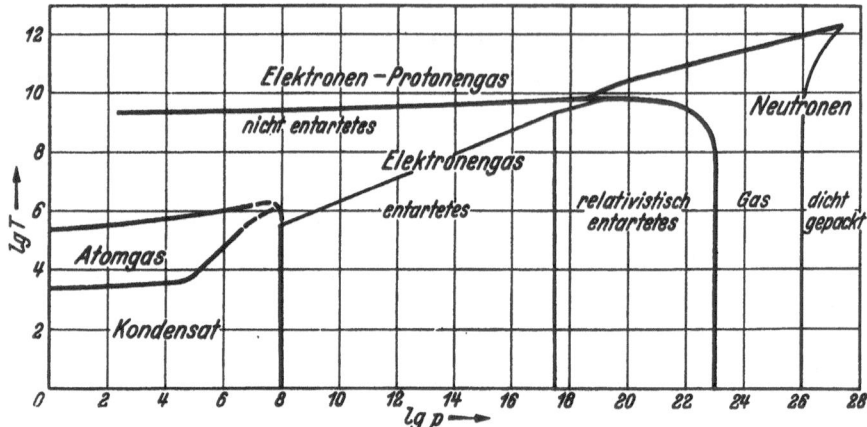

Fig. 4. Phase diagram of matter.

So far we have considered thermodynamic equilibrium without paying attention to the processes which help to produce it. At low pressures we know that nuclei are not transformed; thus, deviations from equilibrium can be maintained practically indefinitely. At the high temperatures and pressures where equilibrium lies on the side of protons and neutrons, all particles have such high energy that transformation processes can occur frequently. Thus it is here that equilibrium is produced.

9. Deviations From Thermodynamic Equilibrium.   At low temperatures and small pressures there is no process that can transform non-radioactive nuclei. Deviations from thermodynamic equilibrium in the composition of nuclei, such as the existence of light nuclei, cannot be eliminated. The reason why light nuclei do not combine to yield heavy nuclei is the COULOMB repulsion between them, which prevents them from coming too close together. Only when the particles obtain very high velocities does this situation change. For a simple model of the nucleus, GAMOW (20) and ATKINSON and HOUTERMANS (2) have calculated the probability that a charged particle penetrates into a heavy nucleus with charge Ze. For the range of velocities considered initially the penetration probability of a proton per collision is given by

$$e^{-2Z\frac{e^2}{\hbar v}} = e^{-\sqrt{2}Z\sqrt{\frac{Me^4/\hbar^2}{\epsilon}}},$$

where v is the velocity of the proton ($\frac{e^2}{\hbar}$ is the atomic velocity unit) and ϵ is its kinetic energy. When this energy reaches $Z^2\frac{M}{m}$ times the unit $\frac{me^4}{\hbar^2}$ or (if the process occurs in temperature equilibrium) when the temperature reaches $Z^2\frac{M}{m}$ times the unit $\frac{me^4}{\hbar^2 k}$ (3×10^{-5} degrees), then the probability approaches one. At a temperature of 4×10^7 degrees (the temperature assumed to exist in the central region of stars) ATKINSON and HOUTERMANS obtained a mean penetration probability (taking account of the MAXWELLian velocity distribution) for a proton per collision with a Li nucleus of 10^{-10}, and for a Ne nucleus a mere 10^{-24}. The fact that a proton has penetrated into a nucleus does not necessarily mean that this will be transformed into a heavier nucleus. However, experimental results concerning the transformation of light elements under bombardment with protons appear to show that the transformation probability is not much smaller than the calculated penetration probability.

At temperatures of about 4×10^7 degrees protons and Li nuclei would only be able to exist simultaneously for any length of time if they were extremely dilute. In the presence of Ne nuclei, protons could exist at densities of the order of $1\,\mathrm{g/cm}^3$ for time periods comparable with the lifetime of a star. We will return later to discuss the significance of this circumstance for the energy content and energy transformation in matter that is not in thermodynamic equilibrium and for the energy generation in stars.

10. Radiation. Any volume, whether it is filled with matter or empty, also contains radiation. In temperature equilibrium the radiation in vacuum has an energy density of

$$\frac{E}{V} = \frac{\pi^2}{15} \frac{(kT)^4}{\hbar^3 c^3} = aT^4.$$

The radiation exerts a pressure of

$$p = \frac{1}{3}\frac{E}{V} = \frac{\pi^2}{45} \frac{(kT)^4}{\hbar^3 c^3} = \frac{a}{3}T^4.$$

The energy of the radiation is distributed over the various frequencies according to PLANCK's law. The maximum energy in a wavelength interval occurs at a wavelength

$$\lambda = \frac{2\pi}{4.965} \frac{\hbar c}{kT},$$

for the temperature $\frac{mc^2}{k}$ (5.9×10^9 degrees), i.e. in the vicinity of the "COMPTON wavelength", $\frac{2\pi\hbar}{mc}$ (2.4×10^{-10} cm).

If the volume considered is filled with matter, then the separation into the energy of the radiation and the energy of the matter can be carried out only when the interaction between the radiation and the matter is weak. If for example the matter is nonabsorbing and its optical behavior can be described by a velocity of light v or an index of refraction $n = \frac{c}{v}$, then we have

$$\frac{E}{V} = \frac{\pi^2}{15} \times \frac{c^3}{v^3} \times \frac{(kT)^4}{\hbar^3 c^3}.$$

In a gas of atoms, the quantity $n^2 - 1$ is of the order of the ratio of the volume occupied by the atoms to the total volume, and thus small. In a gas composed of electrons and nuclei the interaction of radiation and matter is also small, provided the gas is not degenerate and the temperature does not reach the vicinity of $\frac{mc^2}{k}$.

In our p-T diagram (Figs. 2 and 4), which has thus far only given the pressure originating from the matter, let us draw the straight line

$$p = \frac{(kT)^4}{\hbar^3 c^3} .$$

On this line the given pressure of the matter has to be supplemented by a radiation pressure that is about 1/5 of this amount. Below this line the radiation pressure can be neglected (since n varies with T^4 and the matter pressure at most with T). Almost coincident with this line is the boundary between the nondegenerate and the relativistic degenerate electron gas

$$p = \frac{64}{3\pi^2} \frac{(kT)^4}{\hbar^3 c^3} .$$

Here the ratio of radiation pressure to the pressure of matter is given by

$$\frac{p_{rad}}{p_{mat}} = \frac{\pi^4}{15 \times 64} \approx \frac{1}{10} ,$$

if one carries out the calculation as for a vaccuum. Above the straight line

$$p = \frac{(kT)^4}{\hbar^3 c^3}$$

the radiation pressure soon exceeds the pressure of the matter. If in the p-T diagram we give the total pressure, then the diagram is bounded by the straight line

$$p = \frac{\pi^2}{45} \frac{(kT)^4}{\hbar^3 c^3}$$

(Fig. 5), since for a given temperature the total pressure cannot be smaller than this. As one approaches the limit the density of the matter decreases rapidly. But the energy of the radiation also possesses mass and thus a mass density

$$\varrho = \frac{1}{c^2} \frac{E}{V} = \frac{3p}{c^2} .$$

Thus the radiating vacuum has the same equation of state as the relativistic matter. In the region where the heavy particles of matter have velocities that are significantly lower than the velocity of light, the total pressure is

$$p = p_{rad} + p_{mat} = \frac{c^2}{3} \varrho_{rad} + \frac{kT}{\mu M} \varrho_{mat} ;$$

in the region where the heavy particles have a velocity that is almost equal to the speed of light, we have

$$p = p_{rad} + p_{mat} = \frac{c^2}{3} (\varrho_{rad} + \varrho_{mat}) = \frac{c^2}{3} \varrho .$$

In the region where the radiation pressure and matter pressure are comparable, our p-T diagram is very narrow. Thus the problem that arose for strong interaction between radiation and matter, i.e. the fact that we could not say anything straightforward about the energy density of the radiation, turns out not to be very significant.

Following the discovery of the positron, the prediction made from DIRAC's theory of the electron concerning the possibility of the generation of electron-positron pairs became more plausible. The prerequisite for pair creation is radiation of sufficiently high frequency:

$$\hbar\omega \geq 2mc^2 \; ; \qquad \lambda \leq \pi\frac{\hbar}{mc}.$$

Taking into account this transformation possibility, in thermodynamic equilibrium at low temperatures one finds radiation and electrons but no positrons; at higher temperatures pairs of positrons and electrons are created from the radiation. Processes that lead to the thermal equilibrium do indeed exist. The thermodynamic potential Φ of pure radiation is found from the entropy

$$S = \frac{4}{3}aT^3V$$

to be zero. If N_+ and N_- are the numbers of positrons and electrons, and p_+ and p_- are the partial pressures, then for the entire system we have

$$\Phi = N_+\left[mc^2 + kT\log\frac{2^{1/2}\pi^{3/2}\hbar^3 p_+}{m^{3/2}(kT)^{5/2}}\right] +$$
$$+ N_-\left[mc^2 + kT\log\frac{2^{1/2}\pi^{3/2}\hbar^3 p_-}{m^{3/2}(kT)^{5/2}}\right].$$

Thus, in thermodynamic equilibrium

$$p_+p_- = \frac{m^3}{2\pi^3\hbar^6}(kT)^5 e^{-\frac{2mc^2}{kT}},$$

i.e., a significant level of pair creation occurs when

$$\frac{p}{(kT)^{5/2}} > \frac{m^{3/2}}{2^{1/2}\pi^{3/2}\hbar^3}e^{-\frac{mc^2}{kT}}.$$

Let us compare the pressure due to pair creation

$$p_{\text{mat}} = \frac{m^{3/2}}{2^{1/2}\pi^{3/2}\hbar^3}(kT)^{5/2}e^{-\frac{mc^2}{kT}}$$

with the radiation pressure

$$p_{\text{rad}} = \frac{\pi^2(kT)^4}{45\hbar^3 c^3}.$$

The ratio

$$\frac{p_{\text{mat}}}{p_{\text{rad}}} = \frac{45}{2^{1/2}\pi^{7/2}}\left(\frac{mc^2}{kT}\right)^{3/2}e^{-\frac{mc^2}{kT}}$$

has a maximum value of about $1/10$, and so pair creation scarcely changes the overall picture. Some indication that pair creation has a minor effect on the equation of state is reported by CHANDRASEKHAR and ROSENFELD (8).

For a hypothetical process in which positive and negative particles of the proton mass are created we can calculate the equilibrium in a corresponding manner. Such pairs could only be expected in significant numbers when $kT > Mc^2$ is reached.

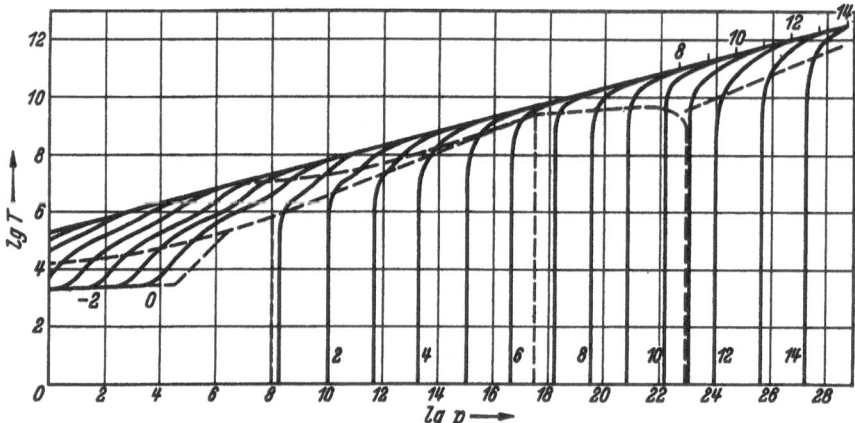

Fig. 5: Density of matter in the p-T diagram (taking account of radiation).

Figure 5 shows the density ϱ (plotted is $\log_{10}$ of the density measured in g/cm^3) as a function of the temperature T and the total pressure p (from matter and radiation pressure) for matter and radiation in thermodynamic equilibrium. With the exception of low pressures, Fig. 5 is valid for all materials; at low pressures we show here values corresponding to iron.

Figure 6 is a perspective representation as a block diagram of the area connecting together the values of p, T, and ϱ which belong together. Let us describe the behavior of matter in general terms by means of *compressibility* and *thermal expansion coefficients*. With the exception of the normal condensed state, one finds that matter is fairly compressible, but in general that it is less compressible than the nondegenerate ideal gas; the thermal expansion over large ranges (the degenerate regions) is virtually zero, and at higher temperatures it is that of the nondegenerate ideal gas; in the region of ionization and of nuclear dissociation it is even higher and where the radiation pressure dominates it becomes very large. It is essentially the compressibility and the thermal expansion that determine the relevance of the individual state regions for the equilibrium in the center of stars.

We can ignore quite a large range of temperature and pressure on account of the fact that the forces between the particles play only a minor role, and so the matter is an ideal gas. The other properties of matter in this region also present no problems (see next sections). This does not mean that the processes that are important in the center of stars are completely explained; indeed these processes can be extremely complicated. However, concerning the basics of these processes in the relevant areas there remains no doubt. The region which remains unknown is at higher pressures and higher temperatures than those found in ordinary stars and also at higher gravitational fields.

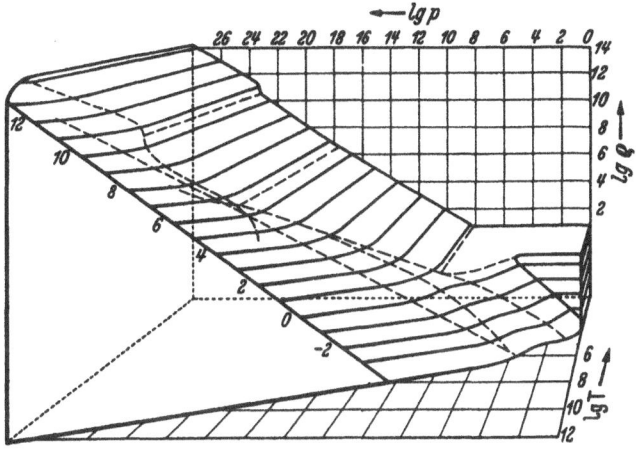

Fig. 6. Relationship between p, T, ϱ.

II. Other Physical Properties.

11. The Energy Content. In addition to the equation of state (relationship between pressure, temperature, and density), a further important factor determining the thermal behavior of matter is its energy content. In a representation of the energy content as a function of pressure and temperature the prominent areas are those in which strong transformations occur in the structure of matter, whereas in the state diagram it is the changes that effect the electrons that are particularly noticeable.

We specify the energy as the energy per unit mass

$$\eta = \frac{E}{\varrho V}$$

(it has dimension c^2). As the zero of our energy scale we take the energy of the condensed state at $p = 0$, $T = 0$. Neglecting the complicated situation in the transition regions, we obtain for the individual simple states the following: the condensed state (provided it is really incompressible according to the DULONG–PETIT rule) has

$$\eta = \frac{3kT}{\mu M} .$$

The gas of atoms has

$$\eta = \frac{\epsilon_1}{\mu M} + \frac{3kT}{2\mu M} \qquad \frac{\eta - \eta_1}{c^2} = \frac{3}{2\mu} \frac{m}{M} \frac{kT}{mc^2},$$

where, for very involatile substances, ϵ_1 has values between 1/10 and 1/5 of the atomic unit $\frac{me^4}{\hbar^2}$ or values of about $\frac{1}{2}10^{-5}$ to 10^{-5} of the unit mc^2. The nonrelativistic nondegenerate electron gas has

$$\eta = \frac{\epsilon_2}{\mu M} + \frac{3kT}{2\mu M},$$

where μ now lies in the vicinity of 2 (except for hydrogen) and ϵ_2 is of the order of $Z^{4/3} \times \frac{m e^4}{\hbar^2}$ [according to MILNE (30) using a statistical treatment of the electron shell with the method of THOMAS and FERMI, and placing a numerical factor of 0.77 in front; in comparison, the evaporation energy ϵ_1 can be neglected]. For the nonrelativistic degenerate electron gas we have

$$\eta = \frac{\epsilon_2}{\mu M} + \frac{3^{7/5}\,\pi^{4/5}}{2 \times 5^{3/5}}\frac{\hbar^{6/5}}{\mu\,M\,m^{3/5}}p^{2/5} \qquad \frac{\eta - \eta_2}{c^2} = \frac{3^{7/5}\,\pi^{4/5}}{2 \times 5^{3/5}}\frac{m}{\mu\,M}\left(\frac{p}{m^4\,c^5/\hbar^3}\right)^{2/5}$$

with the same value of ϵ_2. For the relativistic nondegenerate electron gas

$$\eta = \frac{\epsilon_2}{\mu M} + \frac{3kT}{\mu M} \,,$$

and for the relativistic degenerate electron gas

$$\eta = \frac{\epsilon_2}{\mu M} + \frac{3^{5/4}\,\pi^{1/2}}{2^{3/2}\,\mu}\frac{(\hbar c)^{3/4}}{M}p^{1/4} \qquad \frac{\eta - \eta_2}{c^2} = \frac{3^{5/4}\,\pi^{1/2}}{2^{3/2}\,\mu}\frac{m}{M}\left(\frac{p}{m^4\,c^5/\hbar^3}\right)^{1/4} .$$

For the nonrelativistic neutron gas we obtain

$$\eta = \frac{\epsilon_3}{M} + \frac{3kT}{2M} \qquad \frac{\eta - \eta_3}{c^2} = \frac{3}{2}\frac{kT}{M c^2}$$

and

$$\eta = \frac{\epsilon_3}{M} + \frac{3^{7/5}\,\pi^{4/5}}{2 \times 5^{3/5}}\frac{\hbar^{6/5}}{M m^{3/5}}p^{2/5} \,,$$

where ϵ_3 is our previous $\frac{1}{2}(2R + Q)$, the transformation energy of the nuclei and electrons into neutrons for one neutron (in comparison the quantity ϵ_2 can be neglected). Thus η_3 is of the order of $0.008c^2$. For the nondegenerate proton-electron gas we have correspondingly

$$\eta = \frac{\epsilon_4}{\mu M} + \frac{3kT}{2\mu M} \,,$$

with $\mu = \frac{1}{2}$ and $\epsilon_4 = \frac{1}{2}(2R - Q)$. For fully relativistic matter, as for radiation in a vacuum, we finally have

$$\eta = c^2 .$$

Figure 7 gives the energy per gram as a function of p and T; the numbers represent the values of η in $\log_{10}$, with η measured in $10^7\,\frac{\text{erg}}{\text{g}}$. At the transition from the condensed state to the degenerate electron gas it is not possible to give the exact values of the energies. The idealized equation of state of Sect. 5 would predict no pressure dependence of the energy up to the transition point and then a rapid increase which would finally join onto the curve specified above for the electron gas. The rapid succession of η curves at this point (Fig. 7) results from the artificial use of a logarithmic scale for η.

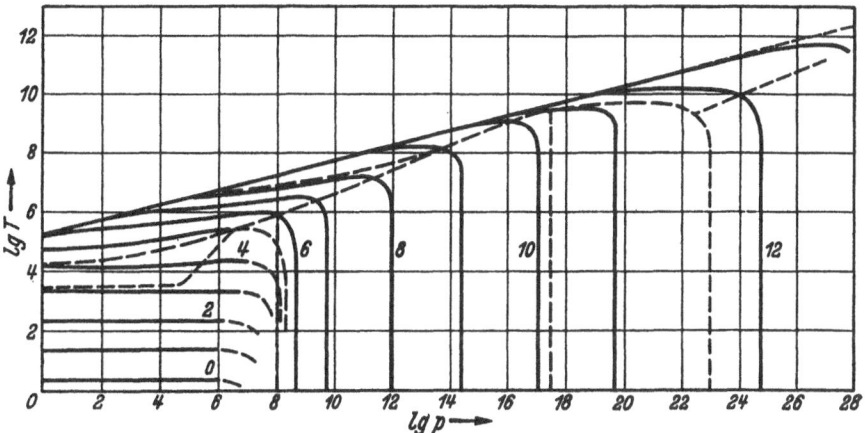

Fig. 7. Energy Content of Matter.

For comparison with these values of the energy per gram, note that the sun radiates almost $2\,\frac{\text{erg}}{\text{g}}$ per second, and thus it requires two months to emit the energy corresponding to the unit used here of $10^7\,\frac{\text{erg}}{\text{g}}$. Thus, if one reduces the numbers given in Fig. 7 by the summand $\log 6 = 0.78$, then one obtains the $\log_{10}$ of the number of years for which the matter, in a given initial state, could continuously emit energy at the same rate as the sun (not including gravitational energy). Note, however, that the figure gives the energy content of the matter in thermodynamic equilibrium. In the presence of light nuclei in a system that deviates from equilibrium the energy content is much larger. Thus, for example, due to their tranformability into heavy nuclei, one finds for protons an energy content per unit mass of $0.008\,c^2$ or $7 \times 10^{11} \times 10^7\,\frac{\text{erg}}{\text{g}}$.

In our overview of the energy content, the rest mass energy of heavy particles does not appear, because in the regions of pressure and temperature considered, there are no tranformations that alter this energy. The largest energy per unit mass that occurred was that of the tranformation of heavy nuclei into protons and neutrons. We have consistently ignored gravitational energy. Indeed its change in ordinary stars is small compared to the other contributions to the energy. This is no longer the case if the star, in the course of its evolution, becomes very small. The absolute magnitude of the (negative) gravitational energy of a whole star can then become comparable with the rest-mass energy of the total stellar material. It is then no longer possible to separate out the gravitational energy in our considerations and we find ourselves in an unknown region.

12. Electrical and Thermal Conductivity. In the state consisting of a gas of atoms, matter is an insulator. In the metallic or electron-gas state it contains mobile electrons which are responsible for the electrical and thermal current. If the electrons were completely free, then we would have an infinitely large conductivity. The electric and thermal resistance arises from the scattering of electrons in the force fields of the nuclei. Our experience with the quantum theory of metallic conductivity indicates that we can ignore the interaction of the electrons with one another, insofar as this is not already included in a statistical force field that is included in the force field of the nuclei. (An exception is the regime of very low temperatures, as shown by the phenomenon of superconductivity.)

We first consider the *electrical conductivity*. This can be written in the form

$$\sigma = \frac{e^2 nl}{mv},$$

where l (for temperatures that are not too low) represents a mean free path for the conduction electrons between two scattering processes, v is the average velocitiy of the electrons, and n is the number of electrons per unit volume. The path length of the electrons is not limited by the cross section πr^2 of the nuclear force fields, otherwise we would have

$$l = \frac{1}{\pi r^2 \, n_K}$$

(n_K is the number of nuclei per unit volume). In fact, a completely regular lattice arrangement of nuclei does not limit the path length at all and also gives infinite conductivity (5, 38). The force fields of the nuclei only effect the electrons when the nuclear positions deviate from the lattice symmetry. For the cross section that limits the path length, one is thus led to take

$$\pi r^2 \frac{d^2}{a^2}, \qquad \frac{1}{a^3} = n_K,$$

where d is the mean deviation from the lattice points and a the separation of the lattice points. If the deviation from the ideal lattice is very large, in other words for completely random positions of the nuclei, then πr^2 should give the correct limit. From this rather crude picture we obtain

$$l = \frac{1}{\pi r^2 \, d^2 \, n_K^{5/3}} \qquad d \le \frac{1}{n_K^{1/3}}.$$

If we have an ideal crystal, where the deviations d of the nuclei from the lattice points are determined purely by thermal vibrations, then we have

$$\frac{1}{2}\overline{\omega^2} M_K \, d^2 = kT,$$

where M_K is the nuclear mass and $\overline{\omega^2}$ is an average value of the square of the frequency ω. Henceforth, neglecting numerical factors, we will write

$$\hbar\omega = k\Theta,$$

and thus we have

$$d^2 = \frac{\hbar^2 \, T}{M_K \, k\Theta^2}.$$

If ϵ is the energy of the electrons, then we choose for r the distance from the nuclei within which the potential energy has a magnitude greater than ϵ. Thus

$$r = \frac{Ze^2}{\epsilon}$$

(here Z is the number of conduction electron per atom, $n = Zn_K$). We thus obtain the estimate

$$l = \frac{M_K \, \epsilon^2 \, k\Theta^2}{Z^{1/3} \, e^4 \, \hbar^2 \, n^{5/3} \, T} \, .$$

Since the exact calculation by BLOCH gives the same result to within numerical factors,[1] one may take our crude approximation to give an intuitive interpretation of the circumstances calculated by BLOCH. The path length which we have thus deduced will be used for the temperatures and pressures for which one can assume an approximate lattice structure for the nuclei, where our calculation yields

$$\frac{d^2}{a^2} < 1.$$

For higher temperatures and pressures we omit this factor and are then in agreement with an estimate by KOTHARI (25). With $\epsilon \approx mv^2$ we thus obtain

$$\sigma = \frac{M_K \, k\Theta^2}{Z^{1/3} \, m^{1/2} \, e^2 \, \hbar^2 \, n^{2/3} \, T} \frac{\epsilon^{3/2}}{} \quad \text{für} \quad n^{2/3} \, kT \leq \frac{Z^{2/3} \, M_K \, (k\Theta)^2}{\hbar^2}$$

$$\sigma = \frac{1}{Z \, m^{1/2} \, e^2} \epsilon^{3/2} \quad \text{für} \quad n^{2/3} \, kT \geq \frac{Z^{2/3} \, M_K \, (k\Theta)^2}{\hbar^2} \, .$$

When we consider that the energy $k\Theta$ of the lattice vibrations behaves, in relation to the atomic energy unit $\frac{me^4}{\hbar^2}$, approximately as $\sqrt{m}$ to $\sqrt{M_K}$, then it follows for the condensed state (insofar as this is metallic) and the transition to the electron gas:

$$\sigma = \frac{m^{5/2} \, e^6}{Z^{1/3} \, \hbar^6 \, n^{2/3} \, kT} \frac{\epsilon^{3/2}}{} \quad \text{für} \quad n^{2/3} \, kT \leq \frac{Z^{2/3} \, m^3 \, e^8}{\hbar^6}$$

$$\sigma = \frac{1}{Zm^{1/2} \, e^2} \epsilon^{3/2} \quad \text{für} \quad n^{2/3} \, kT \geq \frac{Z^{2/3} \, m^3 \, e^8}{\hbar^6} \, .$$

In the nondegenerate gas we naturally have to set $\Theta = 0$; thus here only the second equation is valid. Therefore, for the nondegenerate electron gas with $\epsilon \approx kT$, we finally obtain the electrical conductivity

$$\sigma = \frac{(kT)^{3/2}}{Zm^{1/2} \, e^2},$$

where Z is the number of "free electrons per atom". For the degenerate electron gas with $\epsilon \approx \frac{\hbar^{6/5}}{m^{3/5}} p^{2/5}$, $n \approx \frac{p}{\epsilon}$, the electrical conductivity is

---

[1] In the case of almost free electrons (38)

$$l = \frac{4}{\pi^3} \frac{M_K \epsilon^2 k\Theta^2}{\hbar^2 C^2 n_K T},$$

where C is the mean depth of the potential wells, and thus approximately equal to $Ze^2 n_K^{1/3}$.

$$\varrho = \frac{\hbar^{9/5} \, p^{3/5}}{Z m^{7/5} \, e^2} = \frac{\hbar^3 \, \varrho}{Z \mu \, e^2 \, m^2 \, M}, \quad \text{wenn} \quad p^{2/5} \, kT \geq \frac{Z^{2/3} \, m^{13/5} \, e^8}{\hbar^{26/5}}$$

$$\varrho = \frac{e^6 \, m^{6/5}}{Z^{1/3} \, \hbar^{17/5}} \frac{p^{1/5}}{kT}, \quad \text{wenn} \quad p^{2/5} \, kT \leq \frac{Z^{2/3} \, m^{13/5} \, e^8}{\hbar^{26/5}};$$

for the metal with $\epsilon \approx \frac{\hbar^2}{m} n^{2/3}$ it is

$$\varrho = \frac{e^6 \, m}{Z^{1/3} \, \hbar^3} \frac{n^{1/3}}{kT} \, .$$

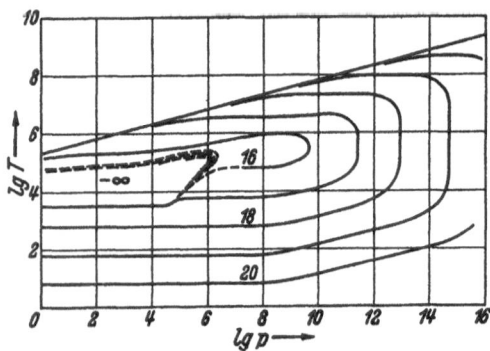

Fig. 8. Electrical conductivity of matter. The condensed state is assumed to be metallic.

We have thereby neglected the transition regions. Thus the function σ that we have calculated in four separate pieces still requires some rounding off. Figure 8 gives this σ as a function of p and T. The *gas of atoms* and, where not metallic, the *condensed state* (with the exception of the neutron gas occurring at very high pressures) are the *only regions where matter does not conduct*. All other states conduct electrical current like a metal or better.

Of course we cannot expect our simplified derivation to yield exact results. Thus our expression for a metal at room temperature gives $\sigma \approx 2 \times 10^{18}$, whereas the best conductors actually have only $\sigma \approx 6 \times 10^{17}$. For the non-degenerate electron gas and the degenerate electron gas with an irregular distribution of nuclei, more exact if still somewhat simplified calculations have been made by KOTHARI (25). The exact values differ from those given here by factors that depend logarithmically on the state variables and which have values of the order of 1 in the regions considered here.

The electrons are also responsible for the conduction of heat. The thermal conductivity can be calculated from the electrical conductivity by using the WIEDEMANN–FRANZ law

$$\lambda \approx \frac{k^2 \, T}{e^2} \sigma \, ;$$

In our approximation the thermal conductivity is constant in a metal, in the nondegenerate electron gas it varies as $T^{5/2}$, and in the degenerate electron gas it varies as $p^{3/5}$ to $p^{1/5}$.

13. Absorption of Light. In general, the gas of atoms allows visible light to pass through it and, in the ultraviolet, it absorbs the frequencies that correspond to the excitation or ionization of the atoms. At temperatures of less than about 10^4 degrees, a gas of atoms scarcely hinders the emission of its own thermal radiation. Only at higher temperatures does it to some extent hold this radiation within itself. The optical behavior of the condensed state is largely dependent upon whether the material is an insulator or a metal.

The electron gas contains electrons that are approximately free, and whose mobility is limited by their interaction with the force fields of the nuclei. This also results in an absorption of radiation. A perfect conductor with infinite conductivity would be transparent. Absorption by an electron gas was calculated by EDDINGTON, MILNE, OPPENHEIMER, GAUNT, KOTHARI, and MAJUMDAR (9, 21, 26, 28, 29, 35) based on a study by KRAMERS of the emission of electrons due to their deceleration in matter (27). This absorption can actually be deduced from the above-calculated conductivity by using quite simple arguments. For the equation of motion of the electrons we assume (as in the dispersion theory of good conductors)

$$e\mathfrak{E} = m\ddot{\mathfrak{r}} + g\dot{\mathfrak{r}},$$

where $\mathfrak{r}$ is the position vector and g is a friction which is related to the conductivity. In place of the equation of motion we can also give a phenomenological relationship between the current density $\mathfrak{s} = \sum e\dot{\mathfrak{r}}$ and the field:

$$ne^2\,\mathfrak{E} = m\dot{\mathfrak{s}} + g\mathfrak{s}.$$

For the stationary case it follows that

$$\mathfrak{s} = \frac{ne^2}{g}\mathfrak{E}, \qquad \sigma = \frac{ne^2}{g}.$$

If, in addition to the conductivity σ we also introduce a quantity $\gamma^2 = \frac{ne^2}{m}$ related to the inertia of the electrons, then

$$\mathfrak{E} = \frac{1}{\gamma^2}\dot{\mathfrak{s}} + \frac{1}{\sigma}\mathfrak{s}$$

becomes the valid "equation of matter" together with the general MAXWELL equations (the relationship between $\mathfrak{E}$ and $\mathfrak{D}$, $\mathfrak{B}$ and $\mathfrak{H}$ is assumed to be that in a vacuum). When an alternating field

$$\mathfrak{E} = \mathfrak{E}_0\,e^{i\omega t}$$

is applied to the matter, then, for the current density,

$$\mathfrak{s} = \mathfrak{s}_0\,e^{i\omega t}$$

it follows that

$$\mathfrak{s} = \frac{\sigma}{1 + i\,\frac{\omega\sigma}{\gamma^2}}\mathfrak{E}.$$

Using the standard calculation from optics one can then find a "refractive index" $\bar{n}$ and an "absorption index" $\bar{k}$:

$$(\bar{n} - i\bar{k})^2 = 1 - \frac{4\pi\gamma^2}{\omega^2 - i\frac{\omega\gamma^2}{\sigma}}.$$

In place of this $\bar{k}$ used in optics, which determines the absorption on the basis of the vacuum wavelength, we will introduce a quantity that determines the absorption of unit mass on the basis of the length unit:

$$\kappa = \frac{\omega}{\varrho c}\bar{k}.$$

The attenuation of the light intensity over a distance x is then given by

$$J \sim e^{-\varrho \kappa x},$$

and we obtain

$$\left(\bar{n} - i\frac{c\varrho}{\omega}\kappa\right)^2 = 1 - \frac{4\pi\gamma^2}{\omega^2 - i\frac{\omega\gamma^2}{\sigma}}.$$

In the simple limiting case $\omega \gg \gamma$, $\omega\sigma \gg \gamma^2$ it follows that

$$\frac{c\varrho}{\omega}\kappa = 2\pi\frac{\gamma^4}{\sigma\omega^3}.$$

We are interested mainly in the absorption of radiation that is in temperature equilibrium with the material, i.e. the "opacity" of the material. Apart from numerical factors, this is given by the absorption of the predominantly radiated frequency

$$\omega \approx \frac{kT}{\hbar}.$$

(Here we ignore numerical factors of order unity.)

For the nondegenerate electron gas one has approximately

$$\sigma = \frac{(kT)^{3/2}}{Zm^{1/2}e^2}.$$

Since, at some distance from the boundary of the degenerate region,

$$\frac{\omega^2}{\gamma^2} = \frac{m(kT)^2}{\hbar^2 e^2 n} \gg 1, \qquad \frac{\sigma\omega}{\gamma^2} = \frac{m^{1/2}(kT)^{5/2}}{Z\hbar e^4 n} \gg 1,$$

it follows that

$$\kappa = \frac{Z\hbar^2 e^6}{c\,m^{3/2}(\mu M)^2}\frac{\varrho}{(kT)^{7/2}}.$$

This is in accordance with the calculations mentioned above (which are also carried out with simplifying assumptions). For the degenerate gas we have, at higher pressures and temperatures,

$$\sigma = \frac{\hbar^3}{Zm^2 e^2}\frac{\sigma}{\mu M}.$$

Here one can assume that the quantity

$$\frac{\sigma\omega}{\gamma^2} = \frac{\hbar^2\,kT}{Zm\,e^4}$$

is large; in contrast

$$\frac{\omega^2}{\gamma^2} = \frac{m(kT)^2}{\hbar^2\,e^2\,n}$$

is not very different from 1. By applying the above-mentioned simple formula for κ we make an error, but one that does not change the order of magnitude of the result. We find

$$\kappa = \frac{Ze^6}{\hbar c\mu M}\frac{1}{(kT)^2}$$

in agreement with the result obtained by MAJUMDAR (28) who also used some simplifications. For the relativistic degenerate gas MAJUMDAR finds a very small constant opacity.

The scattering of light from light, which becomes significant for light of the COMPTON wavelength $\frac{\hbar}{mc}$ at a density of many light quanta in a cube of this sidelength, plays no role at temperatures under $\frac{mc^2}{k}$. Above this, however, it gives a very strong opacity.

The scattering cross section of light quanta of frequency ω is, according to EULER (13), of the order of

$$\left(\frac{e^2}{\hbar c}\right)^4 \cdot \left(\frac{\hbar}{mc}\right)^8 \left(\frac{\omega}{c}\right)^6 = \frac{1}{137^4}\left(\frac{\hbar}{mc}\right)^8 \left(\frac{\omega}{c}\right)^6 .$$

In temperature equilibrium this gives an absorption of the order of

$$\varrho\kappa = \frac{1}{137^4}\frac{mc}{\hbar}\left(\frac{kT}{mc^2}\right)^9 ,$$

which, for temperatures lower than $\frac{mc^2}{k}$, falls rapidly. However, in the vicinity of this temperature, our previous considerations give absorption values of the order of

$$Z\cdot\left(\frac{e^2}{\hbar c}\right)^3 \cdot \frac{mc}{\hbar} = \frac{Z}{137^3}\frac{mc}{\hbar} ,$$

so that at higher temperatures the scattering of light from light soon becomes the dominant process.

14. Energy Transport. To investigate the conditions in the center of a star an important question is whether the transport of energy from the center to the surface takes place by means of movement of material, by thermal conductivity, or by radiation. The effectiveness of the latter two processes can be compared. A temperature gradient produces a thermal current of

$$\mathfrak{Q}_{\text{cal}} = -\lambda_{\text{cal}}\,\text{grad}\,T.$$

However, it also causes a current of radiation energy. When the radiation is approximately in thermal equilibrium, i.e. the temperature gradient so small that within the range of the radiation it plays no significant role, then more energy flows through a unit are oriented perpendicular to the temperature gradient from the side at higher temperature than from the other side. If $\mathfrak{Q}_{\text{rad}}$ is the energy current, then $\varrho\kappa\mathfrak{Q}_{\text{rad}}$ is absorbed per unit volume and thereby exerts a momentum. Thus there is a gradient of radiation pressure

$$-\operatorname{grad} p_{\text{rad}} = \frac{\varrho\kappa}{c}\mathfrak{Q}_{\text{rad}}.$$

Therefore we have

$$\mathfrak{Q}_{\text{rad}} = -\frac{c}{\varrho\kappa}\operatorname{grad}\frac{a}{3}T^4 = -\frac{4acT^3}{3\varrho\kappa}\operatorname{grad}T\,;$$

i.e. the radiation current can be described by a "conductivity"

$$\lambda_{\text{rad}} = \frac{4\pi^2\,k^4\,T^3}{45\hbar^3\,c^2\,\varrho\kappa}.$$

The ratio between the conductivity for heat and that for radiation [KOTHARI (25)] is given, for a nondegenerate gas, (according to the results of Sects. 12 and 13) to within insignificant factors by

$$\frac{\lambda_{\text{cal}}}{\lambda_{\text{rad}}} = \frac{\hbar^5\,ce^2}{m^2}\frac{\varrho^2}{(\mu M)^2\,(kT)^4} = \frac{\hbar^5\,ce^2}{m^2}\frac{p^2}{(kT)^6}$$

$$= \frac{1}{137}\frac{\hbar^6\,c^2}{m^2}\frac{p^2}{(kT)^6} = 137\frac{\hbar^4\,e^2}{m^2}\frac{p^2}{(kT)^6}\,,$$

i.e., at some distance from the boundary of the degenerate region it is small. For a degenerate gas at higher temperatures we have likewise

$$\frac{\lambda_{\text{cal}}}{\lambda_{\text{rad}}} = \frac{\hbar^5\,ce^2}{m^2}\frac{\varrho^2}{(\mu M)^2\,(kT)^4}\,,$$

at some distance from the boundary of the degenerate region this is large. *In the region of the nondegenerate electron gas energy transport by radiation is dominant; in the region of the degenerate electron gas transport by thermal conductivity is dominant.* At lower temperatures the region of thermal conduction extends a small distance over the boundary of the degenerate region. At higher temperatures the same applies for the region of radiation. What is important for the energy transport is the sum of both conductivities; this is shown in Fig. 9. The numbers give the decadic logarithm of the energy conductivity measured in erg/cm sec degree. Neutrons would be able to produce a very strong conductivity for thermal energy (16);

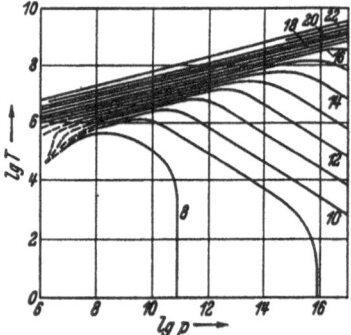

Fig. 9. Conductivity for thermal energy (transport by means of electrons and by means of radiation).

however, except at very high pressures they are not expected to exist in any significant number.

III. The Occurrence of Very High Pressures and Temperatures.

The following sections do not attempt to give a report about the inner structure of celestial bodies;[1] instead we will give a few simple arguments which make some order-of-magnitude predictions about the pressures and temperatures which occur, and which show how the main empirical features of stars are related to the properties of matter that we have been considering.

15. The Planets. In the *laboratory* it is possible to produce temperatures of a few thousand degrees. However, certain effects of much higher temperatures can be produced and investigated by means of experiments with rapidly moving particles (of radioactive origin or accelerated by fields). In the production of high pressures, BRIDGMAN (6) has succeeded in achieving almost 10^5 atm. Thus he is still quite a long way short of the transition from a solid body into the state of the compressibility of the degenerate electron gas.

From the *interior of the earth* we know that the material has no, or only very small, transverse elasticity. Thus we can assume a scalar pressure p in thermal equilibrium with the forces of gravity

$$\frac{dp}{dr} = -G\varrho\frac{\mathfrak{m}}{r^2} \tag{I}$$

$$\frac{d\mathfrak{m}}{dr} = 4\pi\varrho r^2 \tag{II}$$

[r is the distance from the center of the earth, $\mathfrak{m}(r)$ is the mass of the material within radius r, G is the gravitational constant]. Combining these two equations we obtain

$$\frac{d}{dr}\left(\frac{r^2}{\varrho}\frac{dp}{dr}\right) = -4\pi G\varrho r^2. \tag{1}$$

When either $\varrho(r)$ or $\varrho(p)$ is known, this equation has a one-parameter set of solutions (one of the two integration constants is determined by the fact that $\mathfrak{m}(0) = 0$, i.e., by the fact that $\frac{dp}{dr}$ is zero at the center). Let us make an approximate calculation with a constant ϱ. This yields

$$p = \frac{2\pi}{3}G\varrho^2\left(R^2 - r^2\right).$$

With the empirical value $\varrho = 5.5\,\text{g/cm}^3$ one obtains a pressure at the center of the earth of almost 2×10^6 atm. A model based on two shells, an inner shell with a density of $8\,\text{g/cm}^3$ and an outer shell of density $3\,\text{g/cm}^3$, whose ratio is chosen so as to yield the correct mean density, would give a pressure at the center of the earth of 3×10^6 atm. The center of the earth is thus in

---

[1] A report on this subject is given by H. VOGT in Vol.6 of Ergeb. exakt. Naturwiss. (46).

the normal condensed state, a region in which the properties of matter that are determined by electrons do not depend significantly on temperature.

The largest of the planets *Jupiter*, has a mean density of $1.3 \, \text{g/cm}^3$. The rough calculation of the pressure at the center under the assumption of constant density ($R = 7 \times 10^9 \, \text{cm}$) would give $1.2 \times 10^7 \, \text{atm}$. In reality, however, the increase in density as one approaches the center means that the pressure there must be higher. The low density of Jupiter is not compatible with the assumption that the matter is entirely in the condensed state. The matter composing Jupiter lies higher in the diagram of Fig. 5. Then, however, the density ϱ does not have a clear dependence on pressure but is also influenced by the temperature, and thus we cannot integrate the equation describing the structure. Thus to begin with we will be content with a model of Jupiter in which the pressure and density are related in a simple specified manner. We will choose EMDEN's (12) "polytropic" relationship

$$\frac{p}{p_0} = \left(\frac{\varrho}{\varrho_0}\right)^{1+\frac{1}{n}}, \tag{2}$$

where p_0 and ϱ_0 are the pressure and density at the center and n is a fixed number. For a gas in "adiabatic" equilibrium (in which rising and falling gas streams maintain their temperature difference with respect to the environment) such a relationship is valid, in this case with $1 + \frac{1}{n}$ equal to the ratio c_p/c_v of these specific heats. Neglecting dissociation and ionization, n would lie between 3/2 and 3. Within this framework the structure equation (1) can be reduced to a simple equation, which, apart from n, contains only numerical coefficients. This has been solved numerically by EMDEN for a series of n values (12, 31).

Table 2. Results of Model Calculations for the Density and Pressure at the Center of Jupiter

n	$\varrho_0 \, (\text{g/cm}^3)$	$p_0 \, (\text{Atm})$
1.5	8	0.8×10^8
2	14	1.5×10^8
2.5	31	$4 \ \ \times 10^8$
3	71	$11 \ \ \times 10^8$

Table 2 shows the values of the density and pressure at the center of Jupiter for some selected n values.

A glance at Fig. 5 shows that only the highest of these pressure and density values can be realized simultaneously. Thus the matter at the center of Jupiter is in a state that extends into the transition region between atomic gas, condensed state, and degenerate electron gas.

For *the other large planets* we obtain from such an estimate approximately the same central density and somewhat lower central pressures.

16. Ordinary Stars. *The sun* has a mean density of $1.4 \, \text{g/cm}^3$ and a radius of $6.95 \times 10^{10} \, \text{cm}$. The crude estimate of the pressure at the center of the sun with the assumption of constant density would give $1.3 \times 10^9 \, \text{atm}$. This would correspond to a point in the phase diagram (Fig. 5) in the region of the nondegenerate electron gas with a temperature of about 2×10^7 degrees.

More exact estimates make use of assumptions concerning the generation of energy and the energy transport (4, 11, 24, 31, 36, 42, 44, 46). If one neglects energy transports by means of movement of matter, then one can assume for the nondegenerate electron gas that the transport occurs through radiation, approximated by the absorption law

$$\kappa \sim \varrho T^{-7/2}. \tag{3}$$

The two first-order differential equations for the structure (I, II) are joined by a third equation that relates the temperature gradient to the energy current (see Sect. 14):

$$\frac{dT^4}{dr} = -\frac{3}{ac}\varrho\kappa\frac{l}{4\pi r^2}, \tag{4}$$

where l is the amount of energy emitted from a sphere of radius r concentric with the sun. The three equations are not sufficient to determine the four unknown functions p, T, $\mathfrak{m}$, and l (ϱ is determined in conjunction with the equation of state by p and T). Some further statements must be made about the function l. It seems reasonable to suppose that the real situation lies between two limits: the first is the case where only at the center of the sun is more energy emitted than absorbed (point-source model); in the opposite case every gram of matter emits more energy than it absorbs (homogeneous energy generation). With some relatively harmless simplifications, the model of homogeneous energy generation leads to an equation that we have already met. From the relationship [cf. (I) and (4)]

$$\frac{dT^4}{dp} = \frac{3}{4\pi ac}\kappa\frac{l}{\mathfrak{m}} = \frac{3}{4\pi ac}\kappa\frac{L}{\mathfrak{M}},$$

where L and $\mathfrak{M}$ are the total radiation and the total mass, it follows after integration that

$$p \sim T^{17/4} \sim \varrho^{17/13},$$

if we use as a basis the equation of state of the nondegenerate gas

$$p \sim \varrho T$$

and determine the integration constants so that T vanishes with p (since the temperature at the surface of the star is negligible in comparison to that at the center). This, however, gives a "polytropic" equation of state (2) with $n = 3.25$. In equation (1) for the structure again only ϱ_0 and p_0 enter as parameters and we can determine them such that the star concerned has the radius and the mass of the sun. At the center of the sun this yields a pressure of 2.3×10^{11} atm, a density of $120\,\mathrm{g/cm}^3$ and a temperature (with $\mu = 2$) of 4.5×10^7 degrees; this is still in the region of the degenerate electron gas. The point-source model yields values that are not very different from these (4, 11, 42). More exact calculations take into account the fact that κ is somewhat different, since further away from the center the gas is only partially ionized. But even these calculations give essentially similar results (44).

We know from determinations of the ages of rocks that the sun has been radiating almost the same amount of energy for the past 10^9 years. Over this period it has emitted some 10^{18} erg/g. This amount of energy could not possibly have come from its gravitational energy; indeed, even by shrinking to its present size from an infinitely extended cloud far less energy would have been produced. In order to emit quantities of 10^{18} erg/g or $10^{11} \times 10^7$ erg/g in thermodynamic equilibrium from its internal energy, the sun (Fig. 7) must have contained matter at a temperature of some 10^9 degrees or at a pressure of some 10^{19} atm. This, however, would mean that 10^9 years ago the sun would have been in a state very different from that in which it is now. Thus we have to assume that the generation of energy in the sun does not occur in thermodynamic equilibrium. The main source of energy must thus be the transformation of light nuclei into heavy nuclei at temperatures that are significantly lower than those at which this process occurs in thermodynamic equilibrium. The estimate by ATKINSON and HOUTERMANS (2, 22) that were mentioned above show that such processes can take place at a sufficient rate in the relevant temperature range. We will not involve ourselves here in discussing difficulties in interpreting the abundance of various elements. These problems may however be solvable (1, 22). [1]

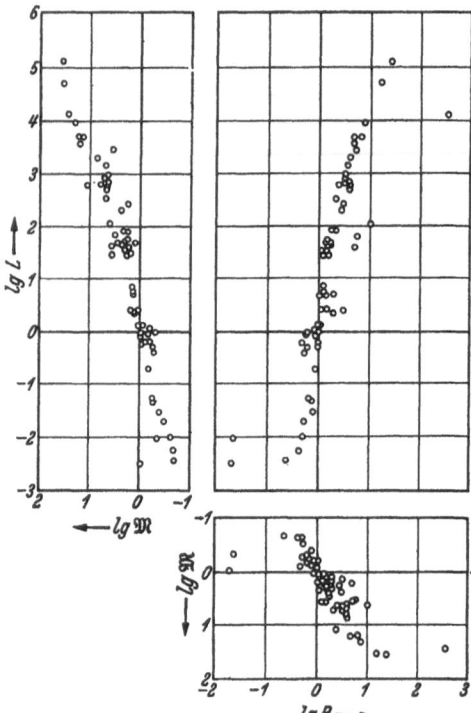

Fig. 10. Distribution of the R , $\mathfrak{M}$, and L values of the stars (numbers taken from PILOWSKI).

Whereas a single star does not hold much interest for the physicist, the *totality of stars* provides a large amount of empirical information and it is conceivable that empirical rules concerning the properties of stars can somehow provide a picture of the behavior of matter at high pressures and temperatures. Neglecting the variable stars, we can assume, due to the similarity with the sun, that the stars are in equilibrium insofar as their properties do not change significantly over very long periods of time. Thus their energy must stem from a supply which, as it is used up, only has a very minor effect on the state variables. The source of energy is probably the transformation of lighter atomic nuclei into heavier nuclei, the transition from a thermodynamically improbable state to a more probable state.

---

[1] A discussion of the energy sources of stars is given by E. FREUNDLICH (19) in Ergeb. exakt. Naturwiss., Vol. 6)

The radius R and radiated energy L are known for many stars and for quite a number of stars, the mass $\mathfrak{M}$ is also known. The measured values are found to have a certain regularity when plotted in R-$\mathfrak{M}$-L coordinate space. Figure 10 shows the positions for the stars for which the three coordinates are known [from PILOWSKI; details in (34)]. For R and L alone we have much more material. The corresponding points do not lie very far from those shown. The values for the stars are found to cluster in the neighborhood of a line in R-$\mathfrak{M}$-L space which can be approximately described by the relationships $L \sim \mathfrak{M}^4$ (at large $\mathfrak{M}$, the increase in $\mathfrak{M}$ is somewhat weaker) and $L \sim R^5$ to $L \sim R^4$. The R-L relation is another expression of the well-known HERTZSPRUNG–RUSSEL diagram of the spectral types and the luminosities. However, there are also numerous stars that do not obey this relationship, in particular the so-called giant stars (on the right in the R-L diagram). The $\mathfrak{M}$-L relation is fairly strictly obeyed and the few giant stars whose mass is known do indeed fit in with it. A few stars with very weak luminosity and very small radius, the so-called white dwarfs (at the lower left of the $\mathfrak{M}$-L and R-L diagram, at the upper left of the R-$\mathfrak{M}$ diagram), fail to obey either the R-L or the $\mathfrak{M}$-L relationship. Thus we will not include them in the ordinary stars.

In summary one can say that for the ordinary stars there appears to be a *general relationship* between $\mathfrak{M}$ and L which is fairly well obeyed (there are signs that for a given value of $\mathfrak{M}$, the value of L has a weak dependence on R, increasing weakly as R increases). *For the large majority* of these stars (the "main sequence stars") *there is a second relationship* which is also quite well obeyed (this can be chosen to be the R-L relation or the R-$\mathfrak{M}$ relation).

Let us see whether our knowledge of the behavior of matter can give us some guidance in understanding such rules. To gain some insight into the nature of the conclusions, we will make some simplifications but will not deviate too far from the actual arguments of EDDINGTON, VOGT et al. (10, 11, 24, 31, 36, 44, 45, 46). The three equations (I, II, III) for the structure and the energy transport

$$\frac{dp}{dr} = -G\varrho\frac{m}{r^2} \tag{I}$$

$$\frac{dm}{dr} = 4\pi\varrho r^2 \tag{II}$$

$$\frac{dT^4}{dp} = \frac{3}{4\pi Gac}\kappa\frac{l}{m} \tag{III}$$

do not determine the structure unambiguously, even when the equation of state $\varrho = \varrho(p, T)$ and the absorption law $\kappa = \kappa(p, T)$ are known, since we only have three equations for four unknown functions. We are lacking a relationship for the generation of energy. Since physics does not tell us much about this relationship, instead of real stars one treats models in which a hypothetical function l is introduced. The following equations describe some of the models that have been suggested or are evidently plausible (f is a function that is taken for all star models, σ and τ are numbers that apply equally to all models, a and b are individual parameters). With (IVa) one obtains a single-parameter family of solutions (T is set to zero at the surface); equations (IVb) and (IVc) yield a two-parameter family, and (IVd) a three-parameter family of solutions.

$$\frac{dl}{d\mathfrak{m}} = f(p, T) \qquad \text{e.g.} \quad f = p^\sigma T^\tau \tag{IV a}$$

$$\frac{dl}{d\mathfrak{m}} = f(a, p, T) \quad \text{e.g.} \quad f = a p^\sigma T^\tau \tag{IV b}$$

$$l = f(a, \mathfrak{m}) \qquad \text{e.g.} \quad l = L \text{ (point-source model)} \tag{IV c}$$

$$l = \frac{L}{\mathfrak{M}} \mathfrak{m} \text{ (homogenous energy generation)}$$

$$l = L \left(\frac{\mathfrak{m}}{\mathfrak{M}}\right)^\sigma$$

$$l = f(a, b, \mathfrak{m}) \quad \text{e.g.} \quad l = \left\{ \begin{array}{ll} \frac{L}{b\mathfrak{M}}\mathfrak{m} & \text{für} \quad \mathfrak{m} \leq b\mathfrak{M} \\ L & \text{für} \quad \mathfrak{m} \geq b\mathfrak{M} \quad (32) \end{array} \right\} \tag{IV d}$$

We restrict ourselves to stellar matter in the state of the nondegenerate electron gas, for which the radiation pressure is not yet significant,

$$p \sim \varrho T$$

and assume an absorption law

$$\kappa \sim \varrho^\alpha T^\beta,$$

e.g., the previously mentioned law

$$\kappa \sim \varrho T^{-7/2}. \tag{3}$$

From a solution to the equations (I), (II), (III) and these two relationships, one can, by means of a simple similarity tranformation, find new solutions, namely, a two-parameter family whose members can be distinguished,e.g. through the $\mathfrak{M}$ and R values. One writes

$$\frac{r'}{r} = \frac{R'}{R}; \quad \frac{\mathfrak{m}'}{\mathfrak{m}} = \frac{\mathfrak{M}'}{\mathfrak{M}}; \quad \frac{\varrho'(r')}{\varrho(r)} = \frac{\mathfrak{M}' R^3}{R'^3 \mathfrak{M}}; \quad \frac{p'(r')}{p(r)} = \frac{\mathfrak{M}'^2 R^4}{R'^4 \mathfrak{M}^2},$$

etc. We will describe the transformations by means of the abbreviated and self-evident notation

$$r \sim R, \quad \mathfrak{m} \sim \mathfrak{M}, \quad \varrho \sim \frac{\mathfrak{M}}{R^3}, \quad p \sim \frac{\mathfrak{M}^2}{R^4}, \quad T \sim \frac{\mathfrak{M}}{R}$$

and

$$L \sim l \sim \frac{\mathfrak{M}^3}{\kappa}, \tag{A}$$

where the transformation of κ is given by the relationship with ϱ and T. With $\kappa \sim \varrho T^{-7/2}$ we obtain $L \sim \mathfrak{M}^{11/2} R^{-1/2}$. Thus from a solution of (I), (II), (III) we clearly do not obtain all solutions, and from a solution of (I), (II), (III) and a solution of equation (IV) we may with the transformation end up with no solution at all of (IV). Let us take the model (IVa) in the particular form given on the right-hand side. By solving this using the transformation given we obtain a new solution if L is transformed as $p^\sigma T^\tau$. From one solution the transformation then yields only a one-parameter family of solutions, but thereby includes all solutions that are possible. For the stars of the model (IVa) in the particular form taken there are two valid R-$\mathfrak{M}$-L relationships; all stars of the model lie on a line in R-$\mathfrak{M}$-L space; one of the two relationships is (A). For the models (IVa) and (IVb) in one of the forms given on the right, we again obtain solutions through the transformation, in this case exactly the two-parameter family of all solutions. All stars of this model lie on a surface in R-$\mathfrak{M}$-L space that is given by (A). The one as yet undetermined factor in (A) can only be found by genuinely solving the equations. In the model (IVd) it is not so simple to use one solution to form new ones.

The R-$\mathfrak{M}$-L relationship

$$L \sim \mathfrak{M}^{11/2} R^{-1/2} \tag{A}$$

which is valid for the models (IVa), (IVb), and (IVc) in the form given on the right, lies in the neighborhood of the empirical relationship $L \sim \mathfrak{M}^4$. More exact calculations with a better κ law, which takes into account the incomplete ionization of the atoms (4, 11), brings the dependence of $\mathfrak{M}$ closer to that which is observed. In contrast, the dependence of $R^{-1/2}$ appears not to correspond to observations; the deviation of the empirical dependence can be attributed to the varying hydrogen content of the stars (and thus various values of μ in the equation of state) (4).

It thus appears that *the empirical $\mathfrak{M}$-L relationship of the ordinary stars expresses a high degree of homogeneity of the stellar material in the equation of state and in the law of energy transport.* The second empirical relationship, which is somewhat less general and not quite so sharp, could then be related to the circumstances of energy generation. Thus the approximately fulfilled relationship $\mathfrak{M} \sim R$ in the above similarity consideration signifies the coincidence of the temperatures at corresponding points of the stars.

17. The Compact Stars. There are just a few stars whose R, $\mathfrak{M}$, L values do not fit the general pattern. These are the *"white dwarfs".* Their characteristic feature is that, despite having almost normal mass (where known), they are very small and have mean densities of about 10^5 g/cm^3 and at their centers even higher densities. Such densities are actually also possible in the state of the nondegenerate electron gas, but the stars would then have to have larger radii. One must thus assume that their material is partially in the state of the degenerate electron gas. It was these stars that drew the attention of physicists to this state as a state of matter under high pressures [FOWLER (17)].

When the majority of the stellar material is in the degenerate state (the ordinary condensed state is also allowed), then the temperature has little bearing on the behavior. Thus, for the structure of a star in equilibrium, energy generation and energy transport are also irrelevant. The equation for the structure (I) admits a one-parameter family of solutions, whose members can be distinguished, for example, by means of the star mass $\mathfrak{M}$. A prerequisite for this, however, is that we really know the equation of state, for which it is sufficient that the star does not contain much hydrogen. The radius of the star is a unique function of the star's mass and for small masses we have of course $R^3 \sim \mathfrak{M}$. For medium masses, where the majority of the material is a nonrelativistic degenerate electron gas, it follows from the equation of state $p \sim \varrho^{5/3}$ and from the equation for the structure by using a similarity transformation that $R^3 \sim \frac{1}{\mathfrak{M}}$. For even larger masses the matter towards the center of the star is relativistic degenerate. By assuming only an equation of state $p \sim \varrho^{4/3}$ one would obtain a singular relationship between $\mathfrak{M}$ and R.

The white dwarfs that are actually observed are even larger still than expected from such estimates. Indeed, their external regions are gaseous, as one can deduce from their high surface temperatures. According to plausible estimates, the companion star of Sirius has a density at the center of somewhat more than 10^6 g/cm^3 and a central pressure of about 10^{17} atm. Thus it just reaches the transition region to the relativistic degenerate electron gas. VAN MAANEN's star is likely to have a significantly higher density and pressure at the center, and thus to consist partly of a relativistic degenerate electron gas. There is no indication for the observed stars that the transformation into neutrons is reached.

CHANDRASEKHAR (7) has calculated the structure of a star, for which the temperature is no longer important, with a more exact equation of state that is valid for the nonrelativistic and the relativistic electron gas. As the mass increases, the radius decreases; for the mass of the sun, the radius is approximately equal to that of the earth, and at even higher masses the radius tends rapidly to zero. The zero radius is reached for a finite mass only slightly larger than the mass of the sun. This last result should not be taken too literally, because, for calculating the equation of state, it was assumed to have unlimited validity for high pressures. This collapse to a zero radius (or to the corresponding value in the general relativity theory) stems from the high compressibility of matter in the state of the relativistic degenerate electron gas. If a star above the limiting mass were to have finite radius, then the pressure of its matter would not be able to support the weight of the matter lying above it; the star would collapse. In this process the pressure would of course increase, but not fast enough to meet the corresponding increase in the weight of the above-lying layers.

The transformation of matter into neutrons at high pressures initially means that, in a certain area, the compressibility is somewhat less than that of the relativistic degenerate electron gas. Thus the limiting mass for finite radius is raised a little. However, when we come to very small radii, the gravitational energy of the particles becomes comparable with their rest-mass energy and the basic assumptions of our calculations lose their validity. From the calculations by CHANDRASEKHAR and other similar calculations that take into account neutron tranformations, we can only draw conclusions provided the stellar radii are not so small that they approach the critical length (Sect. 2) of a few 10^5 cm. Thus we can conclude that stars with sufficient mass can become this small.

One cannot completely exclude that stars, which are at present in the normal state, will expand to become a diffuse cold nebular. But they would have to begin this process before their energy supply from deviations in their chemical composition is exhausted. If they do not undergo this process, then they will finally shrink to become dense stars. Stars with high masses could thereby avoid the above-mentioned small radii by radiating large amounts of the gravitational energy that is set free in the process of contracting. In contracting to the radius of the earth the amount of this energy would be comparable to $\mathfrak{M}c^2$, and thus the star would reduce its mass significantly. It may also be possible for them to lose material as a result of the high radiation pressure at the surface (7). As a *possible final stage in the evolution of stars* we are thus let to expect stars of moderate mass with very high densities.

References[1]

1. ATKINSON, R. D'E.: (The creation of nuclei.) Astrophysic. J. **73**, 250, 308 (1931).
2. — u. F.G. HOUTERMANS: (Penetration of protons into nuclei.) Z. Physik **54**, 656 (1929).
3. BETHE, H.: (Nuclear energies.) Physic. Rev. **47**, 633 (1935).
4. BIERMANN, L.: (Stellar models.) Z. Astrophysik **3**, 116 (1931).
5. BLOCH, F.: (Metallic conductivity.) Z. Physik **52**, 555 (1928).
6. BRIDGMAN, P.W.: Theoretically interesting aspects of high pressure phenomena. Rev. mod. Physics **7**, 1 (1935).
7. CHANDRASEKHAR, S.: (Stars composed of degenerate material.) Monthly not. **91**, 456 (1923); **95**, 207, 226, 676 (1935).
8. — and L. ROSENFELD: (Pair creation.) Nature (Lond.) **135**, 999 (1935)
9. EDDINGTON, A. S.: (Opacity.) Monthly not. **83**, 32, 98 (1922), 431 (1923); **84**, 104, 308 (1924).
10. — ($\mathfrak{M}$-L Relationship.) Monthly not. **84**, 308 (1924).
11. — The internal constitution of the stars. Cambridge 1926. Der innere Aufbau der Sterne. Berlin 1928.
12. EMDEN, R.: Gaskugeln. Leipzig u. Berlin 1907. Thermodynamik der Himmelskörper. Enzyklop. math. Wiss. **6/2**, 373, (1926).
13. EULER, H.: (Scattering of light by light.) Ann. Physik (5) **26**, 398 (1936).
14. FERMI, E.: (Ideal gas.) Z. Physik **36**, 902 (1926).
15. FLEISCHMANN, R. u. W. BOTHE: Künstliche Kernumwandlung. Erg. exakt. Naturwiss. **14**, 1 (1935).
16. FLÜGGE, S.: (Neutrons in stars.) Z. Astrophysik **6**, 272 (1933).
17. FOWLER, R.H.: (Dense material composed of degenerate electron gas.) Monthly not. **87**, 114 (1926).
18. — and E.A. GUGGENHEIM: (Ionization of stellar material.) Monthly not. **85**, 939, 961 (1925).
19. FREUNDLICH, E.: Die Energiequellen der Sterne. Erg. exakt. Naturwiss. **6**, 27 (1927).
20. GAMOW, G.: (Penetration of α particles into nuclei.) Z. Physik **52** , 510 (1928).
21. GAUNT, J.A.: (Radiation of electrons.) Z. Physik **59**, 508 (1930).
22. HOUTERMANS, F.G.: Neuere Arbeiten über Quantentheorie des Atomkerns. Erg. exakt. Naturwiss. **9**, 123 (1930).
23. JEANS, J.H.: (Ionization of stellar material.) Observatory **40**, 43 (1917). (Was not available to me.)
24. — Astronomy and Cosmogony. Cambridge 1928.
25. KOTHARI, D. S.: (Conductivity of an electron gas.) Philosophic. Mag. (7) **13**, 361 (1932). — Monthly not. **93**, 61 (1932).
26. — u. R. C. MAJUMDAR: (Opacity.) Astron. Nachr. **244**, 65 (1931).
27. KRAMERS, H. A.: (Radiation of electrons.) Philosophic. Mag. **46**, 836 (1923).
28. MAJUMDAR, R. C.: (Opacity.) Astron. Nachr. **243**, 5 (1931); **247**, 217 (1932).
29. MILNE, E. A.: (Opacity.) Monthly not. **85**, 750 (1925).
30. — (Energy of the electrons of an atom.) Proc. Cambridge philos. Soc. **23**, 794 (1927).
31. — Thermodynamics of the stars. Handbuch der Astrophysik, Bd 3/1, S.64 1930, insbes. S.183f.
32. — (Stellar models.) Monthly not. **90**, 769; **91**, 1 (1930); **92**, 610 (1922). -Z. Astrophysik **4**, 75; **5**, 337 (1932).
33. MØLLER, C. and S. CHANDRASEKHAR: (Relativistic degenerate gas.) Monthly not. **95**, 673 (1935).

---

[1] The information given in brackets characterizes the relevant subject matter of the work concerned.

34. NERNST, W.: (Stellar evolution.) Z. Physik **97**, 511 (1935).
35. OPPENHEIMER, J. R.: (Radiation of electrons.) Z. Physik. **52**, 725 (1929).
36. ROSSELAND, S.: Astrophysik auf atomtheoretischer Grundlage. Berlin 1931.
37. SAHA, M. N.: (Ionization in stars.) Z. Physik **6**, 40 (1921). — Proc. roy. Soc. A **99**, 135 (1921).
38. SOMMERFELD, A. u. H. BETHE: Elektronentheorie der Metalle. Handbuch der Physik, 2. Aufl., Bd. 24/2, S.333. 1933.
39. SLATER, J. C. and H. M. KRUTTER: (Solid bodies under high pressure.) Physik. Rev. **47**, 559 (1935).
40. STERNE, T. E.: (The transformation of elements in thermodynamic equilibrium.) Monthly not. **93**, 736, 767, 770 (1933).
41. STONER, E. C.: (Degenerate electron gas.) Monthly not. **92**, 651 (1932).
42. STRÖMGREN, B.: (Stellar models.) Z. Astrophysik **2**, 345 (1931).
43. — (Hydrogen content of the stars, opacity.) Z. Astrophysik **4**, 118 (1932); **7**, 222 (1933).
44. — Thermodynamik der Sterne und Pulsationstheorie. Handbuch der Astrophysik, Bd.7, S.121 (1936).
45. VOGT, H.: ($\mathfrak{M}$-L Relationship.) Astron. Nachr. **226**, 301 (1936).
46. — Der innere Aufbau und die Entwicklung der Sterne. Erg. exakt. Naturwiss. **6**, 1 (1927).

References

Abramowitz M. and Stegun I. (1970): Handbook of Mathematical Functions. Dover, New York

Alkofer R. and Reinhardt H. (1992): Fermion Condensation in the Field Strength Approach to Non-Abelian Yang–Mills Theories. *Z. Phys.* **A343**, 79

Anastasio M.R., Müther H. Faessler A., Holinde K. and Machleidt R. (1978): Mesonic and Isobar Degrees of Freedom in the Ground State of the Nuclear Many-Body System. *Phys. Rev.* **C18**, 2416

Anastasio M.R., Celenza L.S., Pong W.S. and Shakin C.M. (1983): Relativistic Nuclear Structure Physics. *Phys. Rep.* **100**, 327

Ando H. and Osaki Y. (1976): *Publ. Astron. Soc. Japan* **27**, 581

Arndt R.A., Hyslop III J.S. and Roper L.D. (1987): Nucleon–Nucleon Partial Wave Analysis to 1100 MeV. *Phys. Rev.* **D35**, 128

Arnett W.D. and Bowers R.L. (1977): A Microscopic Interpretation of Neutron Star Structure. *Astrophys. J. Suppl.* **33**, 415

Baade W. and Zwicky F. (1934): Supernovae and Cosmic Rays. *Phys. Rev.* **45**, 138

Babu S. and Brown G.E. (1973): Quasiparticle Interaction in Liquid ^{3}He. *Ann. of Phys.* **78**, 1

Backer D.C., Kulkarni S.R., Heiles C., Davis M.M. and Goss W.M. (1982): A Millisecond Pulsar. *Nature* **300**, 615

Bahcall, J.N., Lande K., Lanou R.E., Learned J.G., Robertson R.G.H. and Wolfenstein L. (1995): Progress and Prospects in Neutrino Astrophysics. *Nature* **375**, 29

Baym G., Pethick C. and Sutherland P. (1971): The Ground State of Matter at High Densities: Equation of State and Stellar Models. *Astrophys. J.* **170**, 299

Baym G. and Chin S.A. (1976): Can a Neutron Star be a Giant MIT Bag? *Phys. Lett.* **62B**, 241

Bernard V., Meissner U.G. and Zahed I. (1987): Decoupling of the Pion at Finite Temperatures and Densities. *Phys. Rev.* **D36**, 819

Bethe H.A. and Bacher R.F. (1936): Nuclear Physics: Stationary States of Nuclei. *Rev. Mod. Phys.* **8**, 82

Bethe H.A., Brandow B.H. and Petscheck A.G. (1963): Reference Spectrum Method for Nuclear Matter. *Phys. Rev.* **129**, 225

Bina C. R. and Helffrich G. H. (1992): Calculation of Elastic Properties from Thermodynamic Equation of State Principles. *Ann. Rev. Earth Planet. Sci.*, **20**, 527

Bjorken J.D. and Drell S.D. (1964): Relativistic Quantum Mechanics. McGraw-Hill, New York

Blankenbecler R. and Sugar R. (1966): Linear Integral Equation for Relativistic Multichannel Scattering. *Phys. Rev.* **142**, 1051

Blomqvist K. et al. (1995): High-Momentum Components in the $1p$ Orbitals of ^{16}O. *Phys. Lett.* **344B**, 85

Bobeldijk I., Bouwhuis M., Ireland D.G., de Jager C.W., Jans E., de Jonge N., Kasdorp W.-J., Konijn J., Lapikas L., van Leeuwe J.J., van der Meer R.L.J., Nooren G.J.L., Passchier E., Schroevers M., van der Steenhoven G., Steijger J.J.M., Theunissen J.A.P., van Uden M.A., de Vries H., de Vries R., de Witt Huberts P.K.A., Blok H.P., van den Brink H.B., Dodge G.E., Harakeh M.N., Hesselink W.H.A., Kalantar-Nayestanaki N., Pellegrino A., Spaltro C.M., Templon J.A., Hicks R.S., Kelly J.J., Marchand C. (1994): High-Momentum Protons in ^{208}Pb. *Phys. Rev. Lett.* **73**, 2684

Boersma F. and Malfliet R. (1994): From Nuclear Matter to Finite Nuclei. *Phys. Rev.* **C49**, 233 and **C49**, 1495

Bonazzola S., Gourgoulhon E., Salgado M. and Marck J.A. (1993): Axisymmetric Rotating Relativistic Bodies: A New Numerical Approach for "Exact" Solutions. *Astron. Astrophys.* **278**, 421

Bondi H. (1952): On Spherically Symmetrical Accretion. *Mon. Not. Roy. Astron. Soc.* **112**, 195

Born M. and Huang K. (1954): Dynamical Theory of Crystal Lattices. Oxford University Press, London

Bradley P.A. (1995): Asteroseismology of DA White Dwarf Stars. In: *White Dwarfs* (Eds. D. Koester and K. Werner). Lecture Notes in Physics **443**. Springer, Berlin. p. 284

Bradley P.A. and Winget D.E. (1994): An Asteroseismological Determination of the Structure of the DBV White Dwarf GD 358. *Astrophys. J.* **430**, 850

Brockmann R. and Machleidt R. (1984): Nuclear Saturation in a Relativistic Brueckner–Hartree–Fock Approach. *Phys. Lett.* **149B**, 283

Brockmann R. and Machleidt R. (1990): Relativistic Nuclear Structure. *Phys. Rev.* **C42**, 1965

Brown G.E. (1994): The Equation of State of Dense Matter. *Nucl. Phys.* **A574**, 217c

Brown G.E. and Jackson A.D. (1976): The Nucleon–Nucleon Interaction. North-Holland, Amsterdam

Brown G.E. and Weise W. (1976): Pion Condensates. *Phys. Rep.* **27**, 1

Brown G.E., Müther H. and Prakash M. (1990): Effect of Chiral Constraints on Dense Nuclear Matter. *Nucl. Phys.* **A506**, 565

Brown J. M. and McQueen R.G. (1986): Phase Transitions, Grüneisen Parameter and Elasticity for Shocked Iron Between 77 GPa and 400 GPa. *J. Geophys. Res.* **91**, 7485

Brown T.M., Christensen-Dalsgaard J., Dziembowski W.A., Goode P., Gough D.O. and Morrow C.A. (1989): Inferring the Sun's Internal Angular Velocity from Observed p Mode Frequency Splittings. *Astrophys. J.* **343**, 526

Brown T.M. and Gilliland R.L. (1994): Asteroseismology. *Ann. Rev. Astron. Astrophys.* **32**, 37

Bullen K.E. (1956): Seismic Wave Transmission. In: *Encyclopedia of Physics* Vol. XLVII (Ed. S. Flügge). Springer, Berlin

Bukowinski M.S.T. (1994): Quantum Geophysics. *Ann. Rev. Earth Planet. Sci.* **22**, 167

Butterworth E.M. and Ipser J.R. (1976): On the Structure and Stability of Rapidly Rotating Fluid Bodies in General Relativity. I. The Numerical Method for Computing Structure and its Application to Uniformly Rotating Homogeneous Bodies. *Astrophys. J.* **204**, 200

Canuto V. and Chitre S.M. (1974): Cristallization of Dense Neutron Matter. *Phys. Rev. D* **9**, 1587

Ceperly D.M. and Alder B.J. (1980): Ground State of the Electron Gas by a Stochastic Method. *Phys. Rev. Lett.* **45**, 566

Chabrier G., Ashcroft N.W. and DeWitt H.E. (1992): White Dwarfs as Quantum Crystals. *Astrophys. J.* **360**, 48

Chabrier G. (1993): Quantum Effects in Dense Coulombic Matter: Application to the Cooling of White Dwarfs. *Astrophys. J.* **414**, 695

Chandrasekhar S. (1931): The Maximum Mass of White Dwarf Stars. *Astrophys. J.* **74**, 81

Chandrasekhar S. (1939): An Introduction to the Study of Stellar Structure. University of Chicago Press, Chicago

Chanfray G., Aouissat Z., Schuck P. and Nörenberg W. (1991): σ- and ρ-Meson Strength Distribution from in-Medium Corrected $\pi - \pi$ Correlations. *Phys. Lett.* **B256**, 325

Chanmugam G. (1972): Variable White Dwarfs. *Nature Phys. Sci.* **236**, 83

Chinn S.A. (1977): A Relativistic Many-Body Theory of High Density Matter. *Ann. of Phys.* **108**, 301

Christensen-Dalsgaard J., Gough D. and Toomre J. (1985): Seismology of the Sun. *Science* **229**, 923

Christensen-Dalsgaard J., Gough D.O. and Thompson M.J. (1991): The Depth of the Solar Convection Zone. *Astrophys. J.* **378**, 413

Christensen-Dalsgaard J. (1992): Solar Models with Enhanced Energy Transport in the Core. *Astrophys. J.* **385**, 354

Clark J.W. (1979): Variational Theory of Nuclear Matter. *Prog. in Part. and Nucl. Phys.* **2**, 89

Close F.E. (1979): An Introduction to Quarks and Partons. Academic Press, London

Coester F., Cohen S., Day B.D. and Vincent C.M. (1970): Variation in Nuclear Matter Binding Energy with Phase-Shift-Equivalent Two-Body Potentials. *Phys. Rev.* **C1**, 769

Courant R., Hilbert D. (1968): Methoden der Mathematischen Physik I. Springer, Berlin

Cowling T.G. (1941): *Mon. Not. Roy. Astron. Soc.* **101**, 367

Cox J.P. and Giuli R.T. (1968): The Principles of Stellar Structure. Gordon and Breach, New York

Cox J.P. (1984): Rotating, Pulsating Stars. *Pub. Astron. Soc. Pacific* **96**, 577

Cutler C., Lindblom L., Splinter R.J. (1990): Damping Times for Neutron Star Oscillations.

Dahlen F.A. (1972): Elastic Dislocation Theory for a Self-Gravitating Elastic Configuration with an Initial Static Stress Field. *Geophys. J. R. astr. Soc.* **28**, 357

Davies G. F. (1973): Quasi-Harmonic Finite Strain Equations of State of Solids. *J. Phys. Chem. Solids* **34**, 1417

Davies G. F. (1974): Effective Elastic Moduli under Hydrostatic Stress - I. Quasi-Harmonic Theory. *J. Phys. Chem. Solids* **35**, 1513

Day B.D., Coester F. and Goodman A. (1972): Three-Body Correlations in Nuclear Matter. *Phys. Rev.* **C6**, 1992 *Astrophys. J.* **363**, 603

Day B.D. (1981): Nuclear Saturation from Two-Nucleon Potentials. *Phys. Rev. Lett.* **47**, 226

Day B.D. and Wiringa R.B. (1985): Brueckner–Bethe and Variational Calculations of Nuclear Matter. *Phys. Rev.* **32**, 1057

Detweiler S. and Lindblom L. (1985): On the Nonradial Pulsations of General Relativistic Stellar Models. *Astrophys. J.* **292**, 12

Deubner F.-L. (1975): Observations of Low Wavenumber Nonradial Eigenmodes of the Sun. *Astron. Astrophys.* **44**, 371

Deubner F.-L. and Gough D. (1984): Helioseismology: Oscillations as a Diagnostic of the Solar Interior. *Ann. Rev. Astron. Astrophys.* **22**, 593

Dickhoff W.H., Faessler A., Meyer-ter-Vehn J. and Müther H. (1981): Pion Condensation and Realistic Interactions. *Phys. Rev.* **C23**, 1154

Dickhoff W.H., Faessler A. and Müther H. (1982): Multiple Δ (3,3) Excitations and the Binding Energy of Nuclear Matter. *Nucl. Phys.* **A389**, 492

Dickhoff W.H., Faessler A., Müther H. and Wu S.S. (1983): The Screening of the ph Interaction to all Orders. *Nucl. Phys.* **A405**, 534

Dickhoff W.H. and Müther H. (1992): Nucleon Properties in the Nuclear Medium. *Rep. Prog. in Phys.* **55**, 1947

Durso J.W., Saarela M., Brown G.E. and Jackson A.D. (1977): Isobar, Transition Potentials and Short-Range Repulsion in the Nucleon–Nucleon Interaction. *Nucl. Phys.* **A278**, 445

Duvall T.L. (1982): A Dispersion Law for Solar Oscillations. *Nature* **300**, 242

Dziembowski W.A., Goode P.R., Pamyatnykh A.A., Sienkiewicz R. (1994): A Seismic Model of the Sun's Interior. *Astrophys. J.* **432**, 417

Dziewonski A.M. and Anderson D.L. (1981): Preliminary Reference Earth Model. *Phys. Earth Planet. Int.* **25**, 297

Ebeling W., Förster A., Fortov V.E., Polishuk A.Ya. and Gryaznov V.K. (1991): The Physical Properties of Hot Dense Plasmas. Teubner, Leipzig

Elsenhans H., Müther H. and Machleidt R. (1990): Parametrization of the Relativistic Effective Interaction in Nuclear Matter. *Nucl. Phys.* **A515**, 715

Emden R. (1907): Gaskugeln. Teubner, Leipzig

Engvik L., Hjorth-Jensen M., Osnes E., Bao G. and Ostgaard E. (1995): Asymmetric Nuclear Matter and Neutron Star Properties. *Phys. Rev. Lett.* **73**, 2650

Erkelenz K. (1974): Current Status of the Relativistic Two-Nucleon One Boson Exchange Potential. *Phys. Rep.* **13**, 191

Faessler A., Fernandez F., Lübeck G. and Shimizu K. (1983): The NN Interaction and the Role of the [42] Orbital Six-Quark Symmetry. *Nucl. Phys.* **A402**, 555

Fetter A.L. and Walecka J.D. (1971): Quantum Theory of Many Particle Systems. McGraw-Hill, New York

Feynman R.P., Metropolis N., Teller E. (1949): Equations of State of Elements Based on the Generalized Fermi–Thomas Theory. *Phys. Rev.* **75**, 1561

Friedman B. and Pandharipande V.R. (1981): Hot and Cold, Nuclear and Neutron Matter. *Nucl. Phys.* **A361**, 502

Friedman J.L., Ipser J.R. and Parker L. (1986): Rapidly Rotating Neutron Star Models. *Astrophys. J.* **304**, 115

Friedman J.L., Ipser J.R. and Parker L. (1989): Implications of a Half-Millisecond Pulsar. *Phys. Rev. Lett.* **62**, 3015

Friedrich J. and Reinhard P.-G. (1986): Skyrme Force Parametrization: Least Square Fit to Nuclear Ground-State Properties. *Phys. Rev.* **C33**, 335

Fritz R. and Müther H. (1994): NN Correlations and Relativistic Hartree–Fock in Finite Nuclei. *Phys. Rev.* **C49**, 633

Gambhir Y.K., Ring P. and Thimet A. (1990): Relativistic Mean Field Theory for Finite Nuclei. *Ann. of Phys.* **198**, 132

Gautschy A. and Saio H. (1995): Stellar Pulsations across the HR Diagram: Part I. *Ann. Rev. Astron. Astrophys.* **33**, 75

Gell-Mann M. and Low F.E. (1951): Bound States in Quantum Field Theory. *Phys. Rev.* **84**, 350

Gilbert F. and Dziewonski A.M. (1975): An Application of Normal Mode Theory to the Retrieval of Structural Parameters and Source Mechanisms from Seismic Spectra. *Philos. Trans. R. Soc. London* Ser. A **278**, 187

Glendenning N.K., Hecking P. and Ruck V. (1983): Normal and Pion-Condensed States in Neutron Star Matter in a Relativistic Field Theory Constrained by Nuclear Bulk Properties. *Ann. of Phys.* **149**, 22

Glendenning N.K. (1985): Neutron Stars are Giant Hypernuclei? *Astrophys. J.* **293**, 470

Glendenning N.K., Weber F. and Moszkowski S.A. (1992): Neutron Stars in the Derivative Coupling Model. *Phys. Rev. C* **45**, 844

Goldreich P. and Keeley D.A. (1977): Solar Seismology. I. The Stability of Solar p Modes (1977). *Astrophys. J.* **211**, 934

Goldreich P. and Weber S.V. (1980): Homologously Collapsing Stellar Cores. *Astrophys. J.* **238**, 991

Goldreich P., Murray N. and Kumar P. (1994): Excitation of Solar p-Modes. *Astrophys. J.* **424**, 466

Gough D. and Toomre J. (1991): Seismic Observations of the Solar Interior. *Ann. Rev. Astron. Astrophys.* **29**, 627

Graboske H.C., Harwood D.J. and Rogers F.J. (1969): Thermodynamic Properties of Nonideal Gases. I. Free-Energy Minimization Method. *Phys. Rev.* **186**, 210

Green A.M. (1976): Nucleon Resonances in Nuclei. *Rep. Prog. Phys.* **39**, 1109

Gross F. (1969): Three-Dimensional Covariant Integral Equation for Low-Energy Systems. *Phys. Rev.* **186**, 1448

Günter S., Hitschke L. and Röpke G. (1991): Hydrogen Spectral Lines with the Inclusion of Dense-Plasma Effects. *Phys. Rev. A* **44**, 6834

Guenther D.B., Jaffe A., Demarque P. (1989): The Standard Solar Model: Composition, Opacities, and Seismology. *Astrophys. J.* **345**, 1022

Guenther D.B., Demarque P., Kim Y.-C. and Pinsonneault M.H. (1992): Standard Solar Model. *Astrophys. J.* **387**, 372

Guzik J.A. and Cox A.N. (1993): Using Solar p-Modes to Determine the Convection Zone Depth and Constrain Diffusion-Produced Composition Gradients. *Astrophys. J.* **411**, 394

Haftel M.I. and Tabakin F. (1970): Nuclear Saturation and the Smoothness of Nucleon–Nucleon Potentials. *Nucl. Phys.* **A158**, 1

Hamada T. and Johnston I. (1962): A Potential Model Representation of Two-Nucleon Data Below 315 MeV. *Nucl. Phys.* **34**, 382

Hansen C.J. and Kawaler S.D. (1994): Stellar Interiors. Physical Principles, Structure and Evolution. Springer, New York

Harrison B.K., Thorne K.S., Wakano M. and Wheeler J.A. (1965): Gravitation Theory and Gravitational Collapse. University of Chicago Press, Chicago

Hartle, J.B. (1967): Slowly Rotating Relativistic Stars. I. Equations of Structure. *Astrophys. J.* **150**, 1005

Hartle J.B. and Thorne K.S. (1968): Slowly Rotating Relativistic Stars. II. Models for Neutron Stars and Supermassive Stars. *Astrophys. J.* **153**, 807

Harvey J.W., Duvall T.L., and Pomerantz M.A. (1982): Astronomy on Ice. *Sky and Tel.* **64**, 520

Harvey J.W. (1995): Helioseismology. *Physics Today* **48** (No. 10), 32

Henley E.M. and Müther H. (1990): Medium Effects in a Generalized Nambu-Jona-Lasinio Model. *Nucl. Phys.* **A513**, 667

Hernanz M., García-Berro E., Isern J., Mochkovitch R., Segretain L., Chabrier G. (1994): The Influence of Crystallization on the Luminosity Function of White Dwarfs. *Astrophys. J.* **434**, 652

Herold H. and Neugebauer G. (1992): Gravitational Fields of Rapidly Rotating Neutron Stars: Numerical Results. In: *Relativistic Gravity Research* (Eds. J. Ehlers and G. Schäfer). Lecture Notes in Physics **410**. Springer, Berlin. p. 319

Höhler G. and Pietarinen E. (1975): The ρNN Vertex in Vector Dominance Models. *Nucl. Phys.* **B95**, 210

Holinde K., Machleidt R., Anastasio M.R., Faessler A. and Müther H. (1978): Isobar Contributions to the Two-Nucleon Interaction Derived from Noncovariant Perturbation Theory. *Phys. Rev.* **C 18**, 870

Holinde K. (1981): Two-Nucleon Forces and Nuclear Matter. *Phys. Rep.* **68**, 121

Horowitz C.J. and Serot B.D. (1987): The Relativistic Two-Nucleon Problem in Nuclear Matter. *Nucl. Phys.* **A464**, 613

Huebner W.F., Merts A.L., Magee N.H.Jr., Argo M.F. (1977): Astrophysical Opacity Library. *Los Alamos Sci. Lab. Rep.* **LA-6760-M**

Hummer D.G. and Mihalas D. (1988): The Equation of State For Stellar Envelopes. I. An Occupation Probability Formalism for the Truncation of Internal Partition Functions. *Astrophys. J.* **331**, 794

Hund F. (1936): Materie unter sehr hohen Drucken und Temperaturen. *Erg. exakt. Naturwiss.* **15**, 189; see also appendix of this book.

Itzykson C. and Zuber J.-B. (1980): Quantum Field Theory. McGraw-Hill, New York

Jackson A.D., Lande A. and Smith R.A. (1982): Variational and Perturbation Theory Made Planar. *Phys. Rep.* **86**, 55

Jayaraman A. (1983): Diamond Anvil and High-Pressure Physical Investigations. *Rev. Mod. Phys.* **55**, 65

Jeukenne J.-P., Lejeune A. and Mahaux C. (1976): Many-Body Theory of Nuclear Matter. *Phys. Rep.* **25**, 83

Jiang M.F., Kuo T.T.S. and Müther H. (1988): Ring Diagram Nuclear Matter Calculations using Bonn and V_{14} Potentials. *Phys. Rev.* **C38**, 240

Jones P.W., Pesnell W.T., Hansen C.J. and Kawaler S.D.(1989): On the Possibility of Detecting Weak Magnetic Fields in Variable White Dwarfs. *Astrophys. J.* **336**, 403

Kadychevsky V.G. (1968): Quasipotential-Type Equation for the Relativistic Scattering Amplitude. *Nucl. Phys.* **B6**, 125

Kawaler S.D. and Bradley P.A. (1994): Precision Asteroseismology of Pulsating PG 1159 Stars. *Astrophys. J.* **427**, 415

Kawaler S.D., O'Brien M.S., Clemens J.C., Nather R.E., Winget D.E., Watson T.K., Yanagida K., Dixson J.S., Bradley P.A., Wood M.A., Sullivan D.J., Kleinman S.J., Meištas E., Leibowitz E.M., Moskalik P., Zola S., Pajdosz G., Krzesiński J., Solheim J.-E., Bruvold A., O'Donoghue D., Katz M., Vauclair G., Dolez N., Chevreton M., Barstow M. A., Kanaan A., Kepler S.O., Giovannini O., Provencal J.L. and Hansen, C.J. (1995): Whole Earth Telescope Observations and Seismological Analysis of the Pre-white Dwarf PG 2131+066. *Astrophys. J.* **450**, 350

Keilhacker M. and Watkins M.L. (1995): JET Divertor Results. *Europhys. News* **26**, 105

Kepler S.O., Vauclair G., Dolez N., Chevreton M., Barstow M.A., Nather R.E., Winget D.E., Provencal J.L., Clemens J.C. and Fontaine G. (1990): An Observational Limit to the Evolutionary Time Scale of the 13 000 K White Dwarf G117-B15A. *Astrophys. J.* **357**, 204

Kepler S.O., Winget D.E., Nather R.E., Bradley P.A., Grauer A.D., Fontaine G., Bergeron P., Vauclair G., Claver C.F., Marar T.M.K., Seetha S., Ashoka B.N., Mazeh T., Leibowitz E., Dolez N., Chevreton M., Barstow M.A., Clemens J.C., Kleinman S.J., Sansom A.E., Tweedy R.W., Kanaan A., Hine B.P., Provencal J.L., Wesemael F., Wood M.A., Brassard P., Solheim J.-E.and Emanuelsen P.-I. (1991): A Detection of the Evolutionary Time Scale of the DA White Dwarf G117-B15A with the Whole Earth Telescope. *Astrophys. J.* **378**, L45

Kepler S.O. (1993): WET Data on G117-B15A and G226–29. *Baltic Astron.* **2**, 444

Kippenhahn R., and Weigert A. (1990): Stellar Structure and Evolution. Springer, New York

Kleinmann M., Fritz R., Müther H. and Ramos A. (1994): On the Momentum Dependence of the Nucleon–Nucleus Optical Potential. *Nucl. Phys.* **A579**, 85

Klevansky S.P. (1992): The Nambu-Jona-Lasinio Model of Quantum Chromodynamics. *Rev. Mod. Phys.* **64**, 649

Koester D. and Chanmugam G. (1990): Physics of White Dwarf Stars. *Rep. Progr. Phys.* **53**, 837

Kohn W. and Sham L.J. (1965): Self-Consistent Equations Including Exchange and Correlation Effects. *Phys. Rev. A* **140**, 1133

Kraeft W.D., Kremp D., Ebeling W. and Röpke G. (1986): Quantum Statistics of Charged Particle Systems. Plenum, New York

Kramer D., Stephani H., MacCallum M. and Herlt E. (1980): Exact Solutions of Einstein's Field Equations. VEB Deutscher Verlag der Wissenschaften, Berlin

Kudritzki R.P., Gabler R., Kunze D., Pauldrach A.W.A. and Puls J. (1991): Massive Stars in Starbursts. In: *STScI Symposium Series 5* (Eds. C. Leitherer, N.R. Walborn, T.M. Heckman, and C.A. Norman). Cambridge University Press, Cambridge

Kutschera M. and Kotlorz A. (1993): Maximum Quark Core in a Neutron Star for Realistic Equations of State. *Astrophys. J.* **419**, 752

Lacombe M., Loiseaux B., Richard J.M., Vinh Mau R., Côté J., Pirès P. and de Tourreil R. (1980): Parametrization of the Paris NN Potential. *Phys. Rev.* **C21**, 861

Lagaris I.E. and Pandharipande V.J. (1981): Phenomenologic Two-Nucleon Interaction Operator. *Nucl. Phys.* **A359**, 331

Landau L. D. and Lifschitz E. M. (1975): Elastizitätstheorie. Akademie Verlag, Berlin

Landolt A.U. (1968): A New Short-Period Blue Variable. *Astrophys. J.* **153**, 151

Lane J.H. (1870): On the Theoretical Temperature of the Sun, Under the Hypothesis of a Gaseous Mass Maintaining its Volume by its Internal Heat and Depending on the Laws of Gases as Known to Terrestial Experiment. *Amer. J. Sci.* **50**, 57

Lapwood E.R. and Usami T. (1981): Free Oscillations of the Earth. Cambridge University Press, Cambridge

Lasker B.M. and Hesser J.E. (1969): High-Frequency Stellar Oscillations. II. G44–32 A New Short-Period Blue Variable Star. *Astrophys. J.* **158**, L171

Ledoux P. and Walraven Th. (1958): Variable Stars. In: *Encyclopedia of Physics* Vol. LI (Ed. S. Flügge). Springer, Berlin

Leibacher J. and Stein R.F. (1971): A New Description of the Solar Five-Minute Oscillation. *Astrophys. Lett.* **7**, 191

Leibfried G. and Ludwig W. (1961): Theory of Anharmonic Effects in Crystals. In: *Solid State Physics, Vol. 12* (Eds. F. Seitz and D. Turnbull). Academic Press, New York

Lehmann H. (1954): Über Eigenschaften von Ausbreitungsfunktionen und Renormierungskonstanten quantisierter Felder. *Nuovo Cimento* **11**, 342

Leins M. (1994): Nichtradiale Schwingungen von Neutronensternen in der Allgemeinen Relativitätstheorie. Doctoral Dissertation, Universität Tübingen

Leins M., Nollert H.-P. and Soffel M.H. (1993): Nonradial Oscillations of Neutron Stars: A New Branch of Strongly Damped Normal Modes. *Phys. Rev. D* **48**, 3467

Leighton R.B., Nayes R.W. and Simon G.W. (1962): Velocity Fields in the Solar Atmosphere I. Preliminary Report. *Astrophys. J.* **135**, 474

Li G.Q., Machleidt R. and Brockmann R. (1992): Properties of Dense Nuclear and Neutron Matter with Relativistic Nucleon–Nucleon Interactions. *Phys. Rev.* **C45**, 2782

Libbrecht K.G. and Zirin H. (1986): Properties of Intermediate-Degree Solar Oscillation Modes. *Astrophys. J.* **308**, 413

Libbrecht K.G. (1988a): Solar and Stellar Seismology. *Space Sci. Rev.* **47**, 275

Libbrecht K.G. (1988b): Solar p-Mode Phenomenology. *Astrophys. J.* **334**, 510

Libbrecht K.G. and Kaufman J.M. (1988): Frequencies of High-Degree Solar Oscillations. *Astrophys. J.* **324**, 1172

Libbrecht K.G. and Woodard M.F. (1991): Advances in Helioseismology. *Science* **253**, 152

Lindblom L. and Detweiler S.L. (1983): The Quadrupole Oscillations of Neutron Stars. *Astrophys. J. Suppl.* **53**, 73

Machleidt R., Holinde K. and Elster Ch. (1987): The Bonn Meson-Exchange Model for the Nucleon–Nucleon Interaction. *Phys. Rep.* **149**, 1

Machleidt R. (1989): The Meson Theory of Nuclear Forces and Nuclear Structure. *Adv. Nucl. Phys.* **19**, 189

Manzke W. and Gari M. (1978): Resonances in Nuclei — A Fully Self-Consistent Solution of the Coupled Channel Equations for Nuclear Matter (Δ, $\Delta\Delta$). *Nucl. Phys.* **A312**, 457

Mattuck R.D. (1976): A Guide to Feynman Diagrams in the Many-Body Problem. McGraw-Hill, New York

McGraw J.T. (1979): The Physical Properties of the ZZ Ceti Stars and their Pulsations. *Astrophys. J.* **229**, 203

Migdal A.B. (1978): Pion Fields in Nuclear Matter. *Rev. Mod. Phys.* **50**, 10

Mihalas D. (1978): Stellar Atmospheres. Freeman, New York

Mihalas D., Däppen W. and Hummer D.G. (1988): The Equation of State for Stellar Envelopes. II. Algorithm and Selected Results. *Astrophys. J.* **331**, 815

Morse P. M. and Feshbach H. (1953): Methods of Theoretical Physics. McGraw-Hill, New York

Müther H. (1985): Isobar Degrees of Freedom and the Nuclear Many-Body Problem. *Prog. in Part. and Nucl. Phys.* **14**, 123

Müther H. (1986): Nuclear Matter Under Extreme Conditions. *Prog. in Part. and Nucl. Phys.* **17**, 97

Müther H., Prakash M. and Ainsworth T.L. (1987): The Nuclear Symmetry Energy in Relativistic Brueckner–Hartree–Fock Calculations. *Phys. Lett.* **B199**, 469

Müther H., Polls A. and Dickhoff W.H. (1995): Momentum and Energy Distributions of Nucleons in Finite Nuclei due to Short-Range Correlation. *Phys. Rev.* **C51**, 3040

Nambu Y. and Jona-Lasinio G. (1961): Dynamical Model of Elementary Particles Based on an Analogy with Superconductivity. *Phys. Rev.* **122**, 345

Nather R.E., Winget D.E., Clemens J.C., Hansen C.J. and Hine B.P. (1990): The Whole Earth Telescope: A New Astronomical Instrument. *Astrophys. J.* **361** , 309

Nather R.E. and Winget D.E. (1992): Taking the Pulse of White Dwarfs. *Sky & Telescope* **April**, 374

Negele J.W. and Orland H. (1988): Quantum Many-Particle Systems. Addison-Wesley, Redwood City

Neugebauer G. and Herlt E. (1984): Einstein–Maxwell Fields Inside and Outside Rotating Sources as Minimal Surfaces. *Class. Quant. Grav.* **1**, 695

Neugebauer G. and Herold H. (1992): Gravitational Fields of Rapidly Rotating Neutron Stars: Theoretical Foundation. In: *Relativistic Gravity Research* (eds. J. Ehlers and G. Schäfer). Lecture Notes in Physics **410**. Springer, Berlin. p.305

Nollert H.-P. (1996): Visualization of Objects via Four-dimensional Ray Tracing. In: *Relativity and Scientific Computing: Computer Algebra, Numerics, Visualization* (eds. F.W. Hehl et al.). Springer, Berlin (in press)

Nollert H.-P., Ruder H., Herold H. and Kraus U. (1989): About the Relativistic Looks of a Neutron Star. *Astron. Astrophys.* **208**, 153

Nollert H.-P. and Schmidt B.G. (1992): Quasinormal Modes of Schwarzschild Black Holes: Defined and Calculated via Laplace Transformation. *Phys. Rev. D* **45**, 2617

Oppenheimer J.R. and Volkoff G.M. (1939): On Massive Neutron Cores. *Phys. Rev.* **55**, 374

Pandharipande V.R. (1971): Hyperonic Matter. *Nucl. Phys.* **A 178**, 123

Particle Data Group (1994): Review of Particle Properties. *Phys. Rev.* **D50**, 1173

Pekeris C.L. and Jarosch H. (1958): The Free Oscillations of the Earth. In: *Contributions in Geophysics* (ed. H. Benioff). Pergamon Press, London

Perdew J.P. and Zunger A. (1981): Self-Interaction Correlation to Density-Functional Approximations for Many-Electron Systems. *Phys. Rev. B* **23**, 5048

Press F. (1965): Resonant Vibrations of the Earth. *Scientific Am.* **213**, 28

Regge T.R. and Wheeler J.A. (1957): Stability of a Schwarzschild Singularity. *Phys. Rev.* **108**, 1063

Reid R.V. (1968): Local Phenomenological Nucleon–Nucleon Potentials. *Ann. Phys. (N.Y.)* **50**, 411

Reinholz H., Redmer R. and Nagel S. (1995): Thermodynamic and Transport Properties of Dense Hydrogen Plasmas. *Phys. Rev. E* **52**, 5368

Ring P. and Schuck P. (1980): The Nuclear Many-Body Problem. Springer, New York

Rogers F.J, Graboske H.C. and Harwood D.J. (1970): Bound Eigenstates of the Static Screened Coulomb Potential. *Phys. Rev. A* **1**, 1577

Rolfs C.E. and Rodney W.S. (1988): Cauldrons in Cosmos. Univeristy of Chicago Press, Chicago

Salpeter E.E. and Zapolsky H.S. (1967): Theoretical High-Pressure Equations of State Including Correlation Energy. *Phys. Rev.* **158**, 876

Sauer P.U. and Tjon J.A. (1973): Three-Nucleon Calculations Without the Explicit Use of Two-Body Potentials. *Nucl. Phys.* **A216**, 541

Schierholz G. (1972): A Relativistic One-Boson-Exchange Model of the Nucleon–Nucleon Interaction. *Nucl. Phys.* **B40**, 335

Schmid K.W., Müther H. and Machleidt R. (1991): Meson Exchange Potentials and the Problem of Saturation in Finite Nuclei. *Nucl. Phys.* **A530**, 14

Schwarzschild M. (1958): Structure and Evolution of Stars. Princeton University Press, Princeton

Scuflaire R. (1974): The Non-Radial Oscillations of Condensed Polytropes. *Astron. Astrophys.* **36**, 107

Seaton M.J. (1987): Atomic Data for Opacity Calculations: I. General Description. *J. Phys. B: At. Mol. Phys.* **20**, 6767

Segretain L., Chabrier G., Hernanz M., García-Berro E., Isern J. and Mochkovitch R.(1994): Cooling Theory of Crystallized White Dwarfs. *Astrophys. J.* **434**, 641

Serot B.D. and Walecka J.D. (1986): The Relativistic Nuclear Many-Body Problem. *Adv. Nucl. Phys.* **16**, 1

Stevenson D.J. (1982): Interiors of the Giant Planets. *Ann. Rev. Earth Planet Sci.* **10**, 257

Shapiro S.L. and Teukolsky S.A. (1983): Black Holes, White Dwarfs, and Neutron Stars: The Physics of Compact Objects. Wiley, New York

Skyrme T.H.R. (1959): The Effective Nuclear Potential. *Nucl. Phys.* **9**, 615

Smirnova T.V., Tul'bashev S.A. and Boriakoff V. (1994): Statistics of PSR 1133+16 Micropulse Emission Determined at Widely Spaced Frequencies. *Astron. Astrophys.* **286**, 807

Speth J., Werner E. and Wild W. (1977): Theory of Finite Fermi Systems and Applications to the Lead Region. *Phys. Rep.* **33**, 127

Stix, M. (1991): The Sun. Astronomy and Astrophysics Library. Springer, Berlin

Sumiyoshi K. and Toki H. (1994): Relativistic Equation of State of Nuclear Matter for the Supernova Explosion and the Birth of Neutron Stars. *Astrophys. J.* **422**, 700

Tamiya K. and Tamagaki R. (1981): Reaction Matrix Calculations in Neutron Matter with Alternating-Layer-Spin Structure. *Prog. Theor. Phys.* **66**, 948

ter Haar B. and Malfliet R. (1987): Nucleons, Mesons and Deltas in Nuclear Matter. *Phys. Rep.* **149**, 207

Thompson R.H. (1970): Three-Dimensional Bethe–Salpeter Equation Applied to the Nucleon–Nucleon Interaction. *Phys. Rev.* **D1**, 110

t'Hooft G. (1974): A Two-Dimensional Model for Mesons. *Nucl. Phys.* **B75**, 461

Thorne K.S. and Campolattaro A. (1967) Non-Radial Pulsation of General-Relativistic Stellar Models. I. *Astrophys. J.* **149**, 591

Tolstoy I. (1963): The Theory of Waves in Stratified Fluids Including the Effects of Gravity and Rotation. *Rev. of Mod. Phys.* **35**, 207

Trümper J. (1995): Röntgenstrahlung von Neutronensternen. *Phys. Blätter* **51**, 649

Ulrich R.K. (1970): The Five-Minutes Oscillations on the Solar Surface. *Astrophys. J.* **162**, 993

Umeda H., Shibazaki N., Nomoto K. and Tsuruta S. (1993): Thermal Evolution of Neutron Stars with Internal Frictional Heating. *Astrophys. J.* **408**, 186

Unno W., Osaki Y., Ando H. and Shibahashi H. (1989): Nonradial Oscillations of Stars. Univ. Tokyo Press, Tokyo

Van Horn H.M. and Ichimaru S. (Eds.) (1993): Strongly Coupled Plasma Physics. University of Rochester Press, Rochester

von Geramb H.V. (1993): Quantum Inversion Theory and Applications. Springer, Heidelberg

Waldhauser B.M., Maruhn J.A., Stöcker H., and Greiner W. (1988): Nuclear Equation of State from the Nonlinear Relativistic Mean Field Theory. *Phys. Rev.* **C 38**, 1008

von Weizsäcker C.F. (1935): Zur Theorie der Kernmassen. *Z. Phys.* **96**, 431

Walecka J.D. (1974): A Theory of Highly Condensed Matter. *Ann. Phys. (N.Y.)* **83**, 491

Warner B. and Robinson E.L. (1972): Non-Radial Pulsations in White Dwarf Stars. *Nature Phys. Sci.* **239**, 2

Weber F. and Weigel M.K. (1989): Baryon Composition and Macroscopic Properties of Neutron Stars. *Nucl. Phys.* **A505**, 779

Weinberg S. (1972): Gravitation and Cosmology. Wiley, New York

Weise W., Brown G.E. (1975): Equation of State for Neutron Matter in the Presence of a Pion Condensate. *Phys. Lett.* **B 58**, 300

Wick G.C. (1950): The Evaluation of the Collision Matrix. *Phys. Rev.* **80**, 268

Winget D.E., Hansen C.J. and van Horn H.M. (1983): Do Pulsating PG 1159–035 Stars put Constraints on Stellar Evolution? *Nature* **303**, 781

Winget D.E., Kepler S.O., Robinson E.L., Nather R.E. and O'Donoghue D. (1985): A Measurement of Secular Evolution in the Pre-White Dwarf Star PG 1159–035. *Astrophys. J.* **292**, 606

Winget D.E., Nather R.E., Clemens J.C., Provencal J., Kleinman S.J., Bradley P.A., Wood M.A., Claver C.F., Frueh M.L., Grauer A.D., Hine B.P., Hansen C.J., Fontaine G., Achilleos N., Wickramasinghe D.T., Marar T.M.K., Seetha S., Ashoka B.N., O'Donoghue D., Warner B., Kurtz D.W., Buckley D.A., Brickhill J., Vauclair G., Dolez N., Chevreton M., Barstow M.A., Solheim J.E., Kanaan A., Kepler S.O., Henry G.W. and Kawaler S.D.(1991): Astroseismology of the DOV Star PG 1159–035 with the Whole Earth Telescope. *Astrophys. J.* **378**, 326

Winget D.E., Nather R.E., Clemens J.C., Provencal J.L., Kleinman S.J., Bradley P.A., Claver C.F., Dixson J.S., Montgomery M.H., Hansen C.J., Hine B.P., Birch P., Candy M., Marar T.M.K., Seetha S., Ashoka B.N., Leibowitz E.M., O'Donoghue D., Warner B., Buckley D.A.H., Tripe P., Vauclair G., Dolez N., Chevreton M., Serre T., Garrido R., Kepler S.O., Kanaan A., Augusteijn T., Wood M.A., Bergeron P. and Grauer A.D. (1994): Whole Earth Telescope Observations of the DBV White Dwarf GD 358. *Astrophys. J.* **430**, 839

Wiringa R.B., Smith R.A., and Ainsworth T.L. (1984): NN Potential With and Without $\Delta(1232)$ Degrees of Freedom. *Phys. Rev.* **C29**, 1207

Witten E. (1979): Baryons in the 1/N Expansion. *Nucl. Phys.* **B160**, 57

Wood M.A. (1990): PhD Thesis, University of Texas, Austin.

Wood M.A. (1992): Constraints on the Age and Evolution of the Galaxy from the White Dwarf Luminosity Function. *Astrophys. J.* **386**, 539

Wood M.A. (1994): New Model Sequences from the White Dwarf Evolution Code. In: *The Equation of State in Astrophysics* IAU Colloquium **147**. Cambridge Univeristy press, Cambride, p. 612

Wood M.A. (1995): Theoretical White Dwarf Luminosity Functions: DA Models. In: *White Dwarfs* (Eds. D. Koester and K. Werner). Lecture Notes in Physics **443**. Springer, Berlin. p. 41

Wu X., Müther H., Soffel M., Herold H. and Ruder H. (1991): A New Equation of State for Dense Matter and Fast Rotating Pulsars. *Astron. Astrophys.* **246**, 411

Yukawa, H. (1935) *Proc. Phys. Math. Soc. Jpn* **17**, 48

Zamick L., Zheng D.C. and Müther H. (1992): On the Spin-Orbit Splitting for Single-Particle and Single-Hole Energies. *Phys. Rev.* **C45**, 2763

Zerilli F.J. (1970a): Tensor Harmonics in Canonical Form for Gravitational Radiation and Other Applications. *J. of Math. Phys.* **11**, 2203

Zerilli F.J. (1970b): Effective Potential for Even-Parity Regge-Wheeler Gravitational Perturbation Equations. *Phys. Rev. Lett.* **24**, 737

Zhang H., and Bukowinski M.T.S. (1991): Modified Potential-Induced-Breathing Model of Potentials between Close-Shell Ions. *Phys. Rev. B* **44**, 2495

Zienkiewicz O.C. (1977): The Finite Element Method. McGraw-Hill, London

Index

acoustic cutoff frequency, 135
acoustic wave, 136, 142, 146
action integral, 105
adiabatic exponent, 123
adiabatic oscillation, 123
annihilation operator, 29
axial symmetry, 103

bag model, 84
baryon number, 100
– conservation, 100, 163
– density, 163
BBP theorem, 58
β decay, 6, 14, 15, 27
Bethe–Faddeev equations, 62
Bethe–Goldstone equation, 57, 70
Blankenbecler–Sugar equation, 50, 70
Bonn potential, 51, 71
bound-bound transition, 17
bound-free transition, 16
breathing mode, 23, 132
Brueckner–Hartree–Fock, 28, 58
Brunt–Väisälä frequency, 133, 134
bulk modulus, 123, 124, 138

centrifugal potential, 104
Chandrasekhar limit, 101
chemical potential, 13
chiral symmetry, 86
Christoffel symbols, 95, 116
chronological operator, 31
CNO cycle, 14, 15
Coester band, 60, 63, 65, 71, 75
conductive opacity, 18
confinement, 21, 89
conformally flat, 107
constituent quarks, 85
continuity equation, 128
convection, 16, 19, 20
convection zone, 146, 147, 150, 152
cooling rate, 24, 83, 158, 160, 161

creation operator, 29
cutoff frequencies, 136

DAV star, 158, 160
DBV star, 158
Debye length, 12
degenerate electron gas, 6, 8, 13
degenerate gas of fermions, 26
degenerate neutron gas, 6
$\Delta(3,3)$ excitation, 53, 69, 74
density-functional theory, 8
deuterium, 14, 23
Dirac equation, 43, 67
Dirac–Brueckner–Hartree–Fock, 69
Dirac–Hartree–Fock, 68
dispersion relation, 54, 136
displacement vector, 122, 126, 129, 165
Doppler image, 143
Doppler shift, 142
DOV star, 158

Earth model, 138
earthquake, 137
effective mass, 70, 85
eigenfrequencies, 130
Einstein's field equations, 95, 102, 109, 162
Einstein–Hilbert action, 107
embedding diagrams, 115, 116
energy density, 26, 96
– flux, 14
– transport, 16, 18–20
energy-momentum conservation, 97
energy-momentum tensor, 96, 105, 107, 162, 163
equation of state, 5, 9–11, 18, 19, 25–28, 90, 91, 98, 99, 101, 105, 113
– soft, 23
– stiff, 23, 69
ergoregion, 104, 117
evolution operator, 31

f mode, 130, 132, 134, 172
Fermi energy, 6, 25, 38
– momentum, 26
Fermi-hypernetted-chain scheme, 62
Feynman diagram, 29, 34, 41
– propagator, 42
finite element approach, 112
flavors, 84
form factor, 47
four-velocity, 96, 104
free boundary, 125
free energy, 7, 11, 13, 124
free oscillations, 122, 137
free-free opacity, 16
– transition, 16
frequency splitting, 145, 151, 152, 157, 158
fundamental mode, 130, 132, 134, 172

G matrix, 58, 70, 72
g mode, 133–135, 147, 151, 154, 160, 161, 171
G117–B15A, 160
general relativity, 93, 102
gravitational wave damping, 166, 172
Green's function, 29
– many-body, 29
– Lehmann representation, 38
– self-consistent approach (SCGF), 64
– single-particle, 37–40

Hartree–Fock, 23, 40
healing property, 57
heat conduction, 18
Heisenberg scheme, 30
helicity representation, 47
Hertzsprung–Russel diagram, 142, 159, 160
hole-line expansion, 62, 64
Hooke's law, 124
hydrogen burning, 14, 15
hydrostatic equilibrium, 4, 19
hypernuclei, 83
hyperons, 83

ideal gas, 5, 6
– of fermions, 25
interaction scheme, 30
inversion method, 145, 149, 151, 152
irreducible diagrams, 49, 53
irreducible two-body amplitude, 75
isobar degrees of freedom, 74, 76, 82
isotropic matter, 125

Jastrow correlation function, 61

Kaon condensate, 83
κ mechanism, 142
Kepler frequency, 117
Killing vector, 102
Klein–Gordon equation, 66

Lagrangian, 66, 109
Lamé coefficient, 125
Lamb frequency, 134, 135
Landau–Migdal parameter, 80
Lehmann representation, 38
Lense–Thirring frame dragging, 117
light deflection, 117
line broadening, 17
linewidth, 145
linked cluster theorem, 37
linked diagrams, 36
Lippmann–Schwinger equation, 51, 57
local density approximation, 10, 72

m splitting, 145, 151, 152, 157, 158
Mandelstam variable, 50, 54, 82
many-body forces, 24, 66
mass density, 98, 99
– function, 97
– shedding, 113, 116
mass-radius relation, 99, 101
mean molecular weight, 6, 11, 12
metric, 95, 103, 109
millisecond pulsar, 101
minimal-surface formalism, 102
– problem, 109
mixing length, 19, 20, 146
momentum distribution, 39

Nambu and Jona-Lasinio, 84
Nambu–Goldstone, 86
neutrino, 3, 15, 27, 82, 153
neutron drip, 99
neutron matter, 73
neutron stars, 4, 93, 162
– baryon mass, 113
– central density, 101, 113
– equatorial radius, 113
– gravitational mass, 113
– maximum mass, 93, 99, 101
– minimum mass, 101
– moment of inertia, 113
– non-rotating, 94, 110
– outer crust, 99
– rotating, 101, 102
– total mass, 97, 99, 113

Newton's gravitational theory, 93
NN interaction, 43
– local, 43
– medium-range attraction, 52
– non-local, 48
– phenomenological, 21
– realistic, 23, 43, 71
– short-range repulsion, 44, 52
– tensor contribution, 51, 54, 60, 77
non-relativistic electron gas, 28
non-rotating neutron star, 94, 110
normal mode, 101
normal product, 33
nuclear matter, 22, 59, 60, 72
– incompressibility, 22
– saturation density, 56
– saturation point, 22, 69

one-boson exchange, 23, 46
one-pion exchange, 47, 80
opacity, 16–18, 153

p mode, 133–135, 142, 144, 146, 147,
    153, 160, 172
P wave, 126, 129, 132, 137
p–p chain, 14, 15
Paris potential, 55
Parquet technique, 82
partition function, 7, 12
Pauli operator, 57, 63
perfect fluid, 96, 104, 107
period change, 161
PG 1159–035, 155–158, 160
photon scattering, 16
pion condensation, 24, 78, 89
– propagator, 81
– self-energy, 79
Planck spectrum, 17
planetary nebulae, 158
Poisson's equation, 122, 128
polytropic EOS, 7, 133
power spectrum, 137, 144, 156, 157
pressure, 26, 91, 96, 98
– equilibrium, 97, 105
– radiation, 6

QCD, 21, 45
quadrupole moment, 152
quadrupole oscillation, 172
quantumchromodynamics, 21
quark–gluon plasma, 25

quarks, 21, 84
– constituent, 85
quasinormal mode, 166, 168–170

radial perturbation, 99
radiation pressure, 6
radiative diffusion, 16
ray-theory, 122
ray-tracing (4D), 116
Rayleigh wave, 132
redshift, 117
reflection symmetry, 110
Regge–Wheeler gauge, 164, 165
relativistic electron gas, 28
Ricci tensor, 96
Riemann curvature tensor, 96
ring diagram, 63, 76
ROSAT experiment, 83
Rosseland mean opacity, 17
rotating black holes, 117
rotating neutron star, 101, 102
rotation period, 151, 157
RPA approximation, 64, 87

S wave, 126, 129, 132, 137
Saha equation, 12
Schrödinger scheme, 30
Schwarzschild criterion, 19, 133
– metric, 97
– radius, 99
– solution, 110
– – interior, 98
seismology, 136
self-energy, 41, 62, 66–68, 70, 85
shear modulus, 124, 138
short-range repulsion, 44, 52
short-wavelength limit, 134, 146, 148
Skyrme force, 21, 69
Slater determinant, 55
solar neutrinos, 3, 153
– oscillation, 142
– structure, 145, 150, 153
sound speed, 98, 133, 148, 149
spectral function, 39
spherical harmonics, 127, 143, 164
spheroidal mode, 129, 131, 138, 141
spin-orbit term, 73
standard solar model, 146, 147, 153
standing wave, 122
stellar structure, 6, 14, 20, 146
strain tensor, 124
stress tensor, 122–124
Sturm–Liouville problem, 129, 132

tensor force, 51, 54, 60, 77
– operator, 43
– spherical harmonics, 165
thermal conduction, 16
– equilibrium, 7, 17
Thomas–Fermi method, 8
Thomson cross section, 16
three-body force, 65
tidal force, 136
Tolman–Oppenheimer–Volkoff
    equations, 98, 99, 162
toroidal mode, 129, 131, 138, 140
total action, 108

URCA reactions, 82

vector spherical harmonics, 127
visualization, 116
– embedding, 116
– 4D ray-tracing, 116

Walecka model, 22, 66, 69, 72
white dwarfs, 4, 7, 11, 154
white-dwarf cooling, 160, 161
Whole-Earth Telescope, 154, 155
Wick's theorem, 33
WIMP, 153
Wronskian, 170

Zerilli function, 168
– potential, 168

Springer Tracts in Modern Physics

118 **Mechanical Relaxation of Interstitials in Irradiated Metals**
By K.-H. Robrock 1990. 67 figs. VIII, 106 pages

119 **Rigorous Methods in Particle Physics**
Edited by S. Ciulli, F. Scheck, and W. Thirring 1990. 21 figs. X, 220 pages

120 **Nuclear Pion Photoproduction**
By A. Nagl, V. Devanathan, and H. Überall
1991. 53 figs. VIII, 174 pages

121 **Current-Induced Nonequilibrium Phenomena
in Quasi-One-Dimensional Superconductors**
By R. Tidecks 1990. 109 figs. IX, 341 pages

122 **Particle Induced Electron Emission I**
With contributions by M. Rösler, W. Brauer,
and J. Devooght, J.-C. Dehaes, A. Dubus, M. Cailler, J.-P. Ganachaud
1991. 64 figs. X, 130 pages

123 **Particle Induced Electron Emission II**
With contributions by D. Hasselkamp and H. Rothard, K.-O. Groeneveld,
J. Kemmler and P. Varga, H. Winter 1992. 90 figs. IX, 220 pages

124 **Ionization Measurements in High Energy Physics**
By B. Sitar, G. I. Merson, V. A. Chechin, and Yu. A. Budagov
1993. 184 figs. X, 337 pages

125 **Inelastic Scattering of X-Rays with Very High Energy Resolution**
By E. Burkel 1991. 70 figs. XV, 112 pages

126 **Critical Phenomena at Surfaces and Interfaces**
Evanescent X-Ray and Neutron Scattering
By H. Dosch 1992. 69 figs. X, 145 pages

127 **Critical Behavior at Surfaces and Interfaces**
Roughening and Wetting Phenomena
By R. Lipowsky 1996. 80 figs. X, Approx. 180 pages

128 **Surface Scattering Experiments with Conduction Electrons**
By D. Schumacher 1993. 55 figs. IX, 95 pages

129 **Dynamics of Topological Magnetic Solitons**
By V. G. Bar'yakhtar, M. V. Chetkin, B. A. Ivanov, and S. N. Gadetskii
1994. 78 figs. VIII, 179 pages

130 **Time–Resolved Light Scattering from Excitons**
By H. Stolz 1994. 87 figs. XI, 210 pages

131 **Ultrathin Metal Films**
Magnetic and Structural Properties
By M. Wuttig 1996. 103 figs. X, ca. 180 pages

132 **Interaction of Hydrogen Isotopes with Transition Metals and Intermetallic Compounds**
By B. M. Andreev, E. P. Magomedbekov, G. Sicking 1996. 58 figs. VIII, 168 pages

133 **Matter at High Densities in Astrophysics**
Compact Stars and the Equation of State
In Honor of Friedrich Hund's 100th Birthday
By H. Riffert, H. Müther, H. Herold, and H. Ruder 1996. 86 figs. XIV, 274 pages

134 **Fermi Surfaces of Low-Dimensional Organic Metals and Superconductors**
By J. Wosnitza 1996. 88 figs. VIII, 184 pages